Fundamentals of Nanoelectronics

George W. Hanson

Dept. of Electrical Engineering & Computer Science
University of Wisconsin at Milwaukee

Upper Saddle River, New Jersey 07458

Library of Congress Cataloging-in-Publication Data

Hanson, George W.
Fundamentals of Nanoelectronics / George W. Hanson.—1st ed.
p. cm.
Includes bibliographical references and index.
ISBN 978-0-13-195708-4
1. Molecular electronics. 2. Nanoelectronics. I. Title.
TK7874.8.H37 2007
621.382—dc22

2007010007

Vice President and Editorial Director, ECS: *Marcia J. Horton*
Acquisitions Editor: *Michael McDonald*
Associate Editor: *Alice Dworkin*
Executive Managing Editor: *Vince O'Brien*
Managing Editor: *Scott Disanno*
Production Editor: *Craig Little*
Director of Creative Services: *Paul Belfanti*
Creative Director: *Juan Lopez*
Art Director: *Jayne Conte*
Cover Designer: *Bruce Kenselaar*
Managing Editor, AV Management and Production: *Patricia Burns*
Art Editor: *Gregory Dulles*
Photo Research: *Sheila Norman*
Manufacturing Manager: *Alexis Heydt-Long*
Manufacturing Buyer: *Lisa McDowell*
Marketing Manager: *Tim Galligan*
Cover Photo: Colored scanning electron micrograph of two intertwined carbon nanotubes. The image shows approximately 15 microns of tube-length. Dr. Kostas Kostarelos and David McCarthy / Science PhotoLibrary / Photo Researchers, Inc.

Pearson Prentice Hall
Pearson Education, Inc.
Upper Saddle River, NJ 07458

Printed in the United States of America

10 9 8 7 6 5 4 3 2 1

ISBN-10: 0-13-195708-2
ISBN-13: 978-0-13-195708-4

Pearson Education Ltd., *London*
Pearson Education Australia Pty. Ltd., *Sydney*
Pearson Education Singapore, Pte. Ltd.
Pearson Education North Asia Ltd., *Hong Kong*
Pearson Education Canada, Inc., *Toronto*
Pearson Educación de Mexico, S.A. de C.V.
Pearson Education—Japan, *Tokyo*
Pearson Education Malaysia, Pte. Ltd.
Pearson Education, Inc., *Upper Saddle River, New Jersey*

To Sheila

CONTENTS

PREFACE

Nanotechnology refers to any technology that uses nanoscopic objects or devices, i.e., devices on the order of a nanometer (one nanometer is one billionth of a meter). This is roughly the size one would obtain by shrinking a grain of sand by a factor of one thousand, and then again by another factor of one thousand. Thus, it is the technology of the very small.

While there has been considerable hype in the popular media about the promises, and possible perils, of nanotechnology, the field is truly in its infancy. At this time it is impossible to know which aspects of nanotechnology currently under consideration will lead to mature, established fields and to practical applications. It is certain, however, that many as-yet unimagined areas of nanotechnology will be developed in the future. It is also certain that nanotechnology will play an increasingly important role in everyday life, as devices move from the research laboratory to the commercial market.

To develop nanotechnology, scientists and engineers need to understand the fundamental physical principles governing objects having dimensions on the order of nanometers. This is the realm of quantum mechanics, in general, as well as related areas in solid state physics, chemistry, and biology. For example, although there are many proposals to develop nanoscale devices based on electron movement in metals and semiconductors, there is also considerable interest in developing chemical and biological computers, molecular electronics, and other information processing devices outside of the traditional electron physics disciplines.

This book was written to provide an introduction to fundamental concepts of nanoelectronics, including single electron effects and electron transport in nanoscopic systems, for electrical engineers and applied scientists. The intended audience for the book is junior and senior level undergraduate students, although it can also serve as an introduction to the subject for beginning graduate students. Of paramount importance is the idea of understanding quantum dots, quantum wires, and quantum wells, and nanoelectronic applications of these structures. In particular, attention is focused on the quantization of electrical properties, such as conductance quantization and ballistic transport in low-dimensional systems, quantum interference effects arising from the wave nature of electrons, and tunneling phenomena in nanoelectronic devices. Topics were chosen that emphasize, to a large degree, quantum counterparts of classical electronic and electrical devices familiar to junior and senior level undergraduate students, such as transistors and wires. The level of presentation assumes that the reader has some background in basic physics, including fundamental concepts in mechanics, energy, and electromagnetics, and in electrical circuits and traditional electronics. Furthermore, some basic knowledge of the physics of field-effect transistors would be helpful. In the electrical engineering curriculum at most universities, such background material has usually been covered by some point in the junior year.

Although quantum mechanics and solid state physics are treated at an introductory level, the book is not intended to replace discipline specific books or courses in these areas. However, in the ever-increasingly congested undergraduate curriculum at most institutions, a course such as the one this book is intended to accompany can serve as an introduction to the area, and spur interest for further study.

ACKNOWLEDGMENTS

I would like to thank Patricia Whaley and the folks at Quantum Dot Corporation for their helpful comments, and Peter Burke for sharing his expertise, and for his thoughtful considerations of this work. I would also like to thank Richard Sorbello and Carolyn Aita at the University of Wisconsin-Milwaukee for their help, and for their encouragement and interest in this area. Furthermore, Iam grateful to the anonymous reviewers who provided detailed comments and suggestions that significantly improved this work.

Photo Credits

Cover: Colored scanning electron micrograph of two intertwined carbon nanotubes. The image shows approximately 15 microns of tube-length. Dr. Kostas Kostarelos and David McCarthy / Science PhotoLibrary / Photo Researchers, Inc.

Chapter 1: Opener: © Infineon Technologies AG. 1.1: © 2006 Southeastern Universities Research Association. All Rights Reserved. 1.2: © 2004 by David Masse. © MMVII The Porticus Centre. All Rights Reserved. 1.3: Courtesy of the Center for Applied Microtechnology, University of Washington. 1.4: Courtesy of Data Storage Systems Center, Carnegie Mellon University

Chapter 2: Opener: Based on Tonomura, et al., *Amer. Jour. Phys.* 57, 117. 1989. Courtesy of the American Institute of Physics. © Hitachi Ltd. Advanced Research Laboratory. 2.6: Based on Tonomura, et al., *Amer. Jour. Phys.* 57, 117. 1989. Courtesy of the American Institute of Physics. © Hitachi Ltd. Advanced Research Laboratory.

Chapter 3: Opener: Courtesy Almaden Research Center/Research Division/NASA/Media Services. 3.1: Courtesy IBM Research, Almaden Research Center. 3.2: Courtesy IBM Research, Almaden Research Center. Unauthorized use not permitted.

Chapter 4: Opener: From J.M. Zuo, M. Kim, M. O'Keeffe, and J.C.H. Spence, *Nature* 401, 49–52 (1999). Used by permission. 4.12: Courtesy Wikipedia, The Free Encyclopedia. 4.14: Courtesy Carol and Mike Werner/Phototake NYC. 4.19: Reprinted with permission from Roukes, M. L. et al., "Quenching of the Hall Effect in a One-Dimensional Wire," *Phys. Rev. Lett.* 59 (1987): 3011. © 1987, American Physical Society.

Chapter 5: Opener: Image reproduced by permission of IBM Research, Almaden Research Center. Unauthorized use not permitted. 5.12: Based on a figure from S. M. Sze, *Physics of Semiconductor Devices*, John Wiley & Sons, 1969. Used by permission. Page 147 footnote: Reprinted with permission from L. Peterson, P. Laitenberger, E. Laegsgaard, and F. Besenbacher, *Phys. Rev. B*, 58, 7361 1998. Copyright 1998 by the American Physical Society. 5.33: Based on Figure 6.8 (page 55) from *Band Theory and Electronic Properties of Solids* by John Singleton, Oxford University Press, 2001. Reprinted with permission from John Singleton and Oxford University Press. Data from M.D. Sturge, *Phys. Rev.* 127, 768 (1962). 5.35: Based on a figure from Saito, R., G. Dresselhaus, and M. S. Dresselhaus, *Physical Properties of Carbon Nanotubes*, Singapore: World Scientific, 1998. Used by permission. 5.36: Based on a figure in *Physical Properties of Carbon Nanotubes* by R. Saito, G. Dresselhaus, and M.S. Dresselhaus. Singapore: World Scientific Publishing Co. Pte. Ltd., 1998. Used by permission. 5.37: Based on a figure in *Physical Properties of Carbon Nanotubes* by R. Saito, G. Dresselhaus, and M.S. Dresselhaus. Singapore: World Scientific Publishing Co. Pte. Ltd., 1998. Used by permission. 5.38: Based on a figure in *Physical Properties of Carbon Nanotubes* by R. Saito, G. Dresselhaus, and M.S. Dresselhaus. Singapore: World Scientific Publishing Co. Pte. Ltd., 1998. Used by permission. 5.39: Courtesy Shenzhen Nano-Technologies Port Co., Ltd.

Chapter 6: Opener: From Park, J.-Y., et al., "Electrical Cutting and Nicking of Carbon Nanotubes Using an Atomic Force Microscope." *Appl. Phys. Lett.*, 80 (2002): 4446. © 2002, American Institute of Physics. 6.12: From Bonard, J. M. et al., "Field Emission of Individual Carbon Nanotubes in the Scanning Electron Microscope," *Phys. Rev. Lett.* 89 (2002): 197602. © 2002, American Physical Society. 6.17: From Quan, W.-Y., D. M. Kim, and M. K. Cho, "Unified Compact Theory of Tunneling Gate Current in Metal–Oxide-Semiconductor Structures: Quantum and Image Force Barrier Lowering," *J. Appl. Phys.* 92 (2002): 3724. © 2002, American Institute of Physics. 6.20: Courtesy Institut für Allgemeine Physik at the Vienna University of Technology. 6.21: Image reproduced by permission of IBM Research, Almaden Research Center. Unauthorized use not permitted.

Chapter 7: Opener: Reprinted with permission from Hideo Namatsu, Ph.D., *Journal of Vacuum Science & Techology B: Microelectronics and Nanometer Structures*, January 2003, Volume 21, Issue 1, pages 1–5. © 2007, American Institute of Physics. 7.5: From Pépin, A., et al., "Temperature Evolution of Multiple Tunnel Junction Devices Mode with Disordered Two-Dimensional Arrays of Metallic Islands," *Appl. Phys. Lett.* 74 (1999): 3047. © 1999, American Institute of Physics. 7.6: From Pépin, A., et al., "Temperature Evolution of Multiple Tunnel Junction Devices Mode with Disordered Two-Dimensional Arrays of Metallic Islands," *Appl. Phys. Lett.* 74 (1999): 3047. © 1999, American Institute of Physics. 7.7: From Clarke, L., et al., "Room-Temperature Coulomb-Blockade Dominated Transport in Gold Nanocluster Structures," *Semicond. Sci. Technol.* 13 (1998): A111–A114. Courtesy Martin Wybourne. 7.19: Based on a figure in Wilkins, R., E. Ben-Jacob, and R. C. Jaklevic, "Scanning-Tunneling-Microscope Observations of Coulomb Blockade and Oxide Polarization in Small Metal Droplets," *Phys. Rev. Lett.* 63 (1989): 801. © 1989, American Physical Society. 7.27: From Hadley, P., et al., "Single-Electron Effects in Metals and Nanotubes for Nanoscale Circuits," in *Proceedings of the MIOP—The German Wireless Week, 11th Conference on Microwaves, Radio Communication and Electromagnetic Compatibility*, Stuttgart, Germany, 2001. 408–412. Courtesy of Leo Kouwenhoven, Kavli Institute of Nanoscience, Delft University of Technology. 7.28: Based on a figure from Kastner, M. A. "The Single-Electron Transistor," *Rev. Mod. Phys.* 64 (1992): 849. © 1992, American Physical Society. 7.29: Based on a figure from Shirakashi, J.-I., et al., "Single-Electron Changing Effects in Nb/Nb Oxide-Based Single-Electron Transistors at Room Temperature, "*Appl. Phys. Lett.* 72 (1998): 1893. © 1998, American Institute of Physics. 7.32: Based on a figure from Ono, Y., et al., "Si Complementary Single-Electron Inverter with Voltage Gain," *Appl. Phys. Lett.* 76 (2000): 3121. © 2000, American Institute of Physics. 7.33: Based on a figure from Ono, Y., et al., "Si Complementary Single-Electron Inverter with Voltage Gain," *Appl. Phys. Lett.* 76 (2000): 3121. © 2000, American Institute of Physics. 7.34: Based on a figure from Avouris, P., et al., "Carbon Nanotube Electronics," *Proceedings of the IEEE*, 91 (2003): 1772–1784. 7.35: From Avouris, P., et al., "Carbon Nanotube Electronics," *Proceedings of the IEEE*, 91 (2003): 1772–1784. 7.36: Based on a figure from R. Martel, et al., "Single- and Multi-Wall Carbon Nanotube Field-Effect Transistors." *Applied Physics Letters*, 73, 2447 (1998). © 1998, American Institute of Physics. 7.37: Based on a figure from H.R. Shea, R. Martel, T. Hertel, T. Schmidt and PH. Avouris, "Manipulation of Carbon Nanotubes and Properties of Nanotubes Field-Effect Transistors and Rings," *Microelectronic Engineering*, 46, 101–104 (1999). Used by permission from Elsevier. 7.38: Based on a figure from Huang, Y., et al., "Gallium Nitride Nanowire Nanodevices," *Nanoletters*, 2 (2002): 101. 7.39: From De Franceschi, S. and J. A. van Dam, "Single-Electron Tunneling in InP Nanowires," *Appl. Phys. Lett.*, 83 (2003): 344. © 2003, American Institute of Physics. 7.40: From De Franceschi, S. and J. A. van Dam, "Single-Electron Tunneling in InP Nanowires," *Appl. Phys. Lett.*, 83 (2003): 344. © 2003, American Institute of Physics. 7.43: Based on a figure from from Di Ventra, M., S. T. Pantelides, and N. D. Lang, "The Benzene Molecule as a Molecular Resonant-Tunneling Transistor," *Appl. Phys. Lett.*, 76 (2000): 3448. © 2000, American Institute of Physics. 7.44: From Park, J. et al., "Coulomb Blockade and the Kondo Effect in Single-Atom Transistors," *Nature*, 417 (2002): 722–725. © 2002, Macmillan Publishers, Ltd.

Chapter 8: Opener: Courtesy of Prof. Dr. Roland Wiesendanger, Executive Director of the Institute of Applied Physics (IAP) and Interdisciplinary Nanoscience Center Hamburg (INCH).

Chapter 9: Opener: From Harper, J., et al., "Microstructure of GaSb-on-InAs Heterojunction Examined with Cross-Sectional Scanning Tunneling Microscopy," *Appl. Phys. Lett.* 73 (1998): 2805. 9.5: Based on a figure from N. Mayhew, D. Phil. thesis, Oxford University, 1993. (Published in A.M. Fox, "Optoelectronics in Quantum Well Structures," *Contemporary Physics* 37, 11 (1996)). Reprinted with permission from A.M. Fox. 9.8: Based on a figure from Chui, H. C., et al., "Short Wavelength Intersubband Transitions in In GaAs/AlGaAs Quantum Wells Grown on GaAs," *Appl. Phy. Lett.* 64 (1994): 736. © 1994, American Institute of Physics. 9.12: Based on a figure from Saito, R., et al., "Electronic Structure of Chiral Graphene Tubules," *Appl. Phys. Lett.* 60 (1992): 2204. © 1992, American Institute of Physics. 9.14: Reprinted with permission from Quantum Dot Corporation. 9.15: Reprinted with permission from Quantum Dot Corporation. 9.16: From Seydel C., "Quantum Dots Get Wet," *Science* 4 April 2003, 300: 80 [DOI: 10.1126/Science.300.5616.80]. Courtesy Xiaohu Gao and Shuming Nie, Emory University School of Medicine. 9.17: Based on a figure from Link, S. and M. A. El-Sayed, "Spectral Properties and Relaxation Dynamics of Surface Plasmon Electronic Oscillations in Gold and Silver Nanodots and Nanorods," *J. Phys Chem B* 103 (1999): 8410–8426. 9.18: From Sun, Y., B. T. Mayers, and Y. Xia, "Template-Engaged Replacement Reaction: A One-Step Approach to the Large-Scale Synthesis of Metal Nanostructures with Hollow Interiors," *Nano Lett.* 2 (2002): 481. © 2002, American Chemical Society. 9.19: Based on a figure from Li, M., L. Chen, and S. Y. Chou, "Direct Three-Dimensional Patterning Using Nanoimprint Lithography," *Appl. Phys. Lett.* 78 (2001): 3322. © 2001, American Institute of Physics. 9.20: Reused with permission from Li, M., L. Chen, and S. Y. Chou, "Direct Three-Dimensional Patterning Using Nanoimprint Lithography," *Appl. Phys. Lett.* 78 (2001): 3322. © 2001, American Institute of Physics. 9.23: Courtesy of Jonathan Bird and Yuki Takagaki. 9.24: Reprinted with permission from Gilberto Medeiros-Ribeiro at HPL Palo Alto.

Chapter 10: Opener: Ann Bentley & Professor Art Ellis, Materials Science and Engineering Department, University of Wisconsin-Madison. Image courtesy the Nickel Insititute. 10.4: From Toimil Molares, M. E., et al., "Electrical Characterization of Electrochemically Grown Single Copper Nanowires," *Appl. Phys. Lett.* 82 (2003): 2139. © 2003, American Institute of Physics. 10.5: Based on a figure from Toimil Molares, M. E., et al., "Electrical Characterization of Electrochemically Grown Single Copper Nanowires," *Appl. Phys. Lett.* 82 (2003): 2139. © 2003, American Institute of Physics. 10.9: Based on a figure from Thomas, K. J., et al., "Interaction Effects in a One-Dimensional Constriction," *Phys. Rev. B* 58 (1998): 4846. © 1998, American Physical Society. 10.10: Based on Fig. 6 from van Wees, B. J., L.P. Kouwenhoven, E.M.M. Willems, C.J.P.M. Harmans, J.E. Mooij, H. van Houten, C.W.J. Beenakker, J.G. Williamson, and C.T. Foxon, "Quantum Ballistic and Adiabatic Electron Transport Studied with Quantum Point Contacts," *Phys. Rev. B* 43 (1991): 12431. © 1991, American Physical Society. 10.11: Based on a figure from Dreher, M. F., Pauly, J. Heurich, J.C. Cuevas, E. Scheer, and P. Nielaba, "Structure and Conductance Histogram of Atomic-Sized Au Contacts," *Phys. Rev. B* 72 (2005): 075435. © 2005, American Physical Society. 10.12: Based on a figure from Dreher, M. F., Pauly, J. Heurich, J.C. Cuevas, E. Scheer, and P. Nielaba, "Structure and Conductance Histogram of Atomic-Sized Au Contacts," *Phys. Rev. B* 72 (2005): 075435. © 2005, American Physical Society. 10.13: From Tsukagoshi, K., B. W. Alphenaar, and H. Ago, "Coherent Transport of Electron Spin in a Ferro Magnetically Contacted Carbon Nanotube," *Nature* 401 (1999): 572. Courtesy Macmillan Publishers Ltd. and Bruce Alphenaar. 10.14: Based on a figure from S. Li, Z.Y.C. Rutherglen, and P.J. Burke, "Electrical Properties of 0.4 cm Llong Single-Walled Carbon Nanotubes," *Nano Lett.*, 4: 2003–2007, 2004. © 2004 American Chemical Society. 10.15: Based on a figure from Hanson, G.W. "Current on an Infinitely-Long Carbon Nanotube Antenna Excited by a Gap Generator," *IEEE Trans. Antennas Propagat.* 54 (2006): 76–81. © 2006, IEEE. 10.20: Courtesy of WTEC Inc. and Dr. James M. Daughton/NVE Corporation. Based on a graph by Professor Eshel Ben-Jacob.

Part I

FUNDAMENTALS OF NANOSCOPIC PHYSICS

This text is divided into three parts. The first part, Fundamentals of Nanoscopic Physics, introduces some of the basic ideas underlying nanoscopic phenomena, including wave and particle concepts, quantum principles, charge confinement, solid state materials, and the basic ideas of quantum dots, wires, and wells. In essence, this material comprises the basic physics needed to study fairly broad classes of nanoelectronic devices.

The second part, Single-Electron and Few-Electron Phenomena and Devices, introduces the concept of tunneling, and describes related phenomena such as Coulomb blockade. These ideas are then applied to explain the fundamental principles of what are called *single-electron devices*, including the single-electron transistor. These types of devices have generated a lot of excitement in the electronics field, and are expected to play an important role in future electronic and photonic systems. The operation of these single-electron devices relies on the movement of single electrons (more generally, on the movement of relatively small numbers of electrons). This marks a major departure from conventional electronic devices, which use many more electrons in their operation.

The third part, Many Electron Phenomena, describes classical and quantum statistics and related electron phenomena. These concepts are applied to semiconductor quantum wells, wires, and dots, complementing similar material discussed in Part I. Then, ballistic transport is described, where electrons pass through a region of space that is small compared to the mean free path. Thus, ideally, electrons don't scatter in transversing the material, necessitating a vastly different concept of resistance than encountered in conventional, macroscopic circuits.

In an overly broad sense, the first part describes the physical principles needed to understand nanoelectronic devices, the second part describes nanoelectronic devices that have single-electron precision, and the third part considers nanosolids, such as semiconductor quantum dots and nanoscopic wires.

Chapter

1

INTRODUCTION TO NANOELECTRONICS

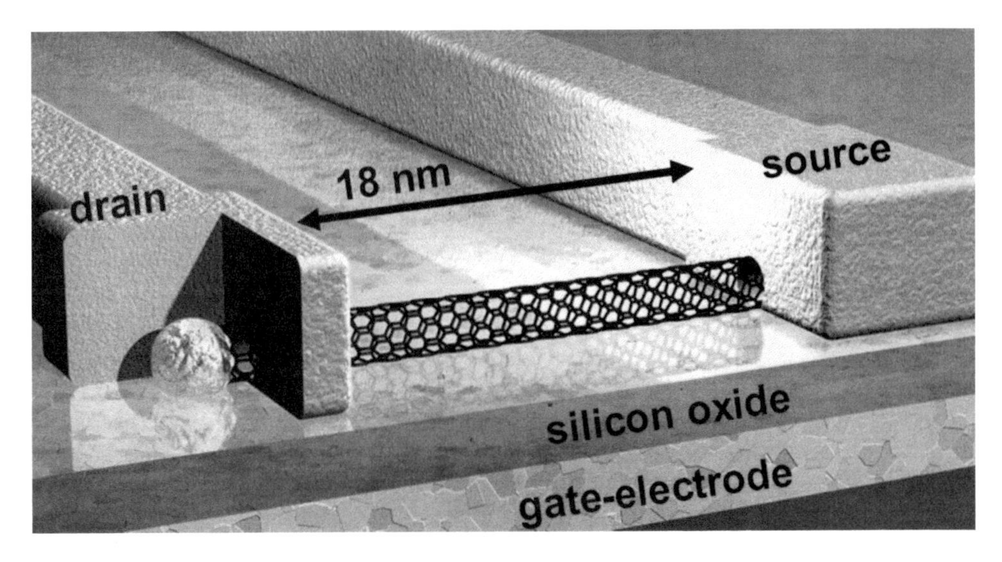

Depiction of a carbon nanotube transistor. (© Infineon Technologies AG.)

Before getting into the details of nanoscale phenomena, it is worthwhile to appreciate the size scale of a nanometer. Typical length units of interest are given in Table 1.1.

Not listed in Table 1.1 is the angstrom, Å, where 1 $\mathring{A} = 10^{-10}$ m, or one-tenth of a nanometer. An angstrom is a typical atomic dimension, so that, roughly speaking, about 10 atoms fit in a nanometer.

A few small objects and their typical sizes are listed in Table 1.2. It can be seen that atoms, DNA, proteins, viruses, and transistors are all in the realm of what can be broadly classified as nanoscale objects (perhaps within a factor of 100 of a nanometer), whereas, for comparison, biological cells are a factor of 10,000 times bigger than a nanometer.

TABLE 1.1 TYPICAL LENGTH UNITS OF INTEREST IN NANOTECHNOLOGY.

Name	Symbol	Size
Meter	m	1
Millimeter	mm	10^{-3} m
Micrometer	μm	10^{-6} m
Nanometer	nm	10^{-9} m
Picometer	pm	10^{-12} m
Femtometer	fm	10^{-15} m
Attometer	am	10^{-18} m

TABLE 1.2 COMMON SMALL OBJECTS AND THEIR TYPICAL DIMENSIONS.

Object	Typical dimension
Atom	1 Å = 0.1 nm
DNA (diameter)	1 nm
Protein	10 nm
Transistor oxide thickness	1.2 nm
Transistor gate length	35 nm
Virus	100 nm
Red blood cell	10 μm = 10,000 nm
Human hair dia.	150 μm = 150,000 nm
Grain of sand	1 mm = 1,000,000 nm

TABLE 1.3 ELECTROMAGNETIC SPECTRUM.

Common name[2]	λ (m)	λ (nm)	f (GHz)
AM radio	10^4	10^{13}	0.00003
FM radio	1	10^9	0.3
Microwaves	10^{-2}	10^7	30
Infrared	10^{-4}	10^5	3,000
Visible—red	7×10^{-7}	700	428,571
Visible—violet	4×10^{-7}	400	750,000
Ultraviolet	10^{-8}	10	30,000,000
X-rays	10^{-10} (1 Å)	0.1	3,000,000,000
γ-rays	10^{-12}	0.001	300,000,000,000

The common name denotes a range of frequencies; in the table a typical value is provided.

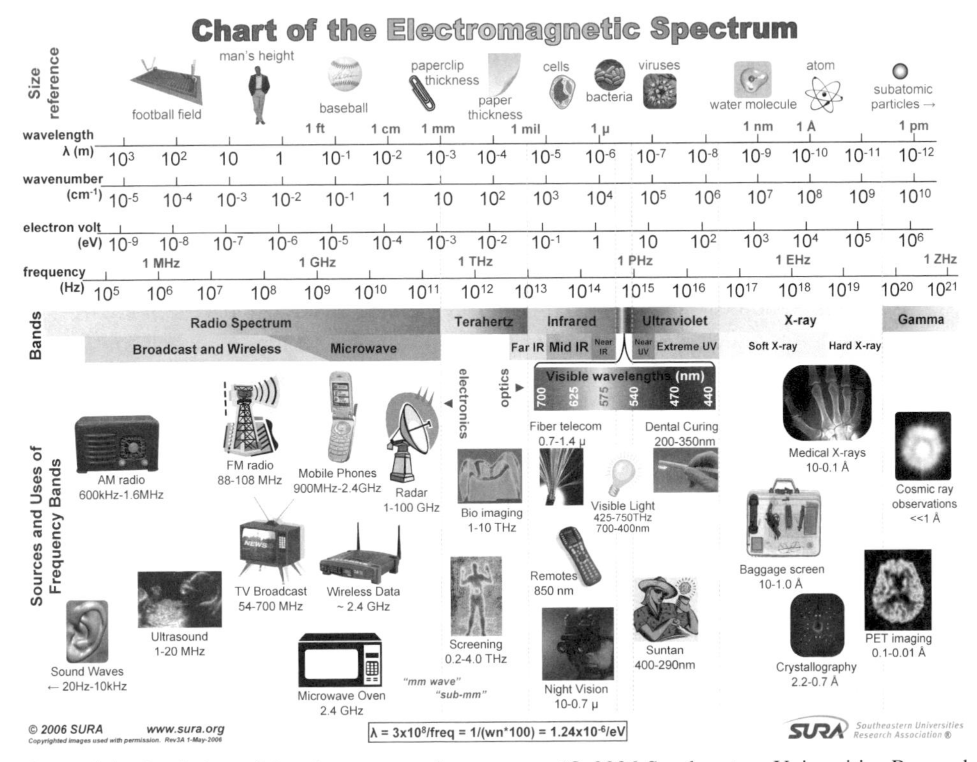

Figure 1.1 Depiction of the electromagnetic spectrum. (© 2006 Southeastern Universities Research Association. All Rights Reserved.)

To further get our bearings as to the size scale involved in nanotechnology, some points in the electromagnetic spectrum are listed in Table 1, and shown in Fig. 1.1, where it can be appreciated that wavelengths between the ultraviolet (UV) and X-ray range are on the general scale of a nanometer.

1.1 THE "TOP-DOWN" APPROACH

Perhaps the biggest single push towards nanoelectronics has been the microelectronics industry, a name that is rapidly becoming outdated. The microelectronics industry was born out of the invention of the transistor in 1947 to 1948 by William Shockley, Walter Brattain, and John Bardeen of Bell Laboratories, and by the invention of the integrated circuit (IC) in 1958 by Jack Kilby of Texas Instruments and Robert Noyce (cofounder) of Fairchild Semiconductor Corporation. Kilby is usually the person mentioned most often in connection with invention of the IC, although both Noyce and Kilby, working independently and without knowledge of each other's efforts, applied for patents in 1959 relating to the IC.

In 1965, Gordon Moore (cofounder of Intel Corporation) observed that the number of transistors per square inch on an IC chip roughly doubled every 12 months. This general rule of thumb is now called *Moore's law*, and has approximately held through the present time, although now the number is taken to double every 18–24 months. Of course, the electronics industry, and the general public, would like this process of miniaturization to continue. However, recently, scientists (even at Intel) predict the end of Moore's law by 2015 to 2018, when manufacturing of transistors with 16 nm feature size is expected to be possible.

The first transistor, shown in Fig. 1.2, was a bipolar device consisting of two sharp metal wires (the emitter and collector) making contact with a germanium substrate, which made contact with a base electrode. By 1960, the minimum feature size of a transistor had shrunk to approximately 100 μm. In 1980, it was close to 1 μm, and in 2001, feature size was on the order of 130 nm. This was followed by 90 nm manufacturing in 2003, and in 2005, 65 nm structures were produced. Forty-five nm technology is scheduled for 2007, with 32 nm technologies being planned for the near future.[†]

Because of the diminishing feature size of transistors and other components, one could say that the electronics industry is already " doing" nanotechnology. However, a fundamental shift seems to be in the works. Transistors on the order of 100 nm still obey, to a certain extent, classical physics, modified by quantum principles.[‡] As feature size is reduced towards a nanometer, more and more purely quantum effects begin to emerge. For example, when the gate oxide thickness of a metal–oxide-semiconductor field-effect transistor (MOSFET) goes below 1–2 nm, significant tunneling through the gate oxide occurs, which is a purely quantum phenomenon. Also, as transistor channel lengths approach a few nanometers, a significant component of electrons may transverse the channel ballistically, i.e., without collisions that characterize ordinary conduction. At last, as channel lengths continue to decrease, direct source-to-drain tunneling can occur. All of these effects serve to drastically alter device performance.

Another very important practical aspect of reducing feature size is that manufacturing processes must also be changed, at great expense. A new integrated circuit manufacturing plant costs upwards of several billion (U.S.) dollars, and costs increase as feature size is reduced. Furthermore, variants of currently used optical lithography can only work to perhaps several tens of nanometers, and other techniques (perhaps not lithography based) need to be developed.

1.1.1 Lithography

The problem of shrinking the size of devices fabricated through optical lithography can be readily understood. In this context, one can broadly define lithography as the process of

[†]The sizes 90 nm, 65 nm, 45 nm, 32 nm, etc., correspond to what are called *technology nodes* in the *Semiconductor Roadmap*. This refers to the International Technology Roadmap for Semiconductors (ITRS), which is an extremely influential cooperative effort of industry manufacturers and suppliers, government organizations, and universities to provide an assessment of the semiconductor industry's technology requirements.

[‡]Although band theory and the concept of effective mass, etc., are central to transistor operation, and are quantum effects, in some sense they serve to modify classical behavior of electrons in semiconductors.

Figure 1.2 The first transistor, invented in 1947 at Bell Laboratories. (© 2004 by David Masse. © MMVII The Porticus Centre. All rights reserved.)

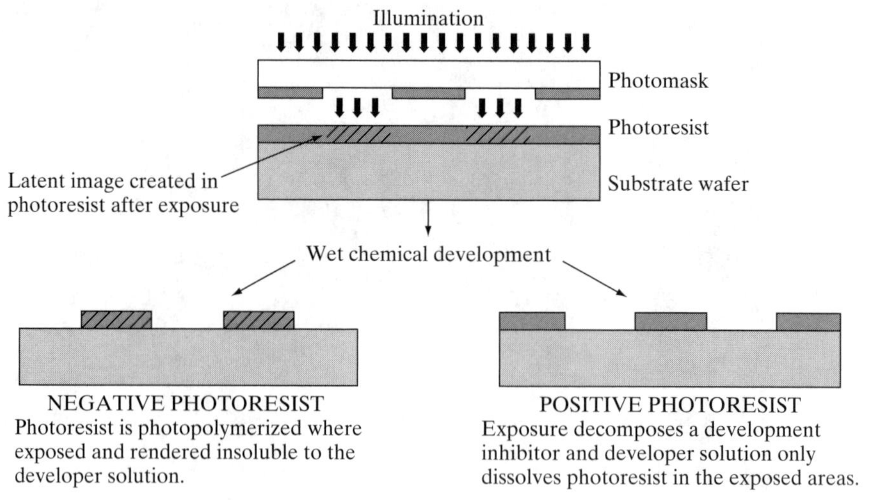

Figure 1.3 Depiction of steps 1–3 of the lithography process. (Courtesy of the Center for Applied Microtechnology, University of Washington.)

using electromagnetic energy to transfer a pattern from a mask to a resist layer deposited on the surface of a substrate, which we'll call the *wafer*, in order to form an electrical circuit. A simplified version of the basic process is described next. (Steps 1–3 are depicted in Fig. 1.3.)

1. A photosensitive emulsion called a *photoresist* is applied to the wafer (in the vast majority of cases the wafer is silicon).
2. Optical energy (light[†]) is directed at a *photomask* containing opaque and transparent regions that correspond to the desired pattern. The light that passes through the photomask reaches the wafer, illuminating the desired pattern on the resist. In projection lithography, which is the most common form in the semiconducting industry, lenses are used to focus the light before and after the photomask.
3. Sections of the photoresist that are exposed to the light coming through the mask undergo chemical reactions.
   (a) For a *negative photoresist*, the resist material is initially soluble (for a particular solvent that will be used in development), and through a chemical reaction when exposed to light, becomes insoluble. When the wafer is later washed with a solvent, the areas that were unexposed (i.e., where the photomask blocked illumination) dissolve, and the exposed areas, corresponding to transparent sections of the photomask, remain.
   (b) In a *positive photoresist*, the resist material is initially insoluble, and through a chemical reaction when exposed to light, becomes soluble. When the wafer is washed with a solvent, the areas that were exposed to the illumination dissolve, and the unexposed areas remain.

[†]Here we use the term "light" in a general way, not necessarily indicating visible light.

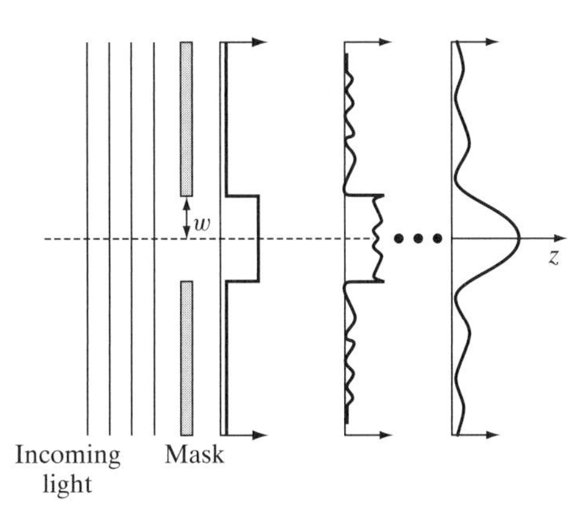

Figure 1.4 Depiction of light diffraction through an aperture in an opaque screen. (Courtesy of Data Storage Systems Center, Carnegie Mellon University.)

4. Different steps, such as the following, may then be performed to transfer the pattern from the resist to the wafer. For example,

(a) Etching may be used to remove substrate material. (See Section 9.4 for a brief description of etching and material deposition.) The photoresist serves to resist the etching and protect sections of the wafer that it covers. After etching the resist is removed, leaving the desired structure;

(b) Material may be deposited, for example, metallization, onto the wafer. Then the photoresist can be removed (the material deposited on the photoresist is also thereby removed, which is known as *lift-off*), leaving the deposited material in areas that were not covered by the resist;

(c) Doping can occur. For instance, a beam of dopant ions can be accelerated towards the wafer. The resist blocks the ions from reaching those regions of the wafer covered by the resist, and thus creates regions of doping in areas not covered by the resist. This is known as *ion implantation*.

Lithography, and other methods of forming nanostructures, is discussed further in Section 9.4.

The resolution R of an optical lithography process describes the ability of an imaging system to resolve two closely spaced objects, and is not actually the smallest feature size of a printed object. The general problem of achieving good resolution can be appreciated by considered the pattern of light (really, any electromagnetic energy) that forms in passing through the transparent regions of the photomask. By a process known as *diffraction*, which is basically the ability of light to "bend" around corners, as light passes through an aperture on the mask, it tends to smear out, as depicted in Fig. 1.4.

Immediately to the right of the mask, the illuminated pattern will have relatively sharp boundaries, but further away from the mask, the pattern becomes as shown.[†] There is an interplay between the aperture size $2w$, wavelength λ, and position z, although, in general, at a fixed position z the smaller the aperture compared to wavelength, the more the beam will

[†]Contact lithography, where the resist is placed immediately after the mask, can, therefore, improve resolution, although there are some technical difficulties to this approach as well.

spread out. Fixing the size of the aperture, one can obtain a sharper pattern using smaller wavelengths. Of course, this ignores the presence of a lens to focus the illumination, but it gives an idea of the role of wavelength in lithographically forming patterns.

The resolution of an optical lithography process incorporating a lens is approximately

$$R = k_1 \frac{\lambda}{NA}, \tag{1.1}$$

where k_1 is a constant, λ is the wavelength of the source, and NA is called the *numerical aperture*. The constant k_1 depends on a variety of factors related to the lithography process, although values of k_1 around 0.8–0.5 are common, with $k_1 = 0.25$ thought to be the limiting value. The numerical aperture is determined by the characteristics of the lens used to focus the incident energy, with typical values being $NA = 0.5$–0.9. When air is used between the lens and the wafer, $NA = 1$ represents the physical limit. For example, an optical lithography system characterized by $\lambda = 193$ nm,[†] $k_1 = 0.25$, and $NA = 0.9$ results in $R = 53.6$ nm.

Although R can be reduced by shrinking λ, this is not necessarily easily accomplished since, for instance, at wavelengths less than 193 nm, light tends to be absorbed by the fused silica lenses that are now used in standard lithographic processes. However, novel exposure techniques such as using phase-shift masks can significantly enhance resolution, and may be able to extend optical lithography into the low tens of nanometers range.

A relatively new method, called *immersion lithography*, is currently being considered, which involves inserting a film of water or another dielectric medium between the lens and the wafer. In this case, resolution is given by

$$R = k_1 \frac{\lambda}{(NA)\,n}, \tag{1.2}$$

where n is the index of refraction of the film. Since water has a refractive index of approximately 1.44 at optical frequencies, immersion lithography can significantly increase resolution. For example, 193 nm immersion lithography would be equivalent to $193/1.44 = 134$ nm lithography in air. Considering the previous example, this would lead to resolution on the order of $R = 37$ nm, down from $R = 53.6$ nm for ordinary lithography. Of course, every technology has advantages and disadvantages, and technical hurdles to overcome. One such technical problem with immersion lithography is maintaining a uniform, bubble-free medium between the lens and the wafer, as different points on the wafer are accessed. Immersion lithography is currently being considered for 32 nm chips, projected to come out in 2009.

Other technologies, such as *extreme ultraviolet* (EUV) and *X-ray* lithography, are now research areas, with EUV being likely to be deployed in the near future. Furthermore, there is *electron beam lithography*, where an electron beam replaces the electromagnetic illumination. In this case, resolution on the order of 1–10 nm is achievable, although currently this is a very high cost, low-throughput (i.e., slow) technology. It is, however, an important tool for nanotechnology research.

[†] 193 nm is the wavelength of an argon fluoride (ArF) laser, which is common in industry.

Furthermore, a method to break the diffraction limit is being investigated using *negative index of refraction* materials,[†] which can focus light much better than ordinary (positive index of refraction) materials. A lens made of negative index material is called a *superlens*. In recent experiments, 365 nm wavelength light was used to write the word NANO onto an organic polymer with a resolution of approximately 60 nm. Because this is an optical lithography method, it is very fast and thus may be a way to extend optical lithography down to a few tens of nanometers. In general, achieving better and better resolution, thus reducing feature size, has been called the "top-down" approach.

1.2 THE "BOTTOM-UP" APPROACH

In 1959, in what can perhaps be called the birth of nanotechnology, influential physicist (and contributor to the development of quantum theory) Richard Feynman, in a speech delivered at Caltech titled "There's Plenty of Room at the Bottom," discussed the technological future of very small devices. In his speech he asked, "What would happen if we could shift atoms, one by one, and arrange them as we wanted?" Instruments have been developed more recently that have allowed scientists to do just that—to see and manipulate nano- and atomic-sized objects. These instruments include the *scanning electron microscope* (SEM), the *scanning tunneling microscope* (STM), the *transmission electron microscope* (TEM), and the *atomic force microscope* (AFM). In what was a remarkable feat at its time, in 1989 scientists at IBM wrote the letters IBM with atoms, by moving individual xenon atoms using STM, which has a resolution of approximately 0.1 nm, although it is limited to scanning conducting surfaces.

In contrast to the "top-down" approach, this nanoscale building is called the " bottom-up" approach, and represents a much more radical technology shift, which is currently being explored in research laboratories. Of course, moving individual atoms one by one is a time-consuming process, and researchers are looking at more efficient methods of building nanoscopic structures. Many avenues are being explored in this regard, including chemical or biological self-assembly of devices, or mechanical assembly of devices by other small devices (called *assemblers*). Since the objects in question are tiny, often electrophoretic forces,[‡] dielectrophoretic forces,[§] and capillary forces can be profitably used. Eric Drexler, in his 1986 book *Engines of Creation: The Coming Era of Nanotechnology*, outlines a possible future for nanotechnology and the "bottom-up" approach that has received a lot of attention.

1.3 WHY NANOELECTRONICS?

Over the course of the previous 50 years, consumers have become used to electronic products becoming simultaneously smaller and cheaper, and yet more powerful (often in the sense of having more features, etc.) This trend is driven in large part by the desire for companies to

[†]Many metals have a negative permittivity at optical wavelengths, but a negative index of refraction material usually must have a simultaneously negative permittivity and permeability.

[‡]The usual force on a charged object; force equals charge multiplied by electric field.

[§]The force on, typically, a dielectric object due to a nonuniform electric field.

be competitive, and reflects the broad wish of consumers for smaller, faster, cheaper, and yet better electronic products. The reduction in product size is due to shrinking the size of individual electronic devices, such as transistors. This reduction also, of course, can lead to improved functionality, as more devices can be packed into a given area.[†] There are direct economic advantages of small device size as well, since the cost of integrated circuit chips is related to the number of chips that can be produced per silicon wafer. Therefore, higher device density leads to more chips per wafer, and reduced cost. Starting with micron-sized transistors in the 1960s, and driven by this trend alone, at some point one must deal with nanoscopic devices. At the time of this writing, that time has come, with current MOSFET dimensions well under 100 nm.

The development of nanoscopic devices includes the possibility of ultrasmall, low-power electronic products, such as communication and computing devices and embedded sensors. Furthermore, as electronics shrink, the possibility of further incorporating electronics with biological systems rapidly expands.[‡] Therefore, there are many factors driving the miniaturization of electronic devices. Thus, the question "Why nanoelectronics?" seems to have an obvious answer.

However, in addition to the benefits of smaller transistors, there are significant problems in shrinking conventional devices to the nanoscale. For example,

- Device fabrication: This has already been discussed—it may be difficult to extend optical lithography into the realm of low tens of nanometers, and other fabrication methods (such as the bottom-up approach) for high-throughput, commercial-level production are not, as yet, mature. Furthermore, for such extremely small devices, random process variations can lead to device characteristics that are unpredictable, and that take on a statistical nature, greatly complicating circuit design.
- Device operation: As device dimensions are reduced, voltage levels also need to be reduced accordingly. This lowers the threshold voltage of MOSFET devices, and makes it difficult to completely turn the device off, wasting power. Furthermore, due to the laws of quantum physics, conventional MOSFETs will behave differently, perhaps radically differently (and often not in a positive sense), when their dimensions shrink to the order of nanometers. Tunneling and ballistic transport are two prominent quantum effects that will be discussed in this text.
- Heat dissipation: As device density increases, the dissipation of heat becomes a major problem, reducing circuit reliability and leading to shorter device lifetimes, or to device failure. With current technologies, if the rate of increasing device density were to continue, microprocessors would soon be producing more heat per square centimeter than the surface of the sun![§]

[†]As an example, one leading chip manufacturer estimates that moving from the 65 nm technology node to the 45 nm node will provide a twofold improvement in transistor density, which can either be used for smaller chip size or for an increased transistor count per chip (and, hence, higher functionality).

[‡]Think of a pacemaker as a traditional example that is still very relevant. However, other small electronic devices are currently in use, or envisioned for near-term use, for implantation in humans to aid hearing, vision, and a host of other medical problems.

[§]Current ICs have power densities in the order of 100 W/cm^2, up from 10 W/cm^2 a decade ago. The power density of a typical hot plate is 10 W/cm^2, whereas the surface of the sun has 7000 W/cm^2.

Given the preceding problems, it can be seen that at some point in time, the electronics industry needs to look beyond shrinking the size of conventional electronic devices. New devices and system architectures will need to be developed, probably based on physical effects only encountered on the nanoscale. For a new electronics technology to supersede the current silicon-based complementary metal–oxide-semiconductor (CMOS) technology, it would need to be more attractive from an economic and technical standpoint (although these are frequently tightly intertwined). It would have to outperform CMOS, be relatively easy and inexpensive to fabricate, have high reproducibility, yield, and low cost, and have good thermal characteristics. At this point, CMOS is still the technology of electronics, and will likely be for the near future. However, some of the new devices that may eventually supersede CMOS, or play a complementary role to CMOS, are described in this text.

1.4 NANOTECHNOLOGY POTENTIAL

The prospects and potential for nanotechnology have not gone unnoticed by government agencies. In the United States, since 2000, the National Nanotechnology Initiative has funded research and development of nanotechnology. In 2001, funding was U.S. $465 million, increasing to U.S. $697 million in 2002, to U.S. $863 million in 2003, to U.S. $989 million in 2004, to U.S. $1,200 million in 2005, and to U.S. 1,300 million in 2006. In Japan, there have been similar levels of funding, and Europe, as well as other regions, has also spent heavily on nanotechnology development. Worldwide government funding on nanotechnology is estimated to be U.S. $7 billion.

There are currently many potential applications of nanotechnology, including

- Nanoparticles—nanoscopic particles for coatings, as catalysts in chemical processing, in organ-specific drug delivery, etc.;
- Nanomaterials—materials with improved strength and weight, perhaps incorporating carbon nanotubes;
- Nanoelectronics—extremely small, low-power electronic devices, such as transistors;
- Nano-optics—forming optical circuits at the nanoscale, often using plasmonic structures;
- Nanomagnetics—using nanoscopic magnetic materials and devices for magnetic storage technologies, logic devices, etc.;
- Nanofluidics—devices based on fluid properties at the nanoscale (leading to what are called chemical and biological laboratories on a chip);
- Nanobioelectronics—combinations of nanodevices and biological structures for, e.g., chemical/bioweapons detection.

In this book, nanoelectronics, and specifically quantum-effect solid state electronics, are emphasized, although the area of molecular electronics is also briefly introduced. The main categories of solid-state nanoelectronic devices are, at present, primarily based on quantum dots, quantum wells and wires, and carbon nanotubes. One of the most fundamental principles of nanoelectronic devices is tunneling, which is a purely quantum effect. In quantum dot and resonant tunneling devices, a small " island" of semiconducting or metallic material is isolated from the rest of the circuit, somewhat confining electrons to a very small region of space (on the order of a few nanometers). Because of tunneling and related phenomena, this confinement is not absolute, giving rise to many interesting effects that may be useful in electronics and photonics applications.

Last, it should be mentioned that many prototype nanoelectronic devices have been developed, and in some cases, their device characteristics are fairly well understood. Some of these devices are, in a sense, the nano analogue of a conventional electronic device, and some represent completely new technologies that rely on purely quantum phenomena. This latter category would include, for example, certain quantum logic devices, electronic devices based on quantum spin, single electron devices, and molecular and biologically based devices. In general, mass fabrication and manufacturing aspects present a great number of challenges that are, for the most part, yet to be overcome. However, there are a considerable number of companies have been formed in the hopes of bringing nanotechnology concepts to market, and there are commercial products available based on nanotechnology. For example, aside from electronics applications, coatings consisting of nano-sized particles embedded in a host medium have been used for a long time. Now, using modern technology, these coatings can be engineered to provide desired effects (e.g., improved strength, color permanence, etc.). As another example, one can purchase tennis rackets made using carbon nanotubes that are said to be five times stiffer than standard carbon racquets, and skis made using nanoparticles that provide good gliding capabilities. Medical and biological applications of quantum dots are also developing rapidly. In the electronics arena, it is generally felt that nanoelectronic devices will allow for fast, low-power chips containing tens of billions or even trillions of devices per chip, with feature size on the order of nanometers. However, it is difficult to say at this point what will be the first commercially available non-CMOS-type nanoelectronic device.

1.5 MAIN POINTS

In this chapter, some general ideas concerning the nanoelectronics field were presented. In particular, after reading this chapter you should

- understand the various size units, and have a general appreciation for the size of micron and nanoscale structures;
- know a bit about the history of microelectronics and the integrated circuit, including Moore's law and what is called the top-down approach;
- understand why optical lithography places a limit on how small electronic components can be made using current technologies, and the pros and cons of possible future lithographic methods;
- understand what is called the bottom-up approach;
- have some understanding of the problems of downscaling current electronics, and using current system architectures;
- have a general sense of the amount of nanotechnology-related products currently available, and of the potential for nanotechnology products in the future.

1.6 PROBLEMS

Note: You will need to use sources other than the text to answer most of the following questions.

1. Determine the current state of the art for transistor feature size using lithography, and briefly describe the lithographic technique (optical, UV, etc., and the wavelengths used).
2. Find a definition of a "quantum dot," one that you can understand, and briefly explain (in one paragraph) an application of quantum dots.
3. From (1.1), it would seem that to enhance resolution, and allow current lithography methods to pattern smaller and smaller structures, one could simply keep reducing the wavelength (i.e., raising the frequency) of sources. Describe the technological difficulties in this approach.
4. Find some information on immersion lithography, and describe in several paragraphs the general idea and its advantages and disadvantages.
5. Repeat Problem 1.4, but on the subject of electron beam and X-ray lithography.
6. Find a company involved in nanotechnology, and briefly describe what it is doing.
7. Find and describe three currently available commercial products that are based on nanotechnology concepts.
8. There are many advocates, and some opponents, of developing nanotechnology. Summarize some of the arguments used by those against nanotechnology development.

Chapter

2

CLASSICAL PARTICLES, CLASSICAL WAVES, AND QUANTUM PARTICLES

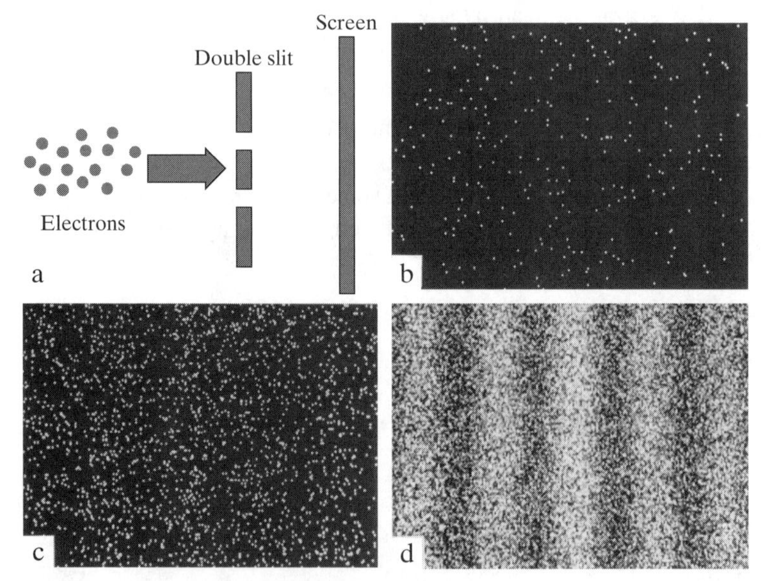

Experimental diffraction pattern of the classic double-slit experiment, applied to electrons. (Based on Tonomura, A., et al., "Demonstration of Single-Electron Build-Up of an Interference Pattern," *Am. J. Phys.* 57 (1989): 117. Courtesy of the American Institute of Physics. © Hitachi Ltd. Advanced Research Laboratory.)

This chapter provides a brief review of a few concepts from classical (Newtonian) physics, and presents some aspects of quantum phenomena that are important in understanding nanoelectronics. A few experiments are discussed that show the inadequacy of classical physics in explaining some aspects of light and matter. However, these experiments are not merely historical. They illustrate some basic properties of what will be called *quantum particles* (a designation that covers both light and electrons), and, in fact, some of these experiments point the way towards novel devices and new technologies.

2.1 COMPARISON OF CLASSICAL AND QUANTUM SYSTEMS

A classical particle is what we think of as an ordinary particle or object, such as a billiard ball, a car, or a bullet. Of course, a classical particle with mass m occupies a definite position in space, $\mathbf{r}(t)$ at a time t (this position indicates, perhaps, the center of mass of the object). For example, in rectangular coordinates, the position of a particle can be given by

$$\mathbf{r}(t) = \mathbf{a}_x x(t) + \mathbf{a}_y y(t) + \mathbf{a}_z z(t), \tag{2.1}$$

where $\mathbf{a}_x$, $\mathbf{a}_y$, and $\mathbf{a}_z$ are unit vectors along the x, y, and z coordinates, respectively. If the particle is moving along a trajectory T, it has a definite velocity, $\mathbf{v} = d\mathbf{r}(t)/dt$, as shown in Fig. 2.1, and the particle has a definite momentum, $\mathbf{p} = m\mathbf{v}$, and definite acceleration, $\mathbf{a} = d^2\mathbf{r}(t)/d^2t$. Note that in this text, vectors are denoted by boldface symbols.

Classical particles obey Newtonian mechanics,

$$\mathbf{F} = m\frac{d^2\mathbf{r}}{dt^2}, \tag{2.2}$$

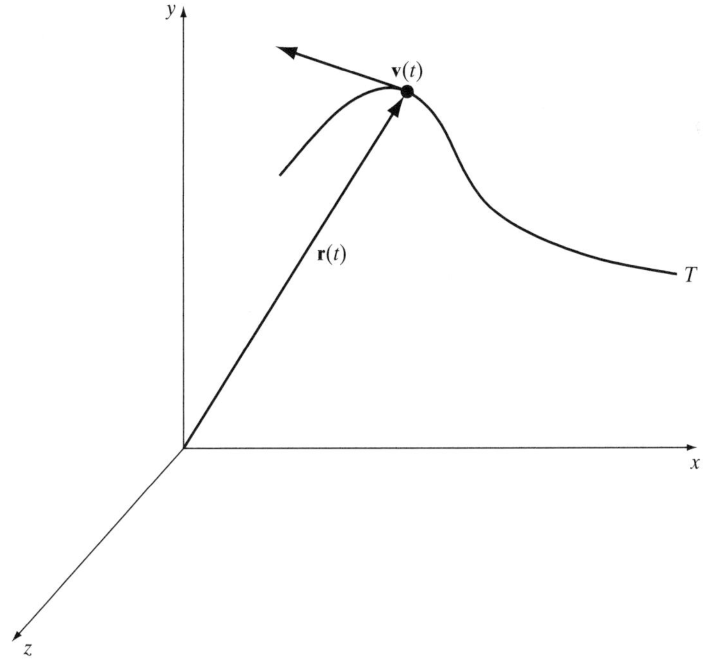

Figure 2.1 A classical particle trajectory T; the particle is located by position $\mathbf{r}(t)$ and has velocity $\mathbf{v}(t)$.

where **F** is an applied force, either mechanical or electrical. In classical physics, physical quantities such as position and momentum can, in principle, be measured with absolute certainty. Using (2.2), we can, in principle, determine with certainty the particle's trajectory for all future times from a set of initial conditions. For systems containing hundreds or perhaps billions of particles, all interacting with each other via collisions and mutual attractions and repulsions, one can still determine, in principle, how each particle will behave in space and time.

The (over)use of the words "in principle" and "definite" in the preceding paragraphs demands some explanation. The point is that the described calculations based on Newton's laws can be, in principle, performed, although it may be extremely difficult to do so. As an example, consider the seemingly simple problem of a ball being thrown through the air. This is actually a very difficult process to model because of the complicated (and often unknown) form of the force term in (2.2). Air molecules, pollution, wind, rain, etc., act as forces on the ball, making its trajectory virtually impossible to determine without simplifying assumptions. For example, perhaps we can assume that friction with air molecules can be ignored, and that there is no wind or rain. In this case, there is, perhaps, only the force of gravity. With such simplifying assumptions, the resulting equation can easily be solved, although whether or not such a simplified problem models physical reality is an important consideration.

In actuality, many of the things that were just mentioned as forces can be thought of as other particles (e.g., air molecules and rain drops), with the result that the force term in Newton's law may be simplified at the expense of keeping track of the mutual interactions among, perhaps, millions or billions of particles. However, given enough computational power, the problem can be solved. On the other hand, for systems having an extremely large number of particles (say, a gas or a liquid), one can obtain information about measurable quantities such as pressure and energy using statistical techniques, which approximate the particle physics in a tractable way.

In the first paragraph of this section, the word "definite" was also used quite a bit, perhaps so much so as to be distracting. However, that was, in fact, the point. Classical physics is characterized by being able to exactly state where a particle will be, and how fast it will be going, at a certain instant of time.† However, the actual computation may be difficult or practically impossible to perform.

Quantum mechanics, on the other hand, indicates a quite different situation. Stated briefly, quantum mechanics states that one can only know the probability of a particle being at a certain position at a certain time. It also indicates that it is impossible to measure precisely both the position and momentum of a particle—not that it is really hard to do, but that it is theoretically impossible! For example, the more accurately position is measured, the less will be known about velocity. Perhaps what is most disturbing is that quantum mechanics indicates that there is no such thing as a particle, in the classical sense. All such objects (billiard balls, bullets, electrons, etc.) exhibit properties usually associated with waves, and properties usually associated with particles, the so-called *particle–wave duality*.

†With this definition, we are somewhat glossing over things like chaotic effects, where often the system is too sensitive on initial conditions to render a definite answer in practical circumstances.

In a mathematical sense, this leads to particles in quantum theory having a phase component, which will be described in the next chapter.

There are many other disturbing (to our intuition) aspects of quantum theory, some of which will be discussed in this book. One aspect of quantum theory that should be appreciated at the utset is that quantum theory is truly a probabilistic theory. This is in contradistinction to much of what one usually considers to be probabilistic systems. Take, for example, the act of rolling a die. If we know the initial position and momentum of the die, properties of the air through which the die travels, and properties of the surface on which the die lands, etc., we could determine, in principle, what side of the die will face up when the die comes to a stop. However, statistical techniques (e.g., the probability of any given side facing up on a six-sided fair die is $1/6$) are used to model this extremely complicated, yet deterministic, problem.[†]

In contrast, quantum mechanics yields quantities that are truly probabilistic. This may indicate a truly probabilistic reality, or perhaps it is merely a probabilistic model of a deterministic reality. As we approach a century of quantum theory, there is still debate on this point; however, it can be safely stated that quantum mechanics is one of the most successful theories ever developed in physics. Philosophy aside, quantum mechanics is currently the only way to model very small (atomic and nanoscopic) objects and devices.

2.2 ORIGINS OF QUANTUM MECHANICS

There are many fine accounts of the development of quantum theory, thanks in part to the fact that quantum theory was developed relatively recently, in the early 1900s. The short and simplified story is that quantum mechanics arose out of experiments performed in the late 1800s and early 1900s that could not be explained by classical physics. In this chapter, the development of quantum theory is not described in any detail, although a few of the basic experiments are described that led to the development of quantum theory, and that, in particular, show the dual wave–particle nature of light, electrons, and, in fact, of all objects. Understanding the wave properties of electrons, that is, that electrons are not small, hard, charged balls but rather (perhaps loosely) localized bundles of energy, is absolutely essential to understanding the fundamental principles of nanoelectronics.

Key to the development of quantum theory were experiments in the 1890s that showed that the specific heat of metals, and thermal blackbody radiation, could not be explained by classical thermodynamics. In addition, in 1887, Heinrich Hertz observed what came to be known as the *photoelectric effect.* Stated briefly, if light is incident on a metal, some energy carried by the light can be transferred to electrons at the metal's surface, and they then may have enough energy to escape from the metal. Classical electromagnetic theory (itself a relatively new discipline at the time) considers light as an electromagnetic wave, and the energy carried by the wave only depends on its amplitude (or intensity), not on its frequency. So if light is indeed a wave phenomenon, experiments

[†]Of course, actually the die is, itself, a quantum object, but we can ignore this fact due to its relatively large mass.

should show that the energy of photo-emitted electrons increases as the intensity of the light increases. However, experiments performed by Philipp Lenard in 1902 showed that this was not the case. Although more electrons were emitted from the metal as the light intensity was increased, the kinetic energy of each emitted electron did not change with intensity. However, when the frequency of the light was increased (i.e., the wavelength was decreased), more energetic electrons were emitted from the metal's surface. Thus, the energy of the emitted electrons was proportional to the incident light's frequency, not to its amplitude.

Furthermore, if light is a wave, then some time should elapse between when the metal is first illuminated and when electrons are first emitted, since a wave continuously transfers energy to the electrons and it should take some time for enough energy to build up to allow the electrons to escape. This was also not found to be the case—sometimes electrons were emitted as soon as the metal was illuminated by the light. There was clearly some problem in how this interaction was being modeled by classical physics.

2.3 LIGHT AS A WAVE, LIGHT AS A PARTICLE

As mentioned above, at the time of the experiments involving the photoelectric effect, light was generally considered to be a wave phenomenon. It is worthwhile to take a brief detour to discuss how this state of knowledge developed.

2.3.1 Light as a Particle, or Perhaps a Wave—The Early Years

Investigations into the nature of light have a long history, not surprisingly. One of the earliest accounts is by Euclid around 300 B.C., who, in his work *Optica*, noted that light travels in straight lines, and described the law of reflection. Jumping far ahead, we find that the 1600s were a particularly important period in the history of light, with such figures as René Descartes, Pierre de Fermat, and Robert Hooke contributing to the field. In 1678, Christiaan Huygens developed a fairly comprehensive wave theory of light, which was very successful in explaining many of the characteristics of light known at the time. Then, in 1704, Isaac Newton, in his work *Opticks*, put forward his view that light is corpuscular in nature, based primarily on the appearance of light traveling in straight lines (since waves can bend around objects to some degree), although he also discussed a wave theory of light. Both the corpuscular and wave theories of light could seemingly be used to explain much of the light phenomena known at the time, such as reflection and refraction.

2.3.2 A Little Later—Light as a Wave

In 1801, Thomas Young performed his famous double-slit experiment, showing evidence for the wave theory of light by demonstrating interference. He was motivated by his earlier work with sound waves, which are known to interfere with each other. In order to grasp the importance of Young's experiment, the concept of interference must first be understood.

Interference. Interference obviously describes the interaction among two or more entities. Let's first consider the familiar example of unit amplitude plane waves, represented as

$$\psi(z) = e^{ikz}, \tag{2.3}$$

where k is a constant called the *wavenumber* (to be discussed later), $i = \sqrt{-1}$ is the imaginary unit, and z is the position coordinate, in this case, the distance that the wave has traveled. We use the symbol ψ to represent the wave, since this notation is generally used in quantum mechanics.

Assume that two plane waves are present, as shown in Fig. 2.2, with the wave represented by ψ_1 emanating from the origin and the wave ψ_2 from the position $z = -L$. Wave ψ_1 travels a distance d to reach the measurement plane, whereas wave ψ_2 must travel a distance $L + d$ to reach the same point. At the position of the measurement plane, the total field ψ_T (i.e., the sum of the waves) is

$$\begin{aligned}\psi_T &= \psi_1 + \psi_2 = e^{ikd} + e^{ik(L+d)} \\ &= e^{ikd}\left(1 + e^{ikL}\right).\end{aligned} \tag{2.4}$$

If

$$kL = 0, 2\pi, 4\pi, \ldots, \qquad |\psi_T| = 2, \tag{2.5}$$

and if

$$kL = \pi, 3\pi, 5\pi\ldots, \qquad |\psi_T| = 0. \tag{2.6}$$

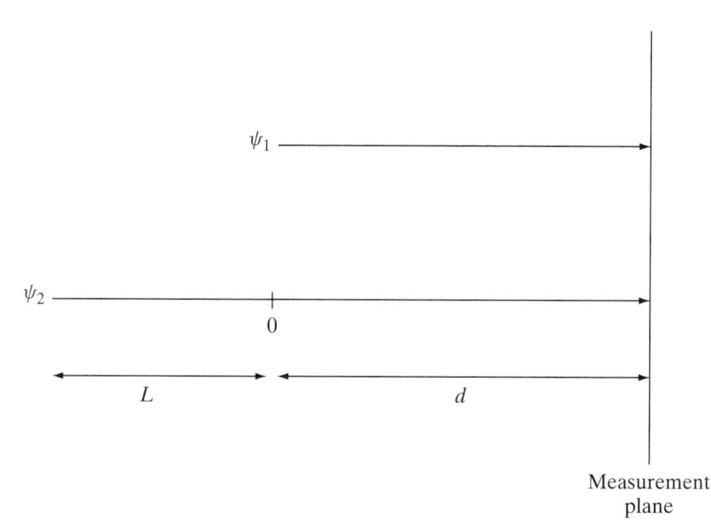

Figure 2.2 Plane waves traveling different distances, resulting in constructive/destructive interference.

That is, at points where the two waves are completely out of phase with each other (i.e., their phase difference is an odd multiple of 180°), they cancel each other out, which is called *destructive interference*. At points where the phase difference is an even multiple of 180°, they add together, doubling their value, which is called *constructive interference*. Therefore, depending of the distance L, the two waves may experience destructive or constructive interference at the measurement plane, or something in between. One can also obtain constructive and destructive interference by combining forward and backward traveling waves,

$$\psi_T = e^{ikz} + e^{-ikz} = 2\cos(kz), \tag{2.7}$$

or

$$\psi_T = e^{ikz} - e^{-ikz} = 2i\sin(kz), \tag{2.8}$$

using Euler's identity ($e^{\pm i\alpha} = \cos\alpha \pm i\sin\alpha$), such as occurs for vibrational waves on guitar strings.

Moreover, more than two plane waves, or other wave types, can be combined to yield constructive and destructive interference. A common example of light interference is a soap film.

The wall thickness of a typical soap film is several microns. The difference in the distance that light must travel in reflecting from the film's top surface, and from the film's bottom surface, is analogous to the difference in distance L depiction in Fig. 2.2. Thus, light reflected from the top surface of the soap film can interfere both constructively and destructively with light reflected from the bottom surface of the film. Since white light is composed of many wavelengths, at any point on the film where wavelengths (i.e., colors) interfere constructively, those colors will be intensified. At points on the film where wavelengths combine destructively, those colors will be suppressed. Thickness variations and other irregularities of the film's surface contribute to the formation of interesting patterns.

Young's Experiment—What to Expect from Waves. Now, getting back to Young's experiment, imagine that two narrow slits are cut into a thin sheet of opaque material, such as metal. A single-frequency (monochromatic) plane wave of light ψ is normally incident on the double slits, as shown in Fig. 2.3.

According to Huygens's principle, each slit reradiates a spherical wave centered on the slit, having the same frequency as the original wave. The waves emanating from slit 1 (the top slit) and slit 2 (the bottom slit) will be denoted by ψ_1 and ψ_2, respectively,

$$\psi_j = A_j \frac{e^{ikr_j}}{r_j}, \tag{2.9}$$

where A_j, $j = 1, 2$, is the amplitude of the wave (if the two slits are equal in size, $A_1 = A_2 = A$), and where r_j is the distance from the center of the jth slit to a given point on the

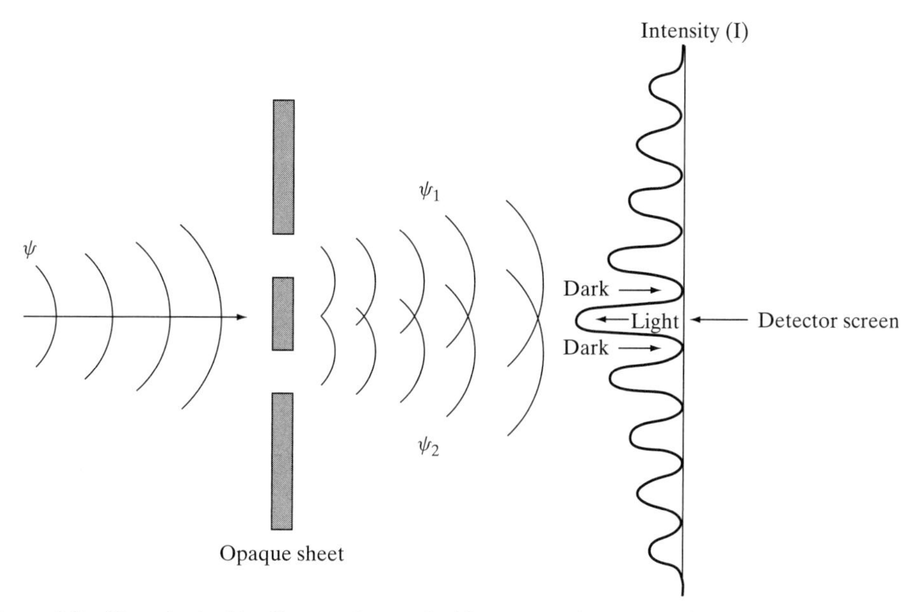

Figure 2.3 Young's double-slit experiment. Incident wave ψ causes spherical waves ψ_1 and ψ_2 to travel toward the detector screen. The resulting intensity pattern I shows interference effects, characteristic of wave behavior.

detector screen. The spherical waves ψ_1 and ψ_2 combine together, such that at any point on the detector screen,

$$\psi_T = A\left(\frac{e^{ikr_1}}{r_1} + \frac{e^{ikr_2}}{r_2}\right), \tag{2.10}$$

forming an interference pattern on the screen (detector) placed behind the slits. That is, at some points on the detector screen the waves ψ_1 and ψ_2 tend to cancel each other out, resulting in destructive interference and a dark patch on the screen. On the other hand, if the waves combine together constructively, a bright patch occurs on the screen.

The location on the screen that lies exactly opposite to the center point of the two slits will be associated with a bright spot, since the two waves travel the same distance from each slit to reach this point. Therefore, they are in phase at this point. Whether or not a particular location on the screen is bright or dark depends on the path difference that the two waves travel to reach the screen. The locations of bright and dark spots can be calculated if the wavelength of the light is known, although the details will be omitted here. However, the resulting bright and dark bands on the screen depicted in Fig. 2.3 show the interference effect, and seemed to be, at the time, conclusive evidence that light is a wave phenomenon. In 1816, Augustin Jean Fresnel presented a mathematically rigorous treatment of the diffraction and interference of light, showing that these phenomena can be explained mathematically by a wave theory of light. Several years later, in 1849, the (seemingly) final death blow to the particle theory of light was dealt by experiments showing that light propagates more slowly through water than though air—the particle theory of light at the time could only account for the law of refraction if light propagated faster through a dense medium, such as water, than through a more rarefied medium, such as air. So, by the mid 1800s, it seemed clear that light was a wave.

Around the same time, pioneers in electromagnetic theory, such as André Marie Ampère, Michael Faraday, and Carl Gauss, were developing the fields of electricity and magnetism, culminating in the development of Maxwell's equations by James Clerk Maxwell in the 1860s. It was found that the speed of an electromagnetic wave is the same as the speed of light, explaining light as an electromagnetic wave. From Maxwell's equations, one can show that the wave nature of light is revealed when the characteristic dimensions of a system are on the order of, or smaller, than the wavelength of light. When the opposite is true, that is, when the system's dimensions are large compared to wavelength, light appears to exhibit ray-like qualities. This explains why the particle versus wave debate went on for such a long time in the development of optics—in some situations, light does act like a particle, or at least a ray, traveling in straight-line trajectories, and in other situations, light does behave as a wave, exhibiting diffraction.

Therefore, by the late 1800s, the matter seemed settled. A comprehensive mathematical theory had been developed (electromagnetics) showing that light was a wave, and explaining light's ray-like behavior at high frequencies. Electromagnetic theory then continued to be developed, and in the 1890s and early 1900s, Heinrich Hertz, Guglielmo Marconi, and others developed communications aspects of electromagnetic waves, leading to today's radio, telephone, television, and radar, and to a host of other applications. However, at about the same time when it was thought that the debate about the nature of light (and all electromagnetic energy) was finally settled, a scientific revolution was about to take place.

2.3.3 Finally, Light as a Quantum Particle

Before describing some problems with the wave theory of light, we need to consider the double-slit experiment in a bit more detail, as well as a similar experiment using particles.

Young's Experiment—One Slit at a Time. Assume as before that a single-frequency plane wave of light, ψ, is incident on the double-slit apparatus, as shown in Fig. 2.4. However, this time we will consider what happens when only one slit is open.

If first only slit 1 is open, the intensity pattern $I_1 = |\psi_1|^2$ is seen on the screen, as shown in the figure. In a similar manner, if only slit 2 is open, then $I_2 = |\psi_2|^2$. However, if both slits are open, we obtain the (interference) intensity pattern $I = |\psi_1 + \psi_2|^2$ shown previously in Fig. 2.3, which is not the same as $I_1 + I_2$. Note that, in general,

$$|\psi_1|^2 + |\psi_2|^2 \neq |\psi_1 + \psi_2|^2 = |\psi_1|^2 + |\psi_2|^2 + 2Re\left(\psi_1^*\psi_2\right), \tag{2.11}$$

where ψ^* is the complex conjugate of ψ and $Re\,(z)$ indicates that we take the real part of the complex quantity z. Thus, when the absolute value of the wavefunction is taken, its phase information is lost. Therefore, since interference effects are a result of phase differences of

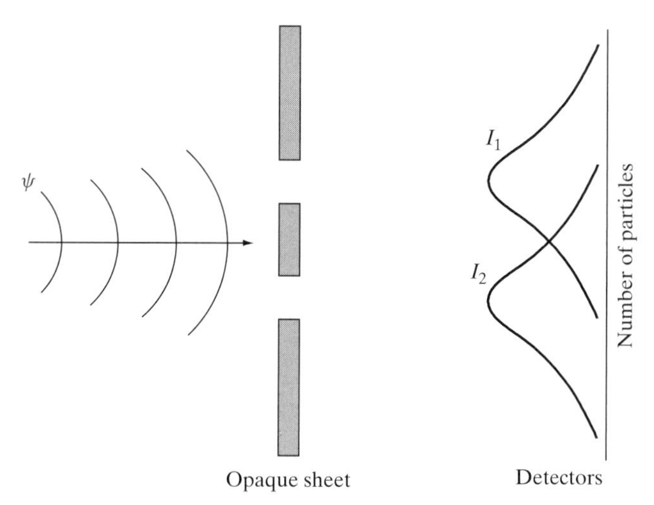

Figure 2.4 Modification of Young's double-slit experiment, when only one slit is open. Intensity pattern I_1 results when slit 1 (the top slit) is open and slit 2 (the bottom slit) is closed, and intensity pattern I_2 results when slit 1 is closed and slit 2 is open.

the two waves involved, ψ_1 and ψ_2 need to be added together before taking the absolute value to obtain the correct result.† So far, all seems well.

Young's Experiment—What to Expect with Classical Particles. Now consider the same experiment, but with classical particles. On the far left of the two slits, a source fires particles at the slits in varying directions, but with the same energy. The screen is now an array of closely spaced particle detectors, and the intensity of the particle beam at any point along the screen will be the number of particles arriving at that point per second. The situation is depicted in Fig. 2.5.

When only one of the slits is open, the intensity pattern I_1 or I_2 is obtained, which resembles those obtained in the similar wave experiment of Fig. 2.4. If both slits are open simultaneously, for classical particles we expect that the intensity pattern is merely $I_1 + I_2$, as shown by the dashed line in Fig. 2.5, since each particle moves along a certain trajectory, either through one slit or the other (or bouncing backwards if they weren't heading for a slit). The fact that either one slit, or both slits, are open makes no difference to any given particle.

The particle beam can be sufficiently sparse, such that only one particle is passing through the slits at any given time, so that they don't interfere (collide) with each other. Again, so far, all seems well.

Young's Experiment and the Concept of Photons—One More Time with Light, but Slowly. Armed with this knowledge, we now reconsider the double-slit experiment with light. Rather than simply watching for bright and dark spots on a screen placed

†As a simple example, consider two plane waves $\psi_1 = e^{ikz}$ and $\psi_2 = e^{-ikz}$. Then, $|\psi_1 + \psi_2| = 2\,|\cos(kz)|$, showing constructive and destructive interference at certain points z. However, $|\psi_1| + |\psi_2| = 1 + 1 = 2$, independent of position.

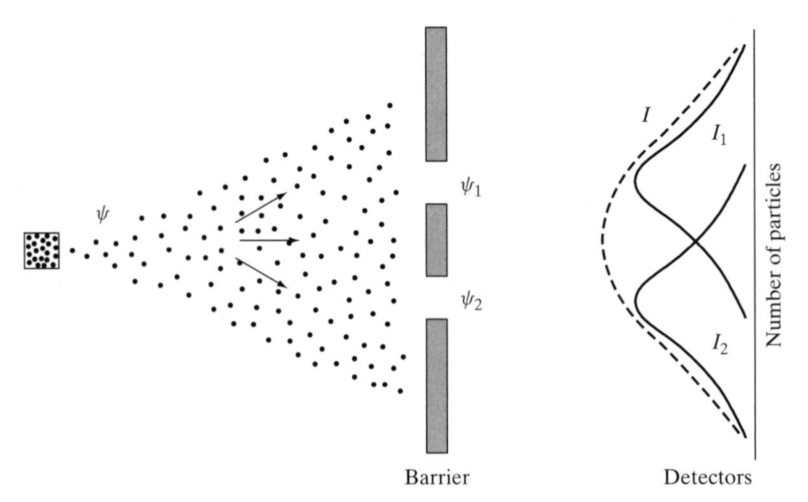

Figure 2.5 Young's double-slit experiment involving classical particles. When one slit is open the individual intensity pattern I_1 or I_2 is obtained. When both slits are simultaneously open, the intensity pattern $I = I_1 + I_2$ results, shown as the dashed line.

behind the slits, we use optical detectors that register the presence of optical energy. With only slit 1 open, we reduce the intensity of the incoming light. If the intensity is reduced enough, so that only one detector is active at any time, we would find that energy is not arriving continuously, but in discrete bursts, pointing to a particle-like nature of light. (It is only because these energy bundles are usually arriving at such a fast rate that we ordinarily don't notice their discrete nature.)

A few years before Lenard's experiment on the energy of photo-emitted electrons, Max Planck had satisfactorily explained blackbody radiation by proposing that the energy emitted by a blackbody should be quantized (i.e., energy should occur in discrete units), such that

$$E = hf = \hbar\omega, \tag{2.12}$$

where h is a constant (now called *Planck's constant*) and f is frequency (ω is radian frequency, $\omega = 2\pi f$), where

$$\begin{aligned} \hbar &= h/2\pi, \\ h &= 6.6261 \times 10^{-34} \text{ Js}, \\ \hbar &= 1.0546 \times 10^{-34} \text{ Js}. \end{aligned} \tag{2.13}$$

In 1905, Albert Einstein explained the photoelectric effect by applying Planck's quantization to all electromagnetic waves, and describing for the first time what is now called a *photon*.

It is interesting that Einstein won the 1921 Nobel prize in physics for his explanation of the photoelectric effect, and not for his theory of relativity, which he also proposed in 1905.

So the discrete nature of light is now recognized as quantized electromagnetics waves, called photons. For the photon single-slit experiment using very low-intensity light (i.e., a very sparse photon stream), over time the intensity pattern I_1 shown in Fig. 2.4 will emerge. We are back to something of a particle theory of light.

Young's Experiment—A Very Strange Result Concerning Interference. Now let both slits be open, and use a light intensity that is so low that only one photon is present at any given time (as in the classical particle experiment). Recording the detector output over time, if photons behaved as classical particles, the resulting intensity pattern should be simply $I_1 + I_2$, the result expected for classical particles shown in Fig. 2.5. Instead, we find that the resulting intensity pattern shows interference effects, depicted in Fig. 2.3, and that the observed pattern corresponds to the pattern expected based on a wave theory! This observation shows that photons are not classical particles. The photons pass through the slits *individually*, but somehow interfere *with themselves* in the process. This is a fairly bad situation, to be sure, from the point of view of understanding light in terms of everyday experiences.

Perhaps the most disturbing thing to consider is that at a point where the interference pattern has a minimum, more photons will arrive at that spot with only one slit open than if both slits are open at the same time. Since the intensity is so low that only one photon is present at a time, opening an additional pathway to the detectors could not possibly reduce the number of photons arriving at a certain spot, if photons were really particles. So we are left with a view that light exhibits wave-like and particle-like behavior, but is clearly *neither* a classical wave nor a classical particle—we call this a *quantum particle*. In a sense, one must accept that light is "its own thing"; light has the properties of, well, light. That is, light† does not have an analogous counterpart to an everyday, familiar object, which makes its interpretation somewhat difficult and unsatisfying. However, it often suffices to know the properties of light, as partially described in the preceding discussion, in order to interpret experiments and predict behavior.

2.4 ELECTRONS AS PARTICLES, ELECTRONS AS WAVES

2.4.1 Electrons as Particles—The Early Years

Although the concept of electrical charge has been around for most of recorded history, it was only in the early 1800s that the existence of atoms was first established. It was, at first, generally thought that an atom was an indivisible particle. The concept of the electron, as a constituent of an atom, was first developed in the late 1800s, due to experiments by

†Recall that all electromagnetic energy (radio waves, light, X-rays, etc.) is the same phenomenon, it is only a matter of different frequencies of oscillation.

John Joseph (J. J.) Thomson. Shortly after, in the early 1900s, Robert Millikan measured the charge of an electron. Therefore, the concept of the electron was fairly new at the time that the photoelectric effect was puzzling scientists.

As the concepts of electrons and atomic structure came about, electrons were naturally thought to be simply very small charged particles. One of the early successes of this model was the prediction of the conductivity of metals, the so-called free *electron gas model.*[†] In fact, this classical model will be used to some degree later on. However, the quantization of light led to the possibility of other phenomena being quantized.

2.4.2 A Little Later—Electrons (and Everything Else) as Quantum Particles

The fact that light is made up of photons lead Louis de Broglie in 1923 to make the radical suggestion that all "particles" having energy E and momentum p should have wavelike properties, too. He proposed that associated with each particle of momentum p is a wave having wavelength (now called the *de Broglie wavelength*)

$$\lambda = \frac{h}{p}, \tag{2.14}$$

so that

$$p = \hbar\frac{2\pi}{\lambda} = \hbar k, \tag{2.15}$$

where k is the wavenumber. When applied to matter, like electrons, these waves are called *matter waves*. De Broglie's prediction was verified for electrons (somewhat accidentally) by Clinton Davission and Lester Germer in 1926 to 1927 using crystal diffraction (not double-slit) techniques. In fact, although double-slit-like experiments with electrons have long been discussed, their successful implementation, using one electron at a time, did not occur until the 1970s. These experiments were repeated with more accuracy in the 1980s. Figure 2.6 shows an experimental demonstration of electron diffraction interference effects using a very sparse stream of electrons. Frames (a)–(d) represent detected electrons at increasingly later times. At first, the detected pattern of electrons seems random. As time increases, the apparent random distribution of electrons begins to form a clear diffraction pattern, in exactly the same way as discussed previously for photons. Since in the experiments electrons pass through the apparatus one at a time, an electron must clearly interfere with itself.

Note that energy quantization, $E = \hbar\omega$, where E is the total (kinetic plus potential) energy, applies to matter waves as well as photons. Since momentum is really a vector quantity, in general,

$$\mathbf{p} = \hbar\mathbf{k}. \tag{2.16}$$

[†]Ohm's law, which states that for many materials the amount of electrical current passing through the material is proportional to the applied voltage, $I = V/R$, where $1/R$ is the proportionality constant, was discovered much earlier, around 1827.

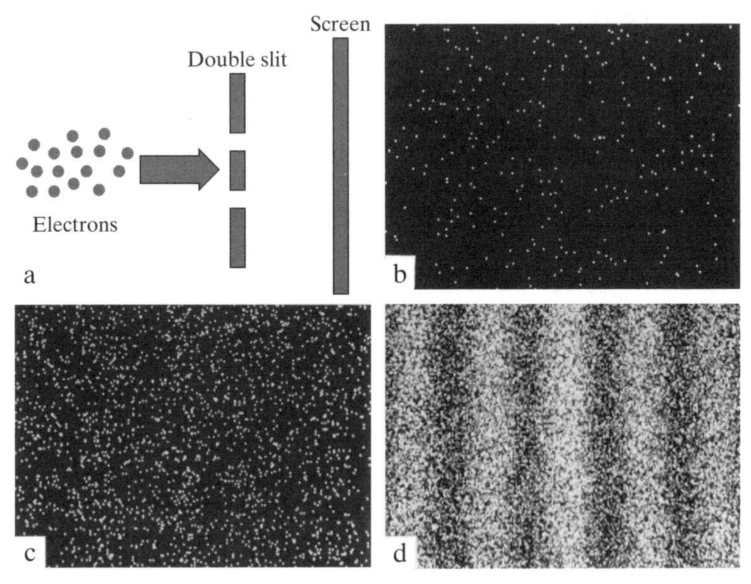

Figure 2.6 Experimental demonstration of interference effects in electron diffraction. Frames (a)–(d) represent increasing time, over which the apparent random distribution of electrons forms a diffraction pattern. (Based on Tonomura, A., et al., "Demonstration of Single-Electron Build-Up of an Interference Pattern," *Am. J. Phys.* 57 (1989): 117. Courtesy of the American Institute of Physics. © Hitachi Ltd. Advanced Research Laboratory.)

The reason we don't see everyday objects behaving as waves is the small numerical value of Planck's constant, given by (2.13). With

$$\lambda = \frac{h}{p} = \frac{h}{mv} \tag{2.17}$$

for objects with mass, it can be seen that, since h is extremely small, m must be very small in order to obtain a wavelength large enough to observe wave-like interactions with physical systems of interest. For example, to see wave effects on the scale of nanometers, we need $\lambda \sim$ nm, which can occur for electrons and other subatomic particles that are extremely light, but generally not for heavier particles. In fact, it is often difficult to observe wave effects for particles heavier than electrons. For example, consider a 1 kg mass moving at 1 m/s. Then, the de Broglie wavelength is

$$\lambda = \frac{h}{p} = \frac{6.6 \times 10^{-34}\ \text{Js}}{(1\ \text{kg})\,(1\ \text{m/s})} = 6.6 \times 10^{-34}\ \frac{\text{Js}^2}{\text{kg m}} = 6.6 \times 10^{-34}\ \text{m}, \tag{2.18}$$

which is much too small to lead to observable wave phenomena.† In particular, when λ is small compared to the distance over which potential energy changes, quantum particles

†As mentioned previously, wave effects such as diffraction are most easily seen when the wavelength is on the order of, or larger than the characteristic dimensions of the system in question.

behave in a classical manner (i.e., quantum effects are too small to notice). In fact, many aspects of classical physics can be obtained by setting $h = 0$ in the various expressions obtained from quantum theory (note that $h = 0$ decouples energy and frequency). Therefore, the 1 kg mass is far too heavy to be able to observe wave effects, and so it acts like a classical particle. However, if the mass was 10^{-31} kg, then $\lambda = 6.6 \times 10^{-3}$ m, and wave effects would be observable in systems on the scale of millimeters. To date, wave effects have been observed with objects ranging from subatomic particles (e.g., electrons) to relatively heavy 2 nm molecules.

As a comparison, the wavelength of a 1 eV photon is[†]

$$\lambda_p = \frac{hc}{E} = 1.24 \ \mu\text{m}, \tag{2.19}$$

since $E = pc$ for photons, whereas the wavelength of a 1 eV electron with only kinetic energy is

$$\lambda_e = \frac{h}{\sqrt{2m_e E}} = 12.3 \ \text{Å}. \tag{2.20}$$

Therefore, these different particles "see" different worlds, and, in particular, electrons generally exhibit wave effects at atomic dimensions, whereas photons show wave effects at micron dimensions.

Furthermore, to get an idea of the large number of photons usually present in light, assume a light wave carrying 1 μW of power (W = J/s) at $\lambda = 600$ nm. Then, each photon carries

$$E_p = \hbar\omega = \hbar\pi \times 10^{15} \ \text{J} \tag{2.21}$$

of energy. Let N be the number of photons per second (so the unit of N is 1/s). The sum of all N photons has power

$$P = NE_p \ (1/\text{s}) \ (\text{J}) = 1\mu \ \text{J/s},$$

so that

$$N = 3 \times 10^{12} \qquad \text{photons/second.}$$

The granularity of this flow is virtually unobservable. This is usually the case, unless great effort is taken to reduce the intensity, and, hence, the number of photons. This is why the discrete nature of light is not easily observed in everyday situations.

[†]eV is *electron volt*—an electron volt is the energy gained by one electron after moving through a potential difference of one volt. Hence, 1 eV= 1.6×10^{-19} J.

2.4.3 Further Development of Quantum Mechanics

In 1925, Erwin Schrödinger gave a talk on de Broglie's work at a seminar in Zürich, after which Peter Debye commented that "to deal properly with waves one had to have a wave equation." While on a vacation in the Swiss Alps one month later, Schrödinger, seemingly motivated by Debye's comment, worked out what came to be known as wave mechanics. This involved a wavefunction ψ that satisfies a wave equation, now called *Schrödinger's equation*.

At first, Schrödinger did not know what the wavefunction actually was, and referred to it by the rather cryptic name of " mechanical field scalar." He at first thought that electrons were not particles at all, but that their particle-like qualities were merely manifestations of a pure wave phenomena involving wavepackets, which are superpositions of waves. In 1927, around the time of Schrödinger's fourth paper on the new wave mechanics, Max Born hypothesized that for each quantum particle there is an associated wave ψ, Schrödinger's wavefunction, the modulus squared ($|\psi|^2$) of which gives the probability of finding the particle at a certain location.

Therefore, in the double-slit experiment, we have the situation where a sparse stream of photons, each with the same energy and momentum, and, hence, the same probability wavefunction ψ, is directed one at a time towards the slits. The wave ψ of each photon interferes *with itself* in passing through the two slits (you could say that the photon passes through both slits at the same time), such that the probability of finding the photon along the row of detectors forms the observed oscillating interference pattern. Of course, a given photon must hit the detectors at some definite point, and so the interference pattern can't be seen by the arrival of one photon. However, over time, the fact that at some locations the probability of finding a photon is small, and at other locations large, leads to the interference pattern ultimately observed. By de Broglie's result, the same is true for electrons, billiard balls, etc. Thus, in a double-slit experiment involving electrons, the electrons are said to pass simultaneously through both slits.

It should be noted that just before Schrödinger's work, in 1925, Werner Heisenberg developed an approach to quantum theory involving sets of complex numbers. Max Born and Pascual Jordan recognized these sets as matrices, and, together with Heisenberg, reformulated the theory as matrix mechanics. In 1926, Schrödinger showed that his method and the method of matrix mechanics were equivalent. In this text, we use Schrödinger's approach because it is a bit more appealing to scientists who are accustomed to dealing with waves. However, perhaps even more important than his matrix mechanics, in 1927, Heisenberg formulated what is now called the *Heisenberg uncertainty principle*, which states that one cannot simultaneously measure the position and momentum of quantum particles with arbitrary precision. Written mathematically, the Heisenberg uncertainty principle is expressed as

$$\Delta p \Delta x \geq \hbar/2. \tag{2.22}$$

This idea is quite important in quantum theory, and it will be used later.

TABLE 2.1 MASS AND CHARGE OF SOME FERMIONS.

Particle	Mass	Charge
electron	$m_e = 9.1095 \times 10^{-31}$ kg	$q_e = -1.6022 \times 10^{-19}$ C $= -e$
proton	$m_p = 1.6726 \times 10^{-27}$ kg	$e = 1.6022 \times 10^{-19}$ C
neutron	$m_n = 1.6750 \times 10^{-27}$ kg	—

Two other key points of quantum theory remain to be discussed, although their historical origins will only be briefly mentioned here. First, in the mid-1920s, it was found that electrons behave in a magnetic field as if they had angular momentum. This phenomenon was initially attributed to the electron spinning about its own axis, and so this quantity was called *spin*. It was quickly determined that the electron was not, in fact, spinning about its own axis, but that spin was an intrinsic quantity associated with the electron. Spin was found to be quantized in multiples of $\hbar$. Particles with integral (in units of $\hbar$) spin are called *bosons*. (Examples are photons and quantized lattice vibrations called *phonons*.) Particles with half-integral spin are called *fermions*. (Examples are electrons, protons, neutrons, quarks, and neutrinos.)[†] Spin has no classical analogue since, although classical particles can possess angular momentum due to orbital motion, or literal spinning (rotating) about an axis, classical objects do not have any intrinsic spin. The typical advice for students learning quantum physics is to accept spin as just another intrinsic property of an object, and not to become bogged down in trying to envision spin in terms of everyday phenomena.

The other key quantum development, also having no classical analogue, was the *Pauli exclusion principle*, proposed in 1925 by Wolfgang Pauli to account for the observed patterns of light emission from atoms. The exclusion principle was quickly generalized to include all quantum particles with nonintegral values of spin, i.e., fermions.

> The Pauli exclusion principle states that two or more identical fermions cannot occupy the same quantum state.

The importance of this seemingly mild statement cannot be overestimated, as this characteristic of fermions is the basis for understanding most materials, including the development of the periodic table, and characteristic properties of insulators, conductors, and semiconductors.

For later reference, the mass and charge of some fermions are listed in Table 2.1.

2.5 WAVEPACKETS AND UNCERTAINTY

Quantum particles (light, electrons, bowling balls, etc.) can be thought of as, in some sense, quantized bundles of energy $E = \hbar\omega$, having wave-like properties (frequency, ω,

[†]Furthermore, any object that is made up of even number of fermions is a (composite) boson, whereas any particle that is made up of an odd number of fermions is a (composite) fermion. Therefore, hydrogen, with one electron and one proton, is a boson. Helium, with two each of electrons, protons, and neutrons, is also a boson.

and wavelength, λ), and particle-like properties (momentum, p) that are interrelated—the so-called wave–particle duality. For example, electrons have a finite and definite charge and mass (in a nonrelativistic sense, the rest mass), which seems like a particle property, although electrons also have a de Broglie wavelength, and can exhibit wave-like diffraction (usually at atomic length scales). Photons have no mass or charge, but can be thought of as "pure energy." As described previously, photons exhibit both wave-like and particle-like behavior.

One attribute of classical particles is that they have a certain trajectory in moving through space that can be described using Newton's laws. For example, in one dimension, a particle occupies a certain position $x(t)$ at a certain time t. On the other hand, consider a typical plane wave

$$\psi(t, x) = Ae^{-i(\omega t - kx)}, \tag{2.23}$$

in one dimension, where A is the amplitude, ω is the radian frequency, and k is the wavenumber ($k = 2\pi/\lambda$). Wave propagation (classical electromagnetic fields, pressure, displacement, etc.) can often be characterized by a plane wave ψ that extends over a region of (or all of) space, rather than being localized to a single point. The field is a function of position x and time t as independent variables (although k and ω are interrelated). Wave–particle duality would seem to imply that quantum particles will not be localized at a single point, like a classical particle, nor spread out over all space, like a classical plane wave, but will be something in between these two cases. For instance, free electrons in space usually have energies that make the de Broglie wavelength very small, such that diffraction and interference effects can often be ignored. Viewed from a distance large compared to its wavelength, an electron appears like a particle. Viewed from a distance small compared to its wavelength, usually atomic dimensions, the " spread" of the electron becomes evident. At sufficiently low energies the wave nature of the electron becomes evident over larger space scales, since momentum becomes smaller and so, by (2.14), wavelength becomes larger.

One way to model this dual behavior is with a *wavepacket*, which is a wave that is both propagating and localized in space and time. If viewed from a sufficiently large distance (relative to wavelength), the wavepacket looks like a particle moving along some trajectory. Viewed from a sufficiently small distance, one can see the " spread" of the wavepacket. Therefore, the wavepacket has attributes of both waves and particles, which is obviously something that is needed for quantum particles. Although wavepackets won't be used explicitly in the detailed examples later in the text, the basic ideas will be presented here as a conceptual tool. In addition, this leads to the important concepts of phase and group velocity.

To understand wavepackets, one first needs to appreciate some aspects of waves. Consider the single frequency plane wave given by (2.23), and recall that frequency and wavenumber are related, that is, $\omega = \omega(k)$ or $k = k(\omega)$. For example, for a photon, the familiar relationship between velocity, wavelength, and frequency,

$$c = \lambda f, \tag{2.24}$$

where c is the speed of light, leads to

$$c = \lambda \frac{\omega}{2\pi} \rightarrow \omega = \frac{2\pi}{\lambda} c = kc,$$

such that

$$\omega(k) = ck. \tag{2.25}$$

For a particle with mass m and only kinetic energy,

$$E = \hbar\omega = \frac{1}{2}mv^2 = \frac{p^2}{2m} = \frac{(\hbar k)^2}{2m}, \tag{2.26}$$

such that

$$\omega(k) = \frac{\hbar k^2}{2m}. \tag{2.27}$$

Relationships between frequency and wavenumber such as (2.25) and (2.27) are called *dispersion relations*.

The *phase velocity* of the plane wave (2.23) is the velocity of a constant phase (and amplitude, in this case) planar wavefront. Therefore, setting the phase term equal to a constant C,

$$\omega t - kx = C, \tag{2.28}$$

and differentiating with respect to time leads to[†]

$$\omega - k\frac{dx}{dt} = \omega - kv_p = 0, \tag{2.29}$$

such that the phase velocity is given by

$$v_p = \frac{\omega}{k}. \tag{2.30}$$

For the plane wave (2.23) this is the only idea of velocity. However, (2.23) describes a wave spread out over all space and time. It certainly has wave behavior, but it does not resemble a particle (localized energy bundle).

[†]We said previously that t and x are independent in the wave picture. This is true, in the manner in which they are used in (2.23). However, once we set the phase term $\omega t - kx$ equal to a constant (to "watch" a specific point on the waveform move), then, as t increases, x must increase to keep $\omega t - kx$ constant. Therefore, $x = x(t)$ for purposes of keeping track of the advancement in time of the constant-phase point.

Now, instead of a single plane wave, consider the quantity

$$\psi(x,t)=\int_{-\infty}^{\infty}a(k)\,e^{-i(\omega(k)t-kx)}dk, \tag{2.31}$$

where the dependence $\omega=\omega(k)$ is explicitly included. It should be first noted that for any value of k, $e^{-i(\omega(k)t-kx)}$ is merely a plane wave having wavenumber k and related frequency $\omega=\omega(k)$, and $a(k)$ is just a number, interpreted as the amplitude of the plane wave. Therefore, the integrand of (2.31) represents plane waves of varying amplitudes and wavenumbers, and the integration is simply a summation of those plane waves.

It can be shown that (2.31) can represent a wave localized in space, i.e., a wavepacket (like a bundle of waves). For example, assume

$$\begin{aligned} a(k) &= 1, \qquad k_0-\Delta k\le k\le k_0+\Delta k, \\ &= 0, \qquad \text{elsewhere,} \end{aligned} \tag{2.32}$$

as shown in Fig. 2.7.

With this form for $a(k)$, one can interpret (2.31) as a summation of waves with wavenumbers within some Δk range of a given value k_0. The integral (2.31) becomes

$$\psi(x,t)=\int_{k_0-\Delta k}^{k_0+\Delta k}e^{-i(\omega(k)t-kx)}dk, \tag{2.33}$$

which can't be evaluated unless the dispersion relation $\omega(k)$ is known.

It is useful to consider several simple cases, beginning with the situation, as for photons in free space, where ω is linear in k, $\omega=ck$. Then, (2.33) becomes

$$\psi(x,t)=\int_{k_0-\Delta k}^{k_0+\Delta k}e^{-ik(ct-x)}dk \tag{2.34}$$

$$=2\Delta k\,e^{-ik_0(ct-x)}\frac{\sin(\Delta k\,(ct-x))}{\Delta k\,(ct-x)}, \tag{2.35}$$

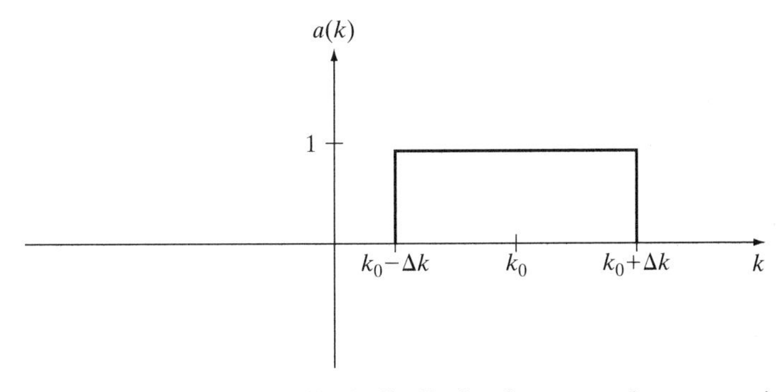

Figure 2.7 Rectangular amplitude distribution for wavepacket construction.

which is a wavepacket moving with velocity c (note that $v_p = \omega/k = c$) and having an envelope proportional to

$$\operatorname{sinc}\left(\Delta k\left(ct - x\right)\right) \equiv \frac{\sin\left(\Delta k\left(ct - x\right)\right)}{\Delta k\left(ct - x\right)}. \tag{2.36}$$

The function $\operatorname{sinc}(x) = \sin(x)/x$ is centered at $x = 0$ and decays away from that point, as shown in Fig. 2.8.

Therefore, the wavepacket (2.35) is concentrated around the point $x = ct$ at a given time t, and moves at the phase velocity $v_p = c$. Referring to the discussion of quantum particles, it seems reasonable to model a quantum particle as being associated with a wavepacket, since a wavepacket exhibits wave-like behavior and, when viewed from a distance large compared to the spread of the envelope (the sinc function), it resembles a particle, since it is somewhat localized in space.

As an interesting aside, the behavior of the sinc function can be used to gain insight into the Heisenberg uncertainty principle. Consider the function $\operatorname{sinc}(\Delta kx)$. If the spread in wavenumbers Δk is very small, then, since $p = \hbar k$, the momentum varies over a small range. Due to the argument of the sinc function, the wavepacket envelope is spread out over a large range of space, x. So a small spread in momentum indicates a large spread in position. As Δk increases, the wavepacket will become more concentrated in space.

In particular, the function $\operatorname{sinc}(\Delta kx)$ is rapidly decreasing outside some range Δx centered at $x = 0$, and one can consider the wave to be "contained" within this region of space. It is useful to choose the width of the wavepacket in space to extend to the points where the amplitude of the wavepacket's envelope decreases to 63 percent of its initial value. That is, the spatial extent of the wavepacket is described by Δx such that

$$\Delta k \Delta x = \frac{\pi}{2}. \tag{2.37}$$

Of course, this choice is somewhat arbitrary, but is one of many reasonable choices. The relation (2.37) makes it clear that increasing Δk causes Δx to decrease, and vice versa.

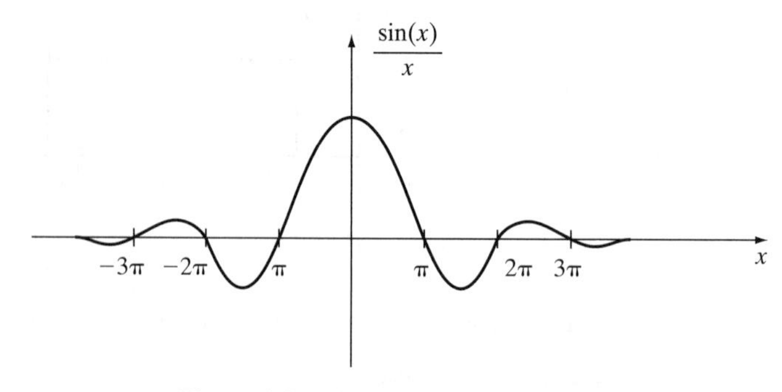

Figure 2.8 Plot of $\sin(x)/x$ versus x.

> Therefore, a wavepacket tightly confined in space is made up of plane waves having a large spread in wavenumbers, and a wavepacket loosely confined in space is made up of plane waves having a small spread in wavenumbers.

If Δk goes to zero, then the wavenumber (and momentum) is known exactly, and then Δx must go to infinity to maintain (2.37), so that the position can be any value. If Δx goes to zero, then the position is known exactly, and, conversely, Δk must go to infinity, so that momentum can take any value. Using $k = p/\hbar$, one obtains

$$\Delta p \Delta x = \frac{\hbar \pi}{2}, \tag{2.38}$$

which, except for the factor of π, is almost the famous Heisenberg uncertainty principle,

$$\Delta p \Delta x \geq \hbar/2. \tag{2.39}$$

The uncertainty principle states that we cannot know both momentum and position to arbitrary accuracy. If we know the position of a particle with great precision (Δx small), then the uncertainty in the particle's velocity is very large. Therefore, since the Heisenberg uncertainty principle applies to quantum particles, and is at least qualitatively consistent with the idea of a wavepacket, we can have some confidence in thinking of quantum particles as wavepackets.

As an example, consider an electron. If we know the position of an electron to nanoscale accuracy, $\Delta x = 10^{-9}$ m, then, from (2.39), $\Delta p \geq 5.25 \times 10^{-25}$ kg m/s, and $\Delta v = \Delta p/m_e \geq 5.77 \times 10^{5}$ m/s, which is a large uncertainty in the electron's velocity. Thus, as expected, at atomic dimensions, the quantum nature of the particle is very important.

If we instead consider a macroscopic object such as a paper clip, the situation is quite different. Assume that we know the position of a paper clip to within 1 mm (10^{-3} m), a reasonable assumption since, perhaps, it is sitting on a table in front of us. Assume that the paper clip has mass 10^{-3} kg. Then, the uncertainty in the paper clip's velocity is

$$\Delta v \geq \frac{\hbar/2}{m \Delta x} = \frac{1.05 \times 10^{-34}}{(2)\, 10^{-3} 10^{-3}} = 5.25 \times 10^{-29} \text{ m/s}, \tag{2.40}$$

which is an extremely small number. Our everyday intuition would lead us to believe that the paper clip is stationary, i.e., that $v = 0$. However, at 5.25×10^{-29} m/s, it would take 1.9×10^{22} seconds (about 6×10^{14} years) for the paper clip to move just 1 μm. Recent observations from the Hubble Space Telescope indicate that the age of the universe is around 14 billion years, which is 14×10^{9} years! Thus, there is no way for us to say that the paper clip sitting on the table in front of us is not really moving, albeit extremely slowly. Thus, the Heisenberg uncertainty principle applies to all objects,

but for familiar macroscopic objects, it leads to results that are consistent with everyday observations.

Notice that the uncertainty principle (2.39) is for a one-dimensional system, and therefore one can say that it relates momentum in the x-coordinate and position along the x-coordinate, i.e.,

$$\Delta p_x \Delta x \geq \hbar/2. \tag{2.41}$$

Generalizing to three dimensions, we have

$$\Delta p_x \Delta x \geq \hbar/2, \tag{2.42}$$

$$\Delta p_y \Delta y \geq \hbar/2, \tag{2.43}$$

$$\Delta p_z \Delta z \geq \hbar/2, \tag{2.44}$$

so that the uncertainty principle does not preclude the simultaneous measurement of, say, momentum p_x and position z. This will be discussed further in Section 3.1.4.

One can gain an intuitive understanding of the position-momentum uncertainty principle for a quantum particle by the following reasoning. If one measures, say, the position of a macroscopic object by shining light on it, the momentum carried by the illuminating photons will not measurably change the momentum of the object, due to its large mass. Thus, the object's position can be measured without changing its momentum. However, this is not the case for an atomic or subatomic particle, due to its extremely small mass.

From (2.39), one can obtain an uncertainty relationship between energy and time. Using $p = \hbar k$, $E = \hbar\omega$, and, specifically for photons, $k = \omega/c$ and $c\Delta t = \Delta x$, then

$$\Delta E \Delta t \geq \hbar/2. \tag{2.45}$$

Although (2.45) was obtained from the quantum uncertainty between momentum and position, it is really a classical uncertainty between time-frequency Fourier transform quantities (i.e., using $E = \hbar\omega$, we obtain $\Delta\omega\Delta t \geq 1/2$).

Dispersion. Returning to the discussion of wavepackets, we find that in the case considered previously, the wavepacket does not change its shape as it propagates. Often, as will be seen in later chapters, one needs to consider a dispersion relation that is more complicated than the simple linear dependence considered above for photons in free space. In general, this leads to the wavepacket changing shape (generally, spreading out) as it propagates. In addition, in this case the phase velocity is not the velocity of primary interest.

To examine this phenomenon, it is convenient to expand $\omega(k)$ in a Taylor's series around the center wavenumber $k = k_0$, obtaining

$$\begin{aligned}\omega(k) &= \omega(k_0) + \left.\frac{\partial\omega}{\partial k}\right|_{k=k_0}(k-k_0) + \frac{1}{2}\left.\frac{\partial^2\omega}{\partial k^2}\right|_{k=k_0}(k-k_0)^2 + \ldots \\ &= \omega_0 + \alpha(k-k_0) + \beta(k-k_0)^2 + \ldots.\end{aligned} \tag{2.46}$$

Now assume that it is sufficient to keep only the first two terms in (2.46). Then,

$$\begin{aligned}\psi(x,t) &\simeq e^{-ik_0(v_p t - x)} \int_{k_0-\Delta k}^{k_0+\Delta k} e^{-i(k-k_0)(\alpha t - x)}dk \\ &= e^{-ik_0(v_p t - x)} 2\Delta k \frac{\sin(\Delta k(\alpha t - x))}{\Delta k(\alpha t - x)},\end{aligned} \tag{2.47}$$

where $v_p = \omega_0/k_0$. The velocity of the envelope is not the phase velocity, but α,

$$\alpha = \left.\frac{\partial\omega}{\partial k}\right|_{k=k_0} = v_g, \tag{2.48}$$

which is called the *group velocity*. Therefore,

$$\psi(x,t) \simeq e^{-ik_0(v_p t - x)} 2\Delta k \frac{\sin\left(\Delta k\left(v_g t - x\right)\right)}{\Delta k\left(v_g t - x\right)}. \tag{2.49}$$

In this case, the wavepacket moves through space and time as a localized bundle of approximate width

$$\Delta k\left(v_g t - x\right) = \frac{\pi}{2} \tag{2.50}$$

that is centered at the point

$$\left(v_g t - x\right) = 0. \tag{2.51}$$

That is, starting at $t = 0$, the wavepacket is centered at $x = 0$, and at a given time t the wavepacket is centered at the point

$$x = v_g t \tag{2.52}$$

and occupies a spatial extent

$$\Delta x = v_g t - \frac{\pi}{2\Delta k}. \tag{2.53}$$

As mentioned above, the group velocity is the velocity of the wavepacket's envelope, rather than the phase velocity of the "center" plane wave that has wavenumber k_0. The phase velocity being associated with the center plane wave is a result of the approximate expansion of the $\omega(k)$ relationship. The wavepacket is actually made up of planes waves, each with an individual wavenumber k and associated phase velocity $v_p = \omega/k$. Each different plane wave moves at a different velocity. Over time, the plane waves will reach a given point at different times, tending to spread out the wavepacket's envelope, the width of which, according to (2.53), grows as time increases. So the center of the wavepacket moves as time changes, indicating propagation (or a particle trajectory), and the wavepacket simultaneously spreads out in space as time increases.

It should be noted that often, rather than the abrupt amplitude function $a(k)$ given by (2.32), a more physically realistic function is used, typically a Gaussian,

$$a(k) = e^{-\frac{(k-k_0)^2}{2\Delta k^2}}. \tag{2.54}$$

However, in this case, the calculations shown previously become a bit more complex than for the simple function (2.32).

2.6 MAIN POINTS

This chapter presented some information on the origins of quantum theory, and compared classical physics and, in particular, classical particles and waves, with quantum physics. After reading this chapter, you should know that

- classical particles have a definite position in space, and definite velocity and momentum;
- quantum mechanics states that we can only know the probability of a particle being at a certain position at a certain time, and that it is impossible to measure precisely both the position and momentum of a particle;
- quantum mechanics arose from experiments in the later 1800s and early 1900s concerning thermal blackbody radiation and the photoelectric effect that could not be explained by classical physics;
- in the quantum picture, energy is quantized as $E = \hbar\omega$;
- all "particles" having energy and momentum have wave-like properties, described by the de Broglie wavelength. When the material space in question is large compared to λ, the particle acts like a classical particle. When the space is small compared to λ, the particle exhibits wave-like properties.

Furthermore, you should

- be familiar with the idea of spin, and the Pauli exclusion principle;
- know what fermions and bosons are, and their relationship to the exclusion principle;
- understand wavepacket concepts, and the idea of phase and group velocity.

2.7 PROBLEMS

1. What is the energy (in J and eV) of a photon having wavelength 650 nm? Repeat for an electron having the same wavelength and only kinetic energy.
2. For light (photons), in classical physics the relation

$$c = \lambda f \tag{2.55}$$

is often used, where c is the speed of light, f is the frequency, and λ is the wavelength. For photons, is the de Broglie wavelength the same as the wavelength in (2.55)? Explain your reasoning. Hint: Use Einstein's formula

$$E = mc^2 = \sqrt{p^2c^2 + m_0c^4}, \tag{2.56}$$

where m_0 is the particle's rest mass, which, for a photon, is zero.

3. Common household electricity in the United States is 60 Hz, a typical microwave oven operates at 2.4×10^9 Hz, and UV light occurs at 30×10^{15} Hz. In each case, determine the energy of the associated photons in J and eV.
4. Assume that a HeNe laser pointer outputs 1 mW of power at 632 nm.
   (a) Determine the energy per photon.
   (b) Determine the number of photons per second, N.
5. Repeat Problem 2.4 if the laser outputs 10 mW of power. How does the number of photons per second scale with power?
6. Calculate the de Broglie wavelength of
   (a) a proton moving at 437, 000 m/s
   (b) a proton with kinetic energy 1,100 eV
   (c) an electron traveling at 10,000 m/s
   (d) an 800 kg car moving at 60 km/h
7. Determine the wavelength of a 150 g baseball traveling 90 m/h. Use this result to explain why baseballs do not seem to diffract around baseball bats.
8. How much would the mass of a ball need to be in order for it to have a de Broglie wavelength of 1 m (at which point its wave properties would be clearly observable)? Assume that the ball is traveling 90 m/h.
9. Determine the momentum carried by a 640 nm photon. Since a photon is massless, does this momentum have the same meaning as the momentum carried by a particle with mass?
10. Consider a 4 eV electron, a 4 eV proton, and a 4 eV photon. For each, compute the de Broglie wavelength, the frequency, and the momentum.
11. Determine the de Broglie wavelength of an electron that has been accelerated from rest through a potential difference of 1.5 V.
12. Calculate the uncertainty in velocity of a 1 kg ball confined to
    (a) a length of 20 μm
    (b) a length of 20 cm
    (c) a length of 20 m

(d) What can you conclude about observing "quantum effects" using 1 kg balls? What kind of objects would you need to use to see quantum effects on these length scales?

13. If we know that the velocity of an electron is 40.23 ± 0.01 m/s, what is the minimum uncertainty in its position? Repeat for a 150 g baseball traveling at the same velocity.

14. If a molecule having mass 2.3×10^{-26} kg is confined to a region 200 nm in length, what is the minimum uncertainty in the molecule's velocity?

15. Determine the minimum uncertainty in the velocity of an electron that has its position specified to within 10 nm.

16. Explain the difference between a fermion and a boson, and give two examples of each.

Chapter

3

QUANTUM MECHANICS OF ELECTRONS

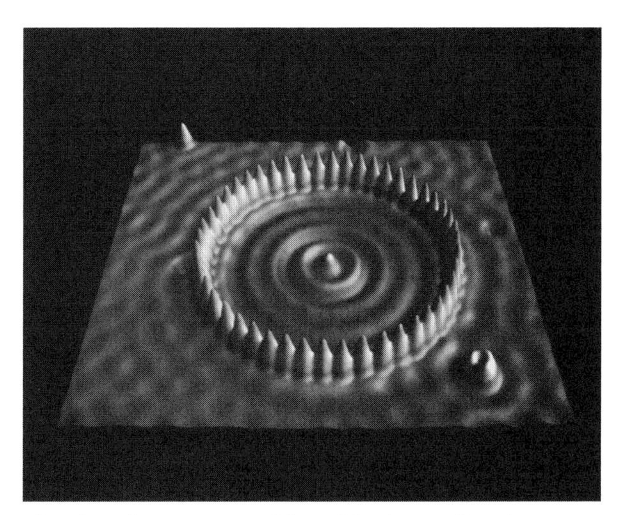

STM image of atoms forming a "quantum corral," resulting in standing electron waves. The diameter of the ring is approximately 14 nm. (Courtesy Almaden Research Center/Research Division/NASA/Media Services.)

As described in the previous chapter, in the early 1900s, it became clear that classical Newtonian mechanics was unable to explain a considerable amount of experimentally observed phenomena. Light was recognized to have a discrete nature, and both light and matter were found to exhibit properties of classical waves and classical particles, and, in addition, to exhibit behavior that was completely unknown to classical physics. In the following discussion, when particles (usually electrons, but perhaps atoms, paper clips, billiard balls, etc.) are referred to, it should be recognized that all such objects, irrespective of size, are really quantum particles. The mathematical description of such particles is given by solutions of Schrödinger's equation.[†]

[†]Schrödinger's equation describes quantum particles having mass. Photons, which do not have mass, obey a quantized version of Maxwell's equations, although we will not consider that development here.

Schrödinger's equation cannot be derived from fundamental principles (the same can be said for Newton's equations), although before presenting Schrödinger's equation formally, a plausible development will be outlined.

From the wave standpoint, a one-dimensional plane wave has the form[†]

$$\psi(x,t) = Ae^{i(kx-\omega t)}, \tag{3.1}$$

where k is the wavenumber and $\omega = \omega(k)$ is the radian frequency. From (3.1), note that

$$\omega = \frac{i}{\psi(x,t)} \frac{\partial \psi(x,t)}{\partial t}, \quad k^2 = -\frac{1}{\psi(x,t)} \frac{\partial^2 \psi(x,t)}{\partial x^2}. \tag{3.2}$$

From the particle perspective, the total energy E of a particle is the sum of kinetic and potential energies,

$$E = \frac{1}{2}mv^2 + V = \frac{p^2}{2m} + V,$$

where p is the particle's momentum, m is the particle's mass, and V is the potential energy seen by the particle. So far, these are classical relations. The quantum nature of the problem is incorporated into the energy expression using quantization of energy, (2.12), and the de Broglie relation, (2.15), leading to

$$\begin{aligned} E = \hbar\omega = \hbar\left(\frac{i}{\psi(x,t)} \frac{\partial \psi(x,t)}{\partial t}\right) &= \frac{(\hbar k)^2}{2m} + V, \\ &= \frac{\hbar^2}{2m}\left(-\frac{1}{\psi(x,t)} \frac{\partial^2 \psi(x,t)}{\partial x^2}\right) + V, \end{aligned} \tag{3.3}$$

such that

$$i\hbar \frac{\partial \psi(x,t)}{\partial t} = \left(-\frac{\hbar^2}{2m} \frac{\partial^2}{\partial x^2} + V\right) \psi(x,t). \tag{3.4}$$

This is Schrödinger's equation in one dimension, although not the most general form. As will be seen, this is the fundamental equation for describing quantum particles with mass, of which, throughout the text, primary consideration will be given to electrons. In the next section, the general postulates of quantum mechanics will be presented, one of which refers to Schrödinger's equation, followed by a discussion of each postulate. In the field of nanoelectronics this constitutes the most difficult material to comprehend; however, it is extremely important.

[†]To obtain the time-domain wave classically, we would take the real part of a time-harmonic phasor,

$$\psi(x,t) = \mathrm{Re}\left\{Ae^{i(kx-\omega t)}\right\} = A\cos(kx - \omega t),$$

assuming the amplitude A is real valued. Here, it is more useful to consider a complex exponential form.

3.1 GENERAL POSTULATES OF QUANTUM MECHANICS

POSTULATE 1. To every quantum system there is a state function, $\Psi(\mathbf{r}, t)$, that contains everything that can be known about the system.†

In the following discussion, a quantum system will be either a single particle, typically an electron, and its environment, or a system of particles. The state function, or *wavefunction*, is complex valued and mathematically well behaved (i.e., it is a finite, single valued, and continuous function), and probabilistic in nature. This latter fact makes the wavefunction fundamentally different from quantities obtained from Newtonian dynamics, such as the trajectory of a particle, or the motion of a wave. In particular, $\rho(\mathbf{r}, t) = |\Psi(\mathbf{r}, t)|^2$ *is the probability density of finding the particle at a particular point in space* $\mathbf{r}$ *at time* t.

That is, the probability of finding the particle in a region of space Ω is

$$P = \int_{\Omega} |\Psi(\mathbf{r}, t)|^2 \, d^3r = \int_{\Omega} \Psi^*(\mathbf{r}, t)\, \Psi(\mathbf{r}, t)\, d^3r. \tag{3.5}$$

In particular, in one dimension, the probability of finding the particle between points $x = a$ and $x = b$ is

$$P = \int_a^b \Psi^*(x, t)\, \Psi(x, t)\, dx. \tag{3.6}$$

By continuity of the wavefunction, the probability of finding the particle in a small volume element $d\Omega$ centered at $\mathbf{r}$ is

$$P = \int_{d\Omega} |\Psi(\mathbf{r}, t)|^2 \, d^3r = |\Psi(\mathbf{r}, t)|^2 \, d\Omega. \tag{3.7}$$

Furthermore, since the particle must be somewhere in space at any time t,

$$\int_{\text{all space}} \Psi^*(\mathbf{r}, t)\, \Psi(\mathbf{r}, t)\, d^3r = 1. \tag{3.8}$$

We frequently use (3.8) to normalize solutions of Schrödinger's equation.

It is worthwhile to reiterate that quantum systems are true probabilistic systems.‡ As mentioned previously, most classical systems we think of as probabilistic, e.g., rolling a die or flipping a coin, are not really probabilistic at a fundamental level.§ Probabilities are

†The vector $\mathbf{r}$ is called the *position vector*, and represents the vector from the origin to some point in space. In rectangular coordinates,

$$\mathbf{r} = \mathbf{a}_x x + \mathbf{a}_y x + \mathbf{a}_z z,$$

where $\mathbf{a}_x$, $\mathbf{a}_y$, and $\mathbf{a}_z$ are unit vectors. The use of the position vector leads to notational convenience, since $\Psi(\mathbf{r}, t)$ simply means $\Psi(x, y, z, t)$.

‡Here we take the view of what is called the *Copenhagen Interpretation* (or *Born interpretation*) of quantum mechanics, which is a mainstream view, although not the only view, in physics.

§Although, if you want to be a stickler, quantum mechanics governs the behavior of the die, too, and so we are back to a probabilistic description. However, the quantum nature of the die is obscured due to its large mass (resulting in a minuscule de Broglie wavelength). Newtonian dynamics, however complicated to apply in this case, models the trajectory of the die very well.

merely assigned to provide a convenient model of the system. Some scientists interpret the probabilistic nature of quantum mechanics to reflect the base probabilistic nature of, well, nature, although others argue that quantum mechanics is simply a probabilistic model of nature, and not the true description of physical reality. However, there is no argument about the accuracy of quantum mechanics—it is considered to be one of the most successful physical models ever developed.

Example

The following chapters will present methods to find the state function for electrons in several relatively simple structures. However, at this point, it is worthwhile to give a concrete example of a state function, since it is a relatively abstract concept (although the mathematical functions involved are usually quite ordinary).

Consider an electron that is confined to a line segment of length L, where the line segment extends from $x = -L/2$ to $x = L/2$ and where, on the line segment, potential energy is zero. As will be seen in the example on page 65 and in the next chapter, it turns out that the electron can be in one of an infinite number of possible states, where each state can be signified by an index[†] n. The possible states are given by wavefunctions $\Psi_n(x, t)$, where

$$\Psi_n(x, t) = \left(\frac{2}{L}\right)^{1/2} \sin\left(\frac{n\pi}{L}x\right) e^{-iE_n t/\hbar}, \tag{3.9}$$

for n even, and

$$\Psi_n(x, t) = \left(\frac{2}{L}\right)^{1/2} \cos\left(\frac{n\pi}{L}x\right) e^{-iE_n t/\hbar}, \tag{3.10}$$

for n odd, where

$$E_n = \frac{\hbar^2}{2m_e}\left(\frac{n\pi}{L}\right)^2 \tag{3.11}$$

is the energy of the electron in the nth state, $n = 1, 2, 3, \ldots$. If the electron is in a state given by an even value of n, then the probability of finding the electron somewhere along the line segment is, from (3.6),

$$P = \int_{-L/2}^{L/2} \Psi_n^*(x, t)\, \Psi_n(x, t)\, dx = \frac{2}{L}\int_{-L/2}^{L/2} \sin^2\left(\frac{n\pi}{L}x\right) dx = 1. \tag{3.12}$$

That is, as expected, the probability of finding the electron somewhere along the line segment is 100 percent. The same result holds if the electron is in a state given by an odd value of n. Furthermore, the probability of finding the electron, say, on the left half of the

[†] In addition, the electron can be in some combination of states, i.e., a sum over n, although this will be discussed later.

line segment is

$$P = \frac{2}{L} \int_{-L/2}^{0} \sin^2 \left(\frac{n\pi}{L} x \right) dx = \frac{1}{2},$$

or 50 percent, and is similar for odd values of n. Note that in a classical model, one would be able to say exactly where the electron is at all times. In the quantum model, one can only talk about probabilities of finding the electron at a certain position at a certain time. For a given state, at time t that probability is high where $|\Psi_n (x, t)|^2$ is large.

POSTULATE 2.†

(a) Every physical observable $\mathcal{O}$ (position, momentum, energy, etc.) is associated with a linear Hermitian operator $\widehat{o}$.

(b) Associated with the operator $\widehat{o}$ is the eigenvalue problem,

$$\widehat{o}\psi_n = \lambda_n \psi_n, \tag{3.13}$$

such that the result of a measurement of an observable is one of the eigenvalues λ_n of the operator $\widehat{o}$.

(c) If a system is in the initial state Ψ, measurement of $\mathcal{O}$ will yield one of the eigenvalues λ_n of $\widehat{o}$ with probability‡

$$P(\lambda_n) = \left| \int \Psi(\mathbf{r}, t)\, \psi_n^* (\mathbf{r})\, d^3 r \right|^2, \tag{3.14}$$

and the state of the system will change from Ψ to ψ_n.

This is a fairly long postulate, and each item will be discussed in some detail.

3.1.1 Operators

First, let us discuss briefly what is meant by an operator. An operator is a mapping from one quantity to another, loosely speaking. For instance, a 3×2 matrix maps 2×1 matrices to 3×1 matrices,

$$\begin{bmatrix} a_{11} & a_{12} \\ a_{21} & a_{22} \\ a_{31} & a_{32} \end{bmatrix} \begin{bmatrix} b_1 \\ b_2 \end{bmatrix} = \begin{bmatrix} c_1 \\ c_2 \\ c_3 \end{bmatrix}. \tag{3.15}$$

†Postulate 2 is quite complicated, and its explanation is rather long and involves difficult concepts. For the specific purpose of understanding the remainder of the text, it is also the least directly useful of the four postulates (beyond this chapter, with the exception of a few examples, we really only need Postulates 1 and 4, although Postulates 2 and 3 are included for completeness). Therefore, if desired the reader may choose to skip this postulate and move on to postulate 4.

‡Here we assume spatial eigenvectors (i.e., those associated with time-independent operators), and it is assumed that the eigenvectors are appropriately normalized to form an orthonormal set. This will be discussed later.

If one considers a set S_1 consisting of all 2×1 matrices, and a set S_2 consisting of all 3×1 matrices, the mapping $M : S_1 \rightarrow S_2$ can be accomplished via 3×2 matrices.

For our purposes, we will often need operators that map one function to another. For instance, when we take the derivative of the sine function, we obtain the cosine function. One can think of the mapping between these two functions as the derivative operator, $\widehat{o} = d/dx : \{\text{sine}\} \rightarrow \{\text{cosine}\}$. That is, apply the derivative operator to the sine function, and obtain the cosine function,

$$\frac{d}{dx} \sin(x) = \cos(x). \tag{3.16}$$

3.1.2 Eigenvalues and Eigenfunctions

Now, with the idea of an operator established, the concepts of eigenvalues and eigenfunctions can be discussed. An *eigenfunction* of a certain operator is a nonzero function such that, when we apply the operator to the eigenfunction, we obtain a multiple of the eigenfunction back again. That is,

$$\widehat{o}\psi_n = \lambda_n \psi_n, \tag{3.17}$$

where $\widehat{o}$ is an operator, ψ_n is an eigenfunction of the operator, and λ_n is called an *eigenvalue* (not to be confused with wavelength). In general, there are many (perhaps infinitely many) eigenfunctions and eigenvalues. For example, consider the operator

$$\widehat{o} = \frac{d^2}{dx^2}, \tag{3.18}$$

the second derivative operator, acting on functions defined over $0 \leq x \leq L$. Every function $\psi_n(x) = \sin(n\pi x/L)$ is an eigenfunction with eigenvalue $\lambda_n = -(n\pi/L)^2$, since

$$\frac{d^2}{dx^2}\left(\sin \frac{n\pi}{L} x\right) = -\left(\frac{n\pi}{L}\right)^2 \left(\sin \frac{n\pi}{L} x\right). \tag{3.19}$$

As an example of a matrix operator,[†] simple matrix multiplication shows that the operator

$$\widehat{o} = \begin{bmatrix} 1 & 1 \\ 2 & -1 \end{bmatrix} \tag{3.20}$$

has eigenvalues and eigenvectors

$$\lambda_1 = \sqrt{3}, \qquad \psi_1 = \begin{bmatrix} \alpha \\ \left(\sqrt{3} - 1\right)\alpha \end{bmatrix}, \tag{3.21}$$

$$\lambda_2 = -\sqrt{3}, \qquad \psi_2 = \begin{bmatrix} \beta \\ \left(-\sqrt{3} - 1\right)\beta \end{bmatrix},$$

[†]Finding the eigenvalues and eigenvectors of a matrix operator is obviously more involved, and will be omitted here, although this topic is discussed in textbooks on linear algebra and applied mathematics.

where $\alpha, \beta \neq 0$. That is,

$$\begin{bmatrix} 1 & 1 \\ 2 & -1 \end{bmatrix}\begin{bmatrix} \alpha \\ \left(\sqrt{3}-1\right)\alpha \end{bmatrix} = \sqrt{3}\begin{bmatrix} \alpha \\ \left(\sqrt{3}-1\right)\alpha \end{bmatrix}, \tag{3.22}$$

and

$$\begin{bmatrix} 1 & 1 \\ 2 & -1 \end{bmatrix}\begin{bmatrix} \beta \\ \left(-\sqrt{3}-1\right)\beta \end{bmatrix} = -\sqrt{3}\begin{bmatrix} \beta \\ \left(-\sqrt{3}-1\right)\beta \end{bmatrix}. \tag{3.23}$$

3.1.3 Hermitian Operators

Hermitian operators form an important special class of operators. Hermitian operators have real eigenvalues, and their eigenfunctions form an *orthogonal*, *complete* set of functions. Orthogonal means that

$$\int \psi_i^* (\mathbf{r})\, \psi_j (\mathbf{r})\, d^3r = 0 \tag{3.24}$$

for all $i \neq j$, which is called an *orthogonality* condition. The subscript indicates a certain eigenfunction. That is, if ψ_1 and ψ_2 are two different eigenfunctions, i.e.,

$$\widehat{o}\psi_1 = \lambda_1\psi_1, \tag{3.25}$$

$$\widehat{o}\psi_2 = \lambda_2\psi_2, \tag{3.26}$$

and $\psi_1 \neq \psi_2$, then

$$\int \psi_1^* (\mathbf{r})\, \psi_2 (\mathbf{r})\, d^3r = 0. \tag{3.27}$$

If the wavefunctions have been normalized such that

$$\int \psi_i^* (\mathbf{r})\, \psi_i (\mathbf{r})\, d^3r = 1, \tag{3.28}$$

then the wavefunctions are said to form an *orthonormal set*, i.e.,

$$\int \psi_i^* (\mathbf{r})\, \psi_j (\mathbf{r})\, d^3r = \begin{cases} 1, & i = j \\ 0, & i \neq j. \end{cases} \tag{3.29}$$

When one says that the eigenfunctions of a Hermitian operator form a *complete set* of functions, it means that any function can be represented as a sum of eigenfunctions,

$$\Psi (\mathbf{r}) = \sum_n a_n \psi_n (\mathbf{r}). \tag{3.30}$$

The expansion coefficients a_n can be obtained using orthonormality. That is, start with (3.30), multiply both sides by ψ_m^*, and integrate.

$$\int \psi_m^* (\mathbf{r})\, \Psi (\mathbf{r})\, d^3r = \sum_n a_n \int \psi_m^* (\mathbf{r})\, \psi_n (\mathbf{r})\, d^3r \tag{3.31}$$

$$= a_m. \tag{3.32}$$

Example

As a familiar example, any periodic function with period 2π can be written as the Fourier series

$$\Psi(x) = \frac{1}{\sqrt{2\pi}} \sum_{n=-\infty}^{\infty} a_n e^{inx}, \tag{3.33}$$

where $\psi_n(x) = e^{inx}/\sqrt{2\pi}$ are normalized eigenfunctions of d^2/dx^2 subject to periodic boundary conditions,

$$\psi_n(-\pi) = \psi_n(\pi), \qquad \psi_n'(-\pi) = \psi_n'(\pi), \tag{3.34}$$

with eigenvalues $\lambda_n = -n^2$, i.e.,

$$\frac{d^2}{dx^2} e^{inx} = -n^2 e^{inx}. \tag{3.35}$$

Note that the same function Ψ has, in general, many different representations, associated with different operators. That is,

$$\Psi(\mathbf{r}) = \sum_n a_n \psi_n(\mathbf{r}) \tag{3.36}$$

$$= \sum_n b_n \phi_n(\mathbf{r}), \tag{3.37}$$

where $\{\psi_n\}$ and $\{\phi_n\}$ are sets of eigenfunctions associated with different operators (say, $\widehat{o}$ and $\widehat{s}$),

$$\widehat{o}\psi_n = \lambda_n \psi_n, \tag{3.38}$$

$$\widehat{s}\phi_n = \upsilon_n \phi_n. \tag{3.39}$$

However, as long as both operators $\widehat{o}$ and $\widehat{s}$ are Hermitian, the expansions (3.36) and (3.37) will be valid.

Example

It is easy to see that the differential operator d^2/dx^2, $\psi_n(0) = \psi_n(L) = 0$, on the segment from $x = 0$ to $x = L$ leads to eigenfunctions

$$\psi_n(x) = \sqrt{\frac{2}{L}} \sin\left(\frac{n\pi x}{L}\right) \tag{3.40}$$

with corresponding eigenvalues $\lambda_n = (n\pi/L)^2$, $n = 1, 2, 3, \ldots$. Functions on this interval can be expanded as

$$\Psi(x) = \sum_{n=1}^{\infty} a_n \sqrt{\frac{2}{L}} \sin\left(\frac{n\pi x}{L}\right). \tag{3.41}$$

For the differential operator d^2/dx^2, $\psi'(0) = \psi'(L) = 0$, the eigenfunctions are found to be

$$\psi_n(x) = \sqrt{\frac{\varepsilon_n}{L}} \cos\left(\frac{n\pi x}{L}\right) \tag{3.42}$$

with corresponding eigenvalues $\lambda_n = (n\pi/L)^2$, $n = 0, 1, 2, \ldots$, where, for notational convenience, we use Neumann's number ε_n, defined as

$$\varepsilon_n \equiv \begin{cases} 1, & n = 0 \\ 2, & n \neq 0. \end{cases} \tag{3.43}$$

Functions on this interval can be expanded as

$$\Psi(x) = \sum_{n=0}^{\infty} b_n \sqrt{\frac{\varepsilon_n}{L}} \cos\left(\frac{n\pi x}{L}\right). \tag{3.44}$$

Since both sets of eigenfunctions form a complete set of functions on the interval $(0, L)$, either representation can be used. For example, the function $\Psi(x)$ that is equal to 0 from $x = 0$ to $x = L/2$, and equal to 1 from $x = L/2$ to $x = L$, can be expressed as (3.41) with

$$a_n = \int_0^L \psi_n^*(x)\, \Psi(x)\, dx \tag{3.45}$$

$$= \sqrt{\frac{2}{L}} \int_0^{L/2} \sin\left(\frac{n\pi x}{L}\right)(0)\, dx + \sqrt{\frac{2}{L}} \int_{L/2}^{L} \sin\left(\frac{n\pi x}{L}\right)(1)\, dx$$

$$= -\frac{\sqrt{2L}}{n\pi}\left(\cos(n\pi) - \cos\left(\frac{n\pi}{2}\right)\right), \tag{3.46}$$

or as (3.44) with

$$b_n = \int_0^L \psi_n^*(x)\, \Psi(x)\, dx \tag{3.47}$$

$$= \int_0^{L/2} \sqrt{\frac{\varepsilon_n}{L}} \cos\left(\frac{n\pi x}{L}\right)(0)\, dx + \int_{L/2}^{L} \sqrt{\frac{\varepsilon_n}{L}} \cos\left(\frac{n\pi x}{L}\right)(1)\, dx$$

$$= \begin{cases} \sqrt{\frac{1}{L}}\frac{L}{2}, & n = 0, \\ -\sqrt{\frac{\varepsilon_n}{L}}\frac{L}{n\pi} \sin\left(\frac{n\pi}{2}\right), & n > 0. \end{cases} \tag{3.48}$$

The only significant difference between the two expansions is the interpretation of the equality at the boundary points $x = 0, L$.

Of course, when one says that *any function* can be expanded as an eigenfunction series, the class or group of functions of interest needs to be specified (e.g., functions that go to zero at $x = a$ and $x = b$, or functions that possess derivatives, etc.) In this text, *any function* means any quantum mechanical state function Ψ, and if we consider the operators encountered in quantum mechanics (as discussed next), we will be able to form the expansion (3.30).

The eigenvalues could alternatively form a continuous, rather than a discrete, set.† In this case, the eigenvalue problem will be denoted as

$$\widehat{o}\psi(\mathbf{r}, \lambda) = \lambda\psi(\mathbf{r}, \lambda), \tag{3.49}$$

and the representation has the form of a continuous summation (integration), rather than a discrete summation,

$$\Psi(\mathbf{r}) = \int c(\lambda)\, \psi(\mathbf{r}, \lambda)\, d\lambda. \tag{3.50}$$

Example

The operator

$$\widehat{o} = -i\frac{\partial}{\partial x} \tag{3.51}$$

on $-\infty < x < \infty$ (i.e., with no "boundary conditions") has $\psi(x, \lambda) = e^{i\lambda x}$ as eigenfunctions,

$$-i\frac{\partial}{\partial x}e^{i\lambda x} = \lambda e^{i\lambda x}, \tag{3.52}$$

where any value of λ is allowed. In one dimension, functions can be represented as the continuous sum of eigenfunctions

$$\Psi(x) = \frac{1}{2\pi}\int_{-\infty}^{\infty} a(\lambda)\, e^{i\lambda x} d\lambda, \tag{3.53}$$

which is merely the familiar Fourier transform representation. The expansion coefficient is the Fourier transform itself,

$$a(\lambda) = \int_{-\infty}^{\infty} \Psi(x)\, e^{-i\lambda x} dx. \tag{3.54}$$

3.1.4 Operators for Quantum Mechanics

Typical operators encountered in quantum mechanics will now be discussed. It is useful to start by examining a plane–wave function of the form

$$\psi(x, t) = Ae^{i(kx - Et/\hbar)}, \tag{3.55}$$

†Such eigenfunctions are sometimes called *improper eigenfunctions*.

which happens to be a solution to Schrödinger's equation in one dimension if $V = 0$ and

$$E = \frac{\hbar^2 k^2}{2m}. \tag{3.56}$$

Momentum Operator. Considering first the spatial part of the function (3.55),

$$\psi(x) = Ae^{ikx}, \tag{3.57}$$

and applying the operator

$$\widehat{o} = -i\hbar\frac{\partial}{\partial x} \tag{3.58}$$

to the function $\psi(x)$ leads to

$$-i\hbar\frac{\partial}{\partial x}\psi(x) = -i\hbar\frac{\partial}{\partial x}Ae^{ikx} = \hbar k\psi(x) = p\psi(x), \tag{3.59}$$

where $p = \hbar k$ is the momentum. Therefore, (3.57) is an eigenfunction of the operator (3.58) with momentum as an eigenvalue. Hence, (3.58) is called the *momentum operator*, written as

$$\widehat{p} = -i\hbar\frac{\partial}{\partial x}. \tag{3.60}$$

Of course, there is nothing special about the coordinate x, and therefore we have momentum operators in rectangular coordinates as

$$\widehat{p}_x = -i\hbar\frac{\partial}{\partial x}, \quad \widehat{p}_y = -i\hbar\frac{\partial}{\partial y}, \quad \widehat{p}_z = -i\hbar\frac{\partial}{\partial z}. \tag{3.61}$$

In three dimensions, the vector momentum operator is

$$\mathbf{p} = \mathbf{a}_x\,\widehat{p}_x + \mathbf{a}_y\,\widehat{p}_y + \mathbf{a}_z\,\widehat{p}_z = -i\hbar\nabla, \tag{3.62}$$

where in rectangular coordinates $\nabla = \mathbf{a}_x\,(\partial/\partial x) + \mathbf{a}_y\,(\partial/\partial y) + \mathbf{a}_z\,(\partial/\partial z)$.

Energy Operator. In the same manner, the operator

$$\widehat{E} = i\hbar\frac{\partial}{\partial t} \tag{3.63}$$

is the *energy operator*. Considering the temporal dependence of (3.55),

$$g(t) = e^{-iEt/\hbar}, \tag{3.64}$$

leads to

$$\widehat{E}g(t) = i\hbar\frac{\partial}{\partial t}g(t) = i\hbar\frac{\partial}{\partial t}e^{-iEt/\hbar} = \hbar\frac{E}{\hbar}g(t) = Eg(t), \tag{3.65}$$

and so (3.64) are eigenfunctions of the energy operator (3.63) with energy as eigenvalues. As will be seen in Postulate 4, the energy operator (3.63) can be related to the Hamiltonian.

Position Operator. Now consider the position operator, $\widehat{x} = x$. Assume that a particle is at position x_α, and denote the position operator by $\widehat{x}$. Then

$$\widehat{x}\psi_\alpha(x) = x_\alpha\psi_\alpha(x) \tag{3.66}$$

must be true, such that

$$x\psi_\alpha(x) = x_\alpha\psi_\alpha(x). \tag{3.67}$$

This should be true for all x, which means that $\psi_\alpha(x) = 0$ for all $x \neq x_\alpha$. The eigenfunction is taken to be the delta function,

$$\psi_\alpha(x) = \psi(x, \alpha) = \delta(x - x_\alpha), \tag{3.68}$$

which, technically, is not a true function, but a distribution. That is, the eigenvalue equation is

$$x\delta(x - x_\alpha) = x_\alpha\delta(x - x_\alpha). \tag{3.69}$$

The delta function has the useful properties

$$\int_a^b \delta(x - x_\alpha) f(x)\, dx = \begin{cases} f(x_\alpha), & a \leq x_\alpha \leq b, \\ 0, & \text{otherwise}, \end{cases} \tag{3.70}$$

$$\int_a^b \delta(x - x_\alpha)\, dx = 1, \quad a < x_\alpha < b, \tag{3.71}$$

$$\int_{-\infty}^{\infty} e^{ikx}\, dx = 2\pi\delta(x). \tag{3.72}$$

Using these properties, we can see that, upon integrating both sides, the eigenvalue equation (3.69) holds.

The position operator $\widehat{x} = x$ is Hermitian, and so its eigenfunctions form a complete (continuous) set. Manipulating (3.70) gives the representation for any good function as

$$\Psi(x) = \int \Psi(x')\, \delta(x - x')\, dx', \tag{3.73}$$

using $x' = x_\alpha$.

Commutation and the Uncertainty Principle. Recall that the Heisenberg uncertainty principle in one dimension,

$$\Delta p \Delta x \geq \hbar/2, \tag{3.74}$$

places a constraint on the product of the uncertainties in momentum and position. On page 38, the generalization to three dimensions was given as

$$\Delta p_x \Delta x \geq \hbar/2, \tag{3.75}$$

$$\Delta p_y \Delta y \geq \hbar/2, \tag{3.76}$$

$$\Delta p_z \Delta z \geq \hbar/2, \tag{3.77}$$

so that, for example, there is no problem with measuring momentum in one direction and position in another. The question then arises as to when it is possible to have simultaneous knowledge of two quantities, and the answer is obtained by considering operators.

Consider two observables, α and β, associated with two operators, $\widehat{\alpha}$ and $\widehat{\beta}$, respectively, and assume that we would like to measure α and β for a given system Ψ. If we are to be able to measure α and β with arbitrary precision, then the measurement of α cannot influence β, and vice versa. Therefore, if this is to be true, the order of the measurements will not be important. Thus,

$$\widehat{\alpha}\widehat{\beta}\Psi = \widehat{\beta}\widehat{\alpha}\Psi, \tag{3.78}$$

and so

$$\left(\widehat{\alpha}\widehat{\beta} - \widehat{\beta}\widehat{\alpha}\right)\Psi = 0. \tag{3.79}$$

The difference operator in (3.79) is called the *commutator*,

$$\left[\widehat{\alpha}, \widehat{\beta}\right] = \left(\widehat{\alpha}\widehat{\beta} - \widehat{\beta}\widehat{\alpha}\right), \tag{3.80}$$

such that when the commutator of two operators is zero (the two operators are then said to *commute*), the corresponding observables can be measured to arbitrary precision.

As an example, consider the position and momentum operators. For $\widehat{x}$ and $\widehat{p}_x$,

$$\left(\widehat{x}\left(-i\hbar\frac{\partial}{\partial x}\right) - \left(-i\hbar\frac{\partial}{\partial x}\right)\widehat{x}\right)\Psi = x\left(-i\hbar\frac{\partial}{\partial x}\right)\Psi - \left(-i\hbar\frac{\partial}{\partial x}\right)x\Psi \tag{3.81}$$

$$= -i\hbar\left(x\frac{\partial}{\partial x}\Psi - \left(x\frac{\partial}{\partial x}\Psi + \Psi\right)\right) \tag{3.82}$$

$$= i\hbar\Psi, \tag{3.83}$$

and, therefore,

$$\left[\widehat{x}, \widehat{p}_x\right] = i\hbar \neq 0. \tag{3.84}$$

So one cannot measure position and momentum along the x-axis with arbitrary precision. However, for the operators $\widehat{x}$ and $\widehat{p}_y$, one obtains

$$\left(\widehat{x}\left(-i\hbar\frac{\partial}{\partial y}\right)-\left(-i\hbar\frac{\partial}{\partial y}\right)\widehat{x}\right)\Psi = x\left(-i\hbar\frac{\partial}{\partial y}\right)\Psi-\left(-i\hbar\frac{\partial}{\partial y}\right)x\Psi \tag{3.85}$$

$$= -i\hbar\left(x\frac{\partial}{\partial y}\Psi-\left(x\frac{\partial}{\partial y}\Psi+0\right)\right) \tag{3.86}$$

$$= 0, \tag{3.87}$$

and so

$$\left[\widehat{x},\widehat{p}_y\right]=0, \tag{3.88}$$

allowing the possibility of measuring position along the x-axis and momentum along the y-axis with arbitrary precision. It can be easily seen that

$$\left[\widehat{\alpha},\widehat{p}_\beta\right]=\left[\widehat{p}_\alpha,\widehat{p}_\beta\right]=\left[\widehat{\alpha},\widehat{\beta}\right]=0 \tag{3.89}$$

for $\alpha,\beta = x, y, z$ with $\alpha \neq \beta$.

3.1.5 Measurement Probability

The last part of Postulate 2 gives the probability of obtaining a certain measurement result (a certain eigenvalue). From (3.14), if the system is already in state ψ_n before measurement ($\Psi = \psi_n$), where ψ_n is an eigenfunction of the measurement operator $\widehat{o}$ (i.e., $\widehat{o}\psi_n = \lambda_n\psi_n$), then (3.14) tells us that, with 100 percent certainty, the result of the measurement will be λ_n, since

$$P(\lambda_n) = \left|\int \Psi(\mathbf{r},t)\,\psi_n^*(\mathbf{r},t)\,d^3r\right|^2 \tag{3.90}$$

$$= \left|\int \psi_n(\mathbf{r},t)\,\psi_n^*(\mathbf{r},t)\,d^3r\right|^2 = 1. \tag{3.91}$$

However, if the initial state of the system is not ψ_n, then the best that can be done is to obtain the probability that the result of a measurement will be a certain λ_n, given by (3.14). For example, assume that the initial state of the particle is

$$\Psi(\mathbf{r},t) = a_1\psi_1(\mathbf{r},t) + a_2\psi_2(\mathbf{r},t). \tag{3.92}$$

Then

$$P(\lambda_1) = \left|\int \Psi(\mathbf{r},t)\,\psi_1^*(\mathbf{r},t)\,d^3r\right|^2 \tag{3.93}$$

$$= \left|\int \left[a_1\psi_1(\mathbf{r},t)+a_2\psi_2(\mathbf{r},t)\right]\psi_1^*(\mathbf{r},t)\,d^3r\right|^2,$$

which, using orthonormality, results in

$$P(\lambda_1) = |a_1|^2 . \tag{3.94}$$

It it obvious that $P(\lambda_2) = |a_2|^2$, and $P(\lambda_n) = 0$ for $n > 2$ (thus, $|a_1|^2 + |a_2|^2 = 1$ must be true).

However, regardless of the initial state, we do know with 100 percent certainty that after a measurement, the system will be in the state ψ_n (the eigenfunction of $\widehat{o}$). So, if we want to obtain a system in a certain state, we can "prepare" this state by doing a measurement.

Example

Coming back to the example of an electron confined to a line segment between $x = -L/2$ and $x = L/2$, we assume that an electron is in the $n = 2$ state. The state function is (3.9),

$$\Psi(x,t) = \Psi_2(x,t) = \left(\frac{2}{L}\right)^{1/2} \sin\left(\frac{2\pi}{L}x\right) e^{-iE_2 t/\hbar}, \tag{3.95}$$

and the probability that the electron is located at some position x_α is

$$P(x_\alpha) = \frac{2}{L}\left|\int_{-L/2}^{L/2} \sin\left(\frac{2\pi}{L}x\right) e^{-iE_2 t/\hbar} \delta(x - x_\alpha)\, dx\right|^2 \tag{3.96}$$

$$= \frac{2}{L}\left|\sin\left(\frac{2\pi}{L}x_\alpha\right)\right|^2 = |\Psi(x,t)|^2, \tag{3.97}$$

consistent with Postulate 1.

Collapse of the State Function—the Measurement Problem. Postulate 2 states that every physically measurable property of a system (any so-called observable) is associated with a linear Hermitian operator, and that the result of every measurement is one of the eigenvalues of the operator associated with that observable. That is, before measurement, the state of the system may not be known, but after measurement (i.e., observation), the system's state will be an eigenstate ψ_n of the operator $\widehat{o}$ associated with the measurement, and that an eigenvalue λ_n must be obtained as the measurement result.

In classical physics, the idea of a measurement is that one is measuring something about the system the instant before the measurement is performed. For example, if the velocity of a car is measured to be 80 km/h, it can be inferred that the instants before and after the measurement was performed, the car was going 80 km/h. Thus, in classical physics, when a measurement is performed, one is "taking a peek" into the system, and it is assumed that the act of measurement does not perturb the system (at least not too much).

The view in the Copenhagen Interpretation of quantum mechanics is quite different. All that we know about the system is its state *after* the measurement is performed. The instant before the measurement, the system is, most generally, in a superposition of states. The act of measuring (observing), no matter how carefully and unobtrusively done, collapses the superposition of states into a single eigenstate. This is called *collapse of the state function*, or the *measurement problem*. This collapse is an instantaneous process. The wave function will stay in the collapsed state until it is perturbed by the outside world, after which, depending on the type of perturbation, it may revert back to a superposition state.

To further consider the collapse of the state function, let's consider consecutive measurements of position and momentum. Let $\delta(x - x_\alpha)$ denote a position eigenfunction, and $e^{ipx/\hbar}$ denote a momentum eigenfunction. Further, note that any momentum eigenfunction can be written as a (continuous) sum of position eigenfunctions,

$$e^{ipx/\hbar} = \int_{-\infty}^{\infty} c(x_\alpha)\, \delta(x - x_\alpha)\, dx_\alpha, \tag{3.98}$$

where $c(x_\alpha) = e^{ipx_\alpha/\hbar}$, and that any position eigenfunction can be written as a (continuous) sum of momentum eigenfunctions,

$$\delta(x - x_\alpha) = \frac{1}{2\pi\hbar} \int_{-\infty}^{\infty} d(p)\, e^{ip(x-x_\alpha)/\hbar} dp, \tag{3.99}$$

where $d(p) = 1$. Therefore, if the system is in a certain momentum state, it is, equivalently, in a superposition of position states, by (3.98). In a similar manner, if the system is in a certain position state, it is, equivalently, in a superposition of momentum states, by (3.99).

Now say that the position of a particle is measured. Before the measurement, assume that the particle wavefunction is a superposition of several position eigenfunctions, as in (3.30), such that

$$\Psi(x) = \int_{-\infty}^{\infty} \Psi(x_\alpha)\, \delta(x - x_\alpha)\, dx_\alpha. \tag{3.100}$$

When the measurement is made, the wavefunction collapses into one of these eigenfunctions, say, $\delta(x - x_\beta)$, the one corresponding to the measured position x_β. That is,

$$\Psi(x) = \int_{-\infty}^{\infty} \Psi(x_\alpha)\, \delta(x - x_\alpha)\, dx_\alpha \xrightarrow{\text{measure position}} \delta(x - x_\beta). \tag{3.101}$$

In fact, position x_β is measured with probability

$$P(x_\beta) = \left| \int \Psi(x)\, \delta(x - x_\beta)\, dx \right|^2 = \left| \Psi(x_\beta) \right|^2. \tag{3.102}$$

If a further position measurement is made immediately afterwards, the wavefunction will still be in the collapsed state, $\delta(x - x_\beta)$, and so the same position (x_β) will be measured. That is,

$$\Psi(x) = \delta(x - x_\beta) \xrightarrow{\text{measure position}} \delta(x - x_\beta). \tag{3.103}$$

However, if immediately after this the momentum of the particle is measured, the wavefunction will collapse to one of the momentum eigenfunctions (recall that the specific position eigenfunction can be written as a continuous sum of momentum eigenfunctions), corresponding to the measured momentum p. That is,

$$\Psi(x) = \delta\left(x - x_\beta\right) = \frac{1}{2\pi\hbar}\int_{-\infty}^{\infty} e^{ip(x-x_\beta)/\hbar}dp \xrightarrow{\text{measure momentum}} e^{ipx/\hbar} \tag{3.104}$$

with probability

$$P(p) = \left|\int_{-\infty}^{\infty} \Psi(x)\, e^{-ipx/\hbar}dx\right|^2 = \left|\int_{-\infty}^{\infty} \delta\left(x - x_\beta\right) e^{-ipx/\hbar}dx\right|^2 = \left|e^{ipx_\beta/\hbar}\right|^2 = 1. \tag{3.105}$$

All values of momentum are equally likely, since the position is known exactly. That is, the uncertainty in position is zero, and so the uncertainty in momentum must be infinite; recall the uncertainty principle (2.39).

Further measurements of momentum will, with 100 percent probability, yield the same value of momentum, and so the system would stay in the momentum state $e^{ipx/\hbar}$,

$$\Psi(x) = e^{ipx/\hbar} \xrightarrow{\text{measure momentum}} e^{ipx/\hbar}, \tag{3.106}$$

until disturbed by an interaction with the "outside world." If a still later measurement of the position is made, because the particle is in a specific momentum state (a superposition of position eigenfunctions), the position recorded by the measurement will once again come down to probability, as the state function collapses to a certain position eigenfunction.

The preceding description follows from the Copenhagen Interpretation, which, obviously, engenders some philosophical problems. Other theories exist, each with their own potential philosophical problems, although the details will not be presented here. However, experiments have shown quantum systems existing in a superposition of states, and so this is not merely a philosophical issue. As an interesting practical application, in quantum cryptography, the collapse of the state function can be used to insure that data is securely delivered to its destination over a network. This is because any detection of information, no matter by what means, constitutes a measurement. This would cause the state function to collapse, indicating that the data had been compromised. Such systems have already been demonstrated in laboratory settings.

As another example, consider the two-slit experiment performed with electrons. Common sense would indicate that the electron passes through one slit or the other to arrive at the detector. However, quantum mechanically, before reaching the detector the electron exists in a superposition of states, and only upon detection does this state function collapse to a single state and a particle-like existence. If one were to try to measure which slit the electron passes through, however carefully the measurement is performed, the double-slit interference pattern would not be observed, since by performing this observation (i.e., measurement), the state function collapses, and the electron acts like a particle.

POSTULATE 3. The mean value of an observable is the expectation value of the corresponding operator.

In classical probability theory, the mean (mathematical expectation) of a function f of a random variable x, $\mu = \langle f \rangle$, is calculated from

$$\langle f \rangle = \int_{-\infty}^{\infty} f(x)\, \rho(x)\, dx, \tag{3.107}$$

where $\rho(x)$ is called the *probability density function* (pdf), such that

$$\int_{-\infty}^{\infty} \rho(x)\, dx = 1. \tag{3.108}$$

Speaking loosely, the state function Ψ is the square root of the probability density function for a particle, and so it might be expected that Ψ^2 will play a role in determining expectation values of quantities. Since the wavefunction is, in general, complex-valued, rather than the square of the function, we actually need the modulus squared,

$$\rho(\mathbf{r}, t) = |\Psi(\mathbf{r}, t)|^2 = \Psi^*(\mathbf{r}, t)\, \Psi(\mathbf{r}, t)\,. \tag{3.109}$$

In general, for any physical observable $\mathcal{O}$, the mean of $\mathcal{O}$ (i.e., the average, taken over many individual measurements under identical conditions, and with the same initial state) is given by

$$\langle \mathcal{O} \rangle = \int \widehat{o}\rho(\mathbf{r}, t)\, d^3r \tag{3.110}$$

$$= \int \Psi^*(\mathbf{r}, t)\, \widehat{o}\, \Psi(\mathbf{r}, t)\, d^3r, \tag{3.111}$$

where $\widehat{o}$ is the operator corresponding to that observable. For example, the expectation value of position ($\mathcal{O} = \mathbf{r}$, $\widehat{o} = \mathbf{r}$) is

$$\langle \mathbf{r} \rangle = \int \Psi^*(\mathbf{r}, t)\, \mathbf{r}\, \Psi(\mathbf{r}, t)\, d^3r, \tag{3.112}$$

and in one dimension,

$$\langle x \rangle = \int \Psi^*(x, t)\, x\, \Psi(x, t)\, dx. \tag{3.113}$$

The expectation value for momentum is

$$\langle p_x \rangle = \int \Psi^*(\mathbf{r}, t) \left(-i\hbar \frac{\partial}{\partial x}\right) \Psi(\mathbf{r}, t)\, d^3r, \tag{3.114}$$

and similarly for p_y and p_z, such that the three-dimensional momentum is obtained from

$$\langle p \rangle = \int \Psi^* (\mathbf{r}, t) (-i\hbar\nabla) \Psi (\mathbf{r}, t) \, d^3r. \tag{3.115}$$

From (3.63), the expectation value of energy is obtained as

$$\langle E \rangle = \int \Psi^* (\mathbf{r}, t) \left(i\hbar \frac{\partial}{\partial t} \right) \Psi (\mathbf{r}, t) \, d^3r. \tag{3.116}$$

In all cases, the wavefunction used in (3.110)–(3.116) is normalized according to (3.8).

Example

Considering again the example of an electron confined to a line segment between $x = -L/2$ and $x = L/2$, with the state function given by (3.9), we find that

$$\Psi (x, t) = \left(\frac{2}{L} \right)^{1/2} \sin \left(\frac{n\pi}{L} x \right) e^{-iE_n t/\hbar}. \tag{3.117}$$

The expectation value of the particle is

$$\langle x \rangle = \int_{-L/2}^{L/2} \Psi^* (x, t) \, x \Psi (x, t) \, dx \tag{3.118}$$

$$= \frac{2}{L} \int_{-L/2}^{L/2} x \sin^2 \left(\frac{n\pi}{L} x \right) dx \tag{3.119}$$

$$= 0. \tag{3.120}$$

That is, the average position of the electron is in the middle of the line segment. Note that this is not necessarily the most likely location to find the electron. For example, for the $n = 2$ state $|\Psi|$ peaks at $x = \pm L/4$, which is the most likely place to find the electron, although the average position is still zero.

Postulate 4. The state function $\Psi (\mathbf{r}, t)$ obeys the Schrödinger equation

$$i\hbar \frac{\partial \Psi (\mathbf{r}, t)}{\partial t} = H \Psi (\mathbf{r}, t), \tag{3.121}$$

where H is the system Hamiltonian (total energy operator).

The Hamiltonian is comprised of kinetic and potential energy terms (and terms corresponding to applied electric or magnetic fields if such fields are present). For a particle with mass† m in the absence of applied electric and magnetic fields,

$$H = \left(-\frac{\hbar^2}{2m} \nabla^2 + V(\mathbf{r}, t) \right), \tag{3.122}$$

where the first term corresponds to momentum and the second term is potential energy. It should be noted that the environment that a particle resides in enters Schrödinger's equation via the potential energy term $V(\mathbf{r}, t)$, and by the boundary conditions enforced on Ψ, which will be discussed shortly.

Example

The state function for an electron confined to a line segment between $x = -L/2$ and $x = L/2$ has been given as (3.9),

$$\Psi(x, t) = \left(\frac{2}{L} \right)^{1/2} \sin\left(\frac{n\pi}{L} x \right) e^{-iE_n t/\hbar}, \tag{3.123}$$

for n even, with zero potential energy on the line segment. In this case, in one dimension

$$H = -\frac{\hbar^2}{2m_e} \frac{d^2}{dx^2}, \tag{3.124}$$

and so it is easy to see that (3.121) is satisfied as long as E_n is defined as in (3.11). That is, plugging (3.123) into (3.121) leads to

$$i\hbar \frac{\partial}{\partial t} \left(\left(\frac{2}{L} \right)^{1/2} \sin\left(\frac{n\pi}{L} x \right) e^{-iE_n t/\hbar} \right) = -\frac{\hbar^2}{2m_e} \frac{d^2}{dx^2} \left(\left(\frac{2}{L} \right)^{1/2} \sin\left(\frac{n\pi}{L} x \right) e^{-iE_n t/\hbar} \right), \tag{3.125}$$

or

$$E_n = \frac{\hbar^2}{2m_e} \left(\frac{n\pi}{L} \right)^2.$$

3.2 TIME-INDEPENDENT SCHRÖDINGER'S EQUATION

Equipped with Schrödinger's equation, and some understanding of the wavefunction, some simple problems involving quantum particles with mass can be formulated and solved. For a particle of mass m in a potential $V(\mathbf{r}, t)$, Schrödinger's equation is

$$i\hbar \frac{\partial \Psi(\mathbf{r}, t)}{\partial t} = \left(-\frac{\hbar^2}{2m} \nabla^2 + V(\mathbf{r}, t) \right) \Psi(\mathbf{r}, t). \tag{3.126}$$

†Photons, which are quantum particles without mass, can be represented as electromagnetic wave oscillators whose states obey (3.121) as well, although with a Hamiltonian different from (3.122). However, in light–material interactions, it is often sufficient to treat the electrons quantum mechanically and the photons as classical electromagnetic fields. We will follow this procedure here.

However, a considerable simplification occurs if the potential energy does not depend on time, which will be the case considered throughout this text. Assuming a product form for the wavefunction,

$$\Psi(\mathbf{r}, t) = \psi(\mathbf{r})\, g(t), \tag{3.127}$$

and substituting into Schrödinger's equation results in

$$\left(-\frac{\hbar^2}{2m}\frac{1}{\psi(\mathbf{r})}\nabla^2\psi(\mathbf{r}) + V(\mathbf{r})\right) = i\hbar\frac{1}{g(t)}\frac{\partial g(t)}{\partial t}. \tag{3.128}$$

The left side of (3.128) is clearly, at most, a function of position $\mathbf{r}$ (but not time), and the right side is, at most, a function of time (but not position). How can a function of position but not time be equal to a function of time but not position, for all times and positions? They clearly cannot be equal unless each side is equal to the same constant, which will be called E (because, although it is not proven here, this constant will represent energy). Therefore,

$$\left(-\frac{\hbar^2}{2m}\nabla^2 + V(\mathbf{r})\right)\psi(\mathbf{r}) = E\psi(\mathbf{r}) \tag{3.129}$$

$$i\hbar\frac{dg(t)}{dt} = Eg(t). \tag{3.130}$$

The method of separating the time and space dependence in (3.126) is called *separation of variables*. Note that (3.129) is really an eigenvalue problem, $\widehat{H}\psi = \lambda\psi$, where the operator is the Hamiltonian (total energy operator),

$$\widehat{H} = \left(-\frac{\hbar^2}{2m}\nabla^2 + V(\mathbf{r})\right), \tag{3.131}$$

and where the eigenvalue is the (usually unknown) energy E. For constant potentials, solutions of the time-independent Schrödinger's equation are eigenfunctions of the Laplacian operator, ∇^2, which for one dimension is d^2/dx^2.

The time-dependent equation (3.130) is easy to solve, leading to

$$g(t) = g_0 e^{-iEt/\hbar}. \tag{3.132}$$

Notice that if (3.132) is compared with a general oscillatory form

$$g(t) = g_0 e^{-i\omega t}, \tag{3.133}$$

then it can be seen that

$$E = \hbar\omega \tag{3.134}$$

makes sense, which is nothing more than the previously stated energy–frequency relation (2.12). Of course, this is not a proof that the constant E in (3.129) is the total energy, but it turns out to be true.

From (3.127), a general time-dependent wavefunction for a particle in a time-independent potential is

$$\Psi(\mathbf{r}, t) = \psi(\mathbf{r})\, e^{-iEt/\hbar}. \tag{3.135}$$

For the time-independent wavefunctions, the orthogonality condition is

$$\int \psi_i^*(\mathbf{r})\, \psi_j(\mathbf{r})\, d^3r = 0 \tag{3.136}$$

for $i \neq j$, where i and j denote different energy eigenvalues,

$$\left(-\frac{\hbar^2}{2m}\nabla^2 + V(\mathbf{r})\right)\psi_i(\mathbf{r}) = E_i \psi(\mathbf{r}), \tag{3.137}$$

$$\left(-\frac{\hbar^2}{2m}\nabla^2 + V(\mathbf{r})\right)\psi_j(\mathbf{r}) = E_j \psi(\mathbf{r}). \tag{3.138}$$

In summary, the time-independent Schrödinger equation is

$$\left(-\frac{\hbar^2}{2m}\nabla^2 + V(\mathbf{r})\right)\psi(\mathbf{r}) = E\psi(\mathbf{r}), \tag{3.139}$$

where E is the total energy of the particle, which has mass m. The time-dependent wavefunction is obtained from the time-independent wavefunction as (3.135). In one dimension, the time-independent Schrödinger equation is

$$\left(-\frac{\hbar^2}{2m}\frac{d^2}{dx^2} + V(x)\right)\psi(x) = E\psi(x), \tag{3.140}$$

where

$$\psi(x, t) = \psi(x)\, e^{-iEt/\hbar}. \tag{3.141}$$

For time-independent potentials, the evolution of the state function is determined uniquely in time upon specification of the initial state $\psi(x, 0)$.

It should be noted that Schrödinger's equation (time dependent, or time independent) is a linear homogeneous equation. As such, a superposition of homogeneous solutions ψ_n is itself a solution. Therefore, the most general state function for the separable case can be written as

$$\Psi(\mathbf{r}, t) = \sum_n a_n \psi_n(\mathbf{r})\, e^{-iE_n t/\hbar}, \tag{3.142}$$

i.e., as an eigenfunction expansion, where ψ_n are eigenfunctions of the Hamiltonian operator and a_n are weighting constants.

For example, a wavepacket can be constructed from a superposition of plane waves, where each plane wave is a solution to Schrödinger's equation having a different wavenumber k. The wavepacket is not associated with a specific energy, but with a range of energies.

This is a general situation—one often cannot say that a particle has a definite energy, only that the particle may have any of the possible energies occurring in the superposition, with a probability proportional to the wavepacket probability, which is related to the amplitude a_n of a particular plane wave having that energy.

3.2.1 Boundary Conditions on the Wavefunction

Before treating some simple problems, notice that Schrödinger's equation is a second-order differential equation, and so boundary conditions need to be imposed on the wavefunction Ψ. The boundary or connection conditions will be obtained in one dimension for simplicity. A typical situation involves considering two adjacent regions, region 1 for $x < a$, where the particle has mass m_1 and "sees" a potential V_1, and region 2 for $x > a$, in which the particle has mass m_2 and "sees" a potential V_2. It will be assumed that the quantity $V_1 - V_2$ is finite.

While it may seem that the particle should have the same mass in both regions, there are many situations (especially for semiconductors, as discussed in Chapter 5) where the particle has an effective mass that is different in different regions of space. By assuming that the transition between regions occurs not abruptly at $x = a$, but over a small range of x values, and then taking the limit that the transition becomes abrupt, it can be shown that

$$\Psi\left(x = a^-\right) = \Psi\left(x = a^+\right), \tag{3.143}$$

$$\frac{1}{m_1}\Psi'\left(x = a^-\right) = \frac{1}{m_2}\Psi'\left(x = a^+\right),$$

where $\Psi\left(x = a^-\right)$ and $\Psi\left(x = a^+\right)$ are the wavefunctions in regions 1 and 2, respectively, infinitesimally close to the transition point $x = a$, and where Ψ' denotes the derivative of Ψ with respect to x.

However, if V has an infinite step discontinuity, say, V_1 is finite and V_2 is infinite, then only the wavefunction, and not its derivative, will be continuous at $x = a$,

$$\Psi\left(x = a^-\right) = \Psi\left(x = a^+\right). \tag{3.144}$$

Example

In previous examples, a one-dimensional space was considered, wherein an electron was constrained to be in the range $-L/2 \leq x \leq L/2$, with $V = 0$. Here we collect the various results concerning this example, and show the derivation of the state function.[†]

Since $V = 0$, Schrödinger's equation (3.140) is

$$-\frac{\hbar^2}{2m_e}\frac{d^2}{dx^2}\psi(x) = E\psi(x), \tag{3.145}$$

[†]The confined region of space considered in this example is often called a *quantum well* or *quantum box*.

which has the solution

$$\psi(x) = \begin{cases} A\sin kx + B\cos kx, & -\frac{L}{2} \leq x \leq \frac{L}{2}, \\ 0, & |x| > \frac{L}{2}, \end{cases} \tag{3.146}$$

where

$$k^2 = \frac{2m_e E}{\hbar^2}. \tag{3.147}$$

It is easy to verify this solution by inserting (3.146) into (3.145). Continuity of the wavefunction at $x = \pm L/2$, i.e., applying (3.144) at $x = \pm L/2$, leads to

$$\psi\left(-\frac{L}{2}\right) = \psi\left(\frac{L}{2}\right) = 0, \tag{3.148}$$

and so it is obvious that either

$$B = 0, \quad \frac{kL}{2} = n\pi, \quad n = 0, \pm 1, \pm 2, \ldots, \tag{3.149}$$

or

$$A = 0, \quad \frac{kL}{2} = \left(n + \frac{1}{2}\right)\pi, \quad n = 0, \pm 1, \pm 2, \ldots. \tag{3.150}$$

The wavefunction for $-L/2 \leq x \leq L/2$ is, therefore,

$$\begin{aligned} \psi_n(x) &= A\sin\frac{n\pi}{L}x, \quad n \text{ even} \\ &= B\cos\frac{n\pi}{L}x, \quad n \text{ odd}, \end{aligned} \tag{3.151}$$

which represents odd and even solutions with respect to the center of the space ($x = 0$). Furthermore, note that (3.145) has the form of an eigenvalue equation $H\psi = E\psi$, and therefore (3.151) are energy eigenfunctions. These wavefunctions clearly exhibit orthogonality,

$$\int \psi_n^*(x)\,\psi_m(x)\,dx = A^*A\int_{-L/2}^{L/2}\left(\sin\frac{n\pi}{L}x\right)\left(\sin\frac{m\pi}{L}x\right)dx = 0 \tag{3.152}$$

$$= B^*B\int_{-L/2}^{L/2}\left(\cos\frac{n\pi}{L}x\right)\left(\cos\frac{m\pi}{L}x\right)dx = 0 \tag{3.153}$$

$$= A^*B\int_{-L/2}^{L/2}\left(\sin\frac{n\pi}{L}x\right)\left(\cos\frac{m\pi}{L}x\right)dx = 0 \tag{3.154}$$

for $m \neq n$ in the first two expressions, and for all m, n in (3.154). The solutions can be normalized according to (3.8),

$$1 = \int_{-L/2}^{L/2} |\psi_n(x)|^2\,dx = \int_{-L/2}^{L/2} A^2\sin^2\frac{n\pi}{L}x\,dx, \tag{3.155}$$

resulting in

$$\psi_n(x) = \left(\frac{2}{L}\right)^{1/2}\sin\frac{n\pi}{L}x, \tag{3.156}$$

or

$$\Psi_n(x,t) = \left(\frac{2}{L}\right)^{1/2} \sin\left(\frac{n\pi}{L}x\right) e^{-iEt/\hbar}, \tag{3.157}$$

for the odd solution (n even). The even solution is

$$\Psi_n(x,t) = \left(\frac{2}{L}\right)^{1/2} \cos\left(\frac{n\pi}{L}x\right) e^{-iEt/\hbar}, \tag{3.158}$$

for n odd. These are the functions provided in the original example on page 46.

Since k is discrete, the energy of the particle (in this case, the energy eigenvalue) is found to be discrete,

$$E = \frac{\hbar^2 k^2}{2m_e} = \frac{\hbar^2}{2m_e}\left(\frac{n\pi}{L}\right)^2 = E_n. \tag{3.159}$$

The momentum associated with any given state is

$$p = \frac{h}{\lambda} = \frac{hk}{2\pi} = \frac{n\pi\hbar}{L} = p_n, \tag{3.160}$$

where the relation $k = 2\pi/\lambda$ was used.

From (3.113), the expectation values of the particle's position and momentum are, by Postulate 3,

$$\langle x \rangle = \int_{-L/2}^{L/2} \psi^*(x,t)\, x\, \psi(x,t)\, dx \tag{3.161}$$

$$= \frac{2}{L}\int_{-L/2}^{L/2} x \sin^2\left(\frac{n\pi}{L}x\right) dx = 0, \tag{3.162}$$

$$\langle p_x \rangle = \int \Psi^*(x,t)\left(-i\hbar\frac{\partial}{\partial x}\right)\Psi(x,t)\, dx \tag{3.163}$$

$$= -i\hbar\frac{2}{L}\frac{n\pi}{L}\int_{-L/2}^{L/2} \sin\left(\frac{n\pi}{L}x\right)\cos\left(\frac{n\pi}{L}x\right) dx = 0. \tag{3.164}$$

Furthermore, for example, the odd eigenfunction (sine) can be written as

$$\psi_n(x) = \frac{1}{\sqrt{2}}\left[\frac{1}{i\sqrt{L}}e^{i\frac{p_n}{\hbar}x} - \frac{1}{i\sqrt{L}}e^{-i\frac{p_n}{\hbar}x}\right] \tag{3.165}$$

$$= \psi_n^+ - \psi_n^-.$$

(See problem 3.8.) The term ψ_n^+ (ψ_n^-) is a plane wave propagating with positive (negative) momentum. Thus, any particular state described by the sine function (eigenfunctions of the Hamilton) can be thought of as representing a superposition of positive and negative momentum states (eigenfunctions of the momentum operator). See the comment about different representations of the state function on page 50. The same comment naturally applies to the cosine eigenfunction.

The previous example is one of the simplest quantum mechanical problems that can be solved. This problem will be revisited in considerable detail in the next chapter. It shows that if a particle is confined to a finite region of space, its wavefunction forms standing waves and the possible energy levels that it can occupy are discrete. In a two-dimensional space, one can consider an analogous situation of a circular boundary at $r = a$ and solve Schrödinger's equation in a circular region $r < a$, subject to $\psi = 0$ for $r \geq a$. The solution involves Bessel functions and will not be discussed here, although the result is a standing wave pattern in the radial direction. Assuming that the particle in question is an electron, we can consider these to be standing electron waves, which have been experimentally observed in a structure called a *quantum corral*. The corral is formed by moving atoms on the surface of a material to form a boundary, setting up an electron standing wave pattern, as shown in the figure at the beginning of this chapter. In Fig. 3.1, the various panels depict assembling the corral from iron atoms on a copper surface, and the resulting electron standing waves are obtained by radial interference. Imaging is done using STM. In Fig. 3.2, a similar structure called a *stadium corral* is shown, and other shapes are possible.

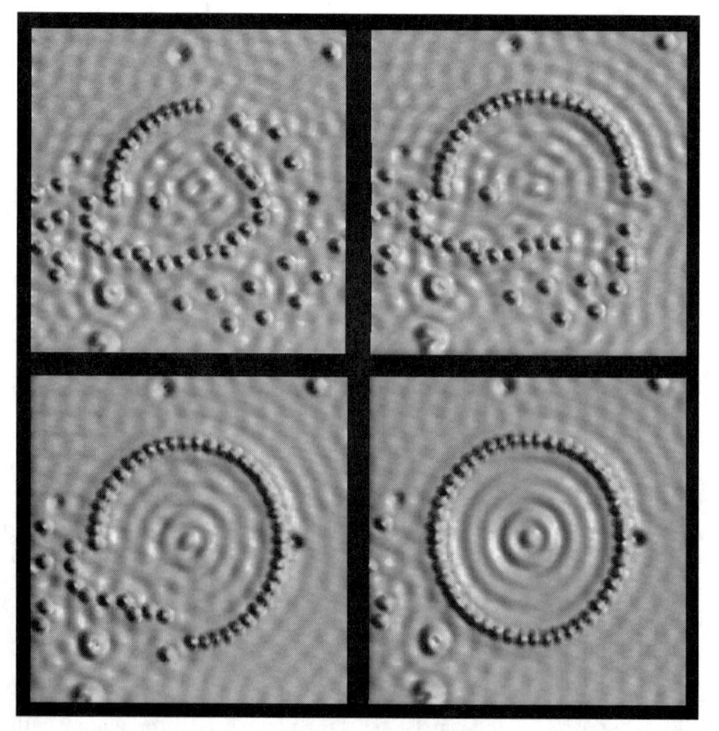

Figure 3.1 Evidence for the existence of electron waves—electron waves on the surface of copper, imaged by STM. The waves are trapped by a ring of iron atoms, which were manipulated into place using an atomic force microscope. (Reprinted with permission from Crommie, M. F., C. P. Lutz, and D. M. Eigler, "Confinement of Electrons to Quantum Corrals on a Metal Surface," *Science* 262 (1993): 218–220. Courtesy IBM Research, Almaden Research Center. Unauthorized use not permitted.)

Figure 3.2 Stadium corral made from iron atoms on a copper surface. Ripples are due to the standing wave patterns of the electron density distribution of quantum states. (Courtesy IBM Research, Almaden Research Center. Unauthorized use not permitted.)

3.3 ANALOGIES BETWEEN QUANTUM MECHANICS AND CLASSICAL ELECTROMAGNETICS

For students already familiar with classical electromagnetic theory, it is worthwhile to point out briefly some analogies between Schrödinger's equation and the equations describing electromagnetic waves.

Maxwell's equations describe classical electromagnetics waves, and for a vacuum, Maxwell's equations are

$$\begin{aligned}
\nabla \cdot \mathbf{E}(\mathbf{r}, t) &= \rho_e(\mathbf{r}, t)/\varepsilon_0, \\
\nabla \cdot \mathbf{B}(\mathbf{r}, t) &= 0, \\
\nabla \times \mathbf{E}(\mathbf{r}, t) &= -\frac{\partial}{\partial t}\mathbf{B}(\mathbf{r}, t), \\
\nabla \times \mathbf{B}(\mathbf{r}, t) &= \mu_0\varepsilon_0\frac{\partial}{\partial t}\mathbf{E}(\mathbf{r}, t) + \mu_0\mathbf{J}_e(\mathbf{r}, t),
\end{aligned} \tag{3.166}$$

where $\mathbf{E}$ is the electric field intensity (V/m), $\mathbf{B}$ is the magnetic flux density (Wb/m^2), ρ_e is the electric charge density (C/m^3), $\mathbf{J}_e$ is the electric current density (A/m^2), ϵ_0 is the permittivity of vacuum ($\varepsilon_0 \simeq 8.85 \times 10^{-12}$ F/m), and μ_0 is the permeability of vacuum ($\mu_0 \simeq 4\pi \times 10^{-7}$ H/m), and where V stands for volts, C for coulombs, Wb for webers, A for amperes, F for farads, H for henrys, and m for meters. For dimensional analysis, C = A · s = F · V and Wb = V · s = H · A, where s stands for seconds. The equations are known, respectively, as *Gauss's law*, the *magnetic Gauss's law*, *Faraday's law*, and *Ampère's law*.

Wave phenomena are predicted from Maxwell's equations by suitable manipulation. For example, taking the curl of Faraday's law, and using Ampère's law, we have

$$\nabla \times \nabla \times \mathbf{E}(\mathbf{r}, t) = -\left(\mu_0 \varepsilon_0 \frac{\partial^2}{\partial t^2} \mathbf{E}(\mathbf{r}, t) + \mu_0 \frac{\partial}{\partial t} \mathbf{J}_e(\mathbf{r}, t) \right). \tag{3.167}$$

If we assume that for any position and time of interest the electric and magnetic currents and charges (the sources of the field) are not present, then

$$\frac{\partial^2}{\partial t^2} \mathbf{E}(\mathbf{r}, t) = \frac{1}{\mu_0 \varepsilon_0} \nabla^2 \mathbf{E}(\mathbf{r}, t), \tag{3.168}$$

where the vector identity $\nabla \times \nabla \times \mathbf{A} = \nabla (\nabla \cdot \mathbf{A}) - \nabla^2 \mathbf{A}$, for $\mathbf{A}$, a general vector, was used.

There is obviously some similarity between (3.168) and Schrödinger's equation for a particle of mass m in the absence of a potential,

$$i\hbar \frac{\partial \Psi (\mathbf{r}, t)}{\partial t} = -\frac{\hbar^2}{2m} \nabla^2 \Psi (\mathbf{r}, t), \tag{3.169}$$

although the presence of the second time derivative in (3.168) is actually quite an important distinction. However, if an oscillatory time variation $e^{-i\omega t}$ is assumed for the electric field (corresponding to a radian frequency ω) and for the wavefunction (corresponding to the form (3.135) with $E = \hbar\omega$), then

$$\nabla^2 \mathbf{E}(\mathbf{r}, t) = -k_{ME}^2 \mathbf{E}(\mathbf{r}, t) \tag{3.170}$$

for the electromagnetic field. Also,

$$\nabla^2 \Psi (\mathbf{r}, t) = -k_{SE}^2 \Psi (\mathbf{r}, t), \tag{3.171}$$

for Schrödinger's equation, where

$$k_{ME}^2 = \omega^2 \mu \varepsilon, \quad k_{SE}^2 = \frac{2mE}{\hbar^2}, \tag{3.172}$$

are in each case simply constants. (E on the right side of the equation for k_{SE} is energy, not electric field.) In this case, the equations governing the electric field and the Schrödinger wavefunction have the same form.

If a particle is bound to a certain region of space Ω, then outside of Ω, the wavefunction will be zero, and at the boundary S of the region $\Psi (S) = 0$. Thus, one would solve (3.171) subject to $\Psi (S) = 0$, leading to discrete values of k_{SE}, and, hence, discrete energy levels E_n, as obtained in the example of page 65. In an analogous manner, if a perfectly conducting electromagnetic cavity were considered, then $\mathbf{E}_{\tan} (S) = 0$, where $\mathbf{E}_{\tan}$ is the tangential electric field. Solving (3.170) subject to $\mathbf{E}_{\tan}(S) = 0$ leads to discrete values of k_{ME}, and, hence, discrete resonant frequencies[†] ω_n. Aside from the difference arising from the vector nature of $\mathbf{E}$ and the scalar nature of Ψ, the mathematical solutions for Ψ and for $\mathbf{E}$ will be the same.

Furthermore, in quantum mechanics $|\Psi|^2$ is a probability density, and provides the likelihood of finding a particle in a certain location at a certain time. If $|\Psi(\mathbf{r}, t)|^2$ is large at a certain $(\mathbf{r}, t)$, then it is more likely to find the particle there. In electromagnetics, $|\mathbf{E}(\mathbf{r}, t)|^2$ is related to the intensity of the field, and, viewed from a photon standpoint, where $|\mathbf{E}(\mathbf{r}, t)|^2$ is large, there are a large number of photons present at $(\mathbf{r}, t)$. Viewed quantum mechanically, there is *likely* to be a large number of photons present at $(\mathbf{r}, t)$ when $|\mathbf{E}(\mathbf{r}, t)|^2$ is large.

Many more analogies are possible, although these will not be discussed in detail here. For example, one analogous situation will arise later when quantum wires are considered. These structures lead to discrete modal solutions of Schrödinger's equation, similar to electromagnetic modes of a conducting or dielectric waveguide in classical electromagnetics. In fact, these structures are often called *electron waveguides*.

3.4 PROBABILISTIC CURRENT DENSITY

In electromagnetic theory, $\mathbf{J}_e$ in Maxwell's equations (3.166) represents electric current density. Current I flowing through a conductor is classically obtained from the current density as

$$I = \int_S \mathbf{J}_e \cdot \widehat{\mathbf{n}}\, dS, \tag{3.173}$$

where S is the cross-sectional surface through which we want to determine the current (charge movement), and $\widehat{\mathbf{n}}$ is a unit amplitude vector normal to the surface (such that $\mathbf{J}_e \cdot \widehat{\mathbf{n}}$ is the component of current crossing the surface, as depicted in Fig. 3.3.)

Recall also that electrical current is the flow of electrical charge q (often, but not necessarily, electrons),

$$I = \frac{\partial q(t)}{\partial t}. \tag{3.174}$$

An important equation that demonstrates that charge conservation is embedded in (3.166) is known as the *continuity equation*. Taking the divergence of Ampère's law, we have

$$0 = \nabla \cdot \nabla \times \mathbf{H} = \nabla \cdot \mathbf{J}_e + \nabla \cdot \frac{\partial \mathbf{D}}{\partial t}, \tag{3.175}$$

noting the vector identity $\nabla \cdot \nabla \times \mathbf{A} = 0$ for any vector $\mathbf{A}$. Upon interchanging the spatial and temporal derivatives, and invoking Gauss's law, we obtain the continuity equation

$$\nabla \cdot \mathbf{J}_e(\mathbf{r}, t) = -\frac{\partial \rho_e(\mathbf{r}, t)}{\partial t}. \tag{3.176}$$

[†]In fact, this really relates to photons having discrete energy via $E_n = \hbar\omega_n$.

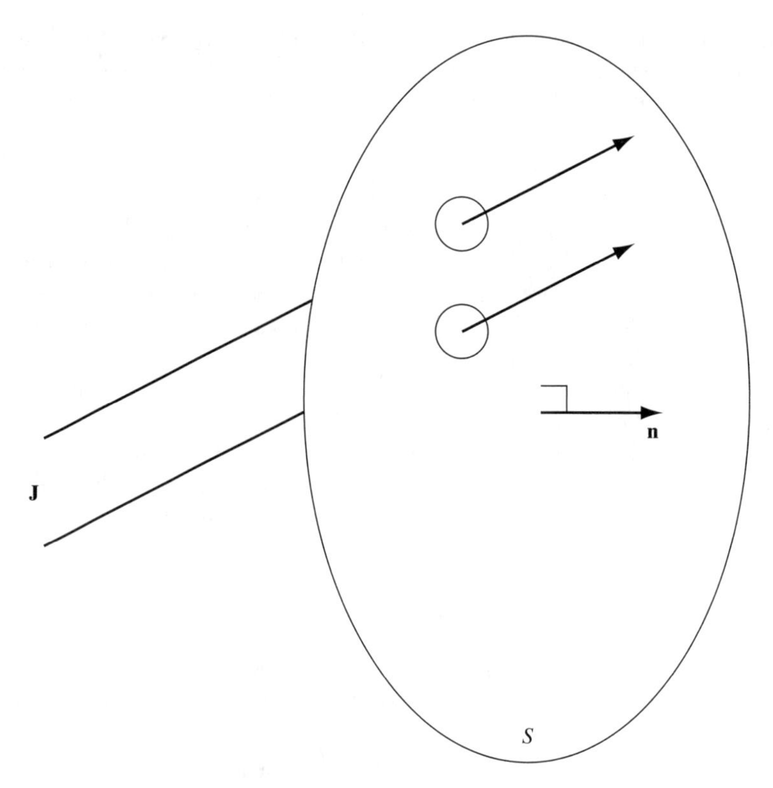

Figure 3.3 Current density incident on a surface S.

Often it is useful to take the volume integral of (3.176), leading to

$$\int_{\Omega} \nabla \cdot \mathbf{J}_e(\mathbf{r}, t)\, d^3r = \oint_S \mathbf{J}_e \cdot \widehat{\mathbf{n}}\, d^2r = -\int_{\Omega} \frac{\partial \rho_e(\mathbf{r}, t)}{\partial t} d^3r = -\frac{d}{dt} Q(t), \tag{3.177}$$

where the divergence theorem was used,

$$\int_{\Omega} \nabla \cdot \mathbf{A}\, d^3r = \oint_S \widehat{\mathbf{n}} \cdot \mathbf{A}\, d^2r, \tag{3.178}$$

where S is the surface of Ω and $\widehat{\mathbf{n}}$ is a unit vector normal to the surface. The physical interpretation of (3.177) is that the net outward flux of current from a volume Ω bounded by a closed surface S must equal the time rate of decrease of charge Q within the surface.

In quantum mechanics, we are often concerned with the location and movement of particles, which are described by the wavefunction. The flow of quantum particles can be considered using the following development, which will result in a relationship analogous to (3.176).

It is easy to see that

$$\frac{\partial}{\partial t}\left(\Psi^{*}(\mathbf{r},t)\,\Psi(\mathbf{r},t)\right)=\Psi^{*}(\mathbf{r},t)\,\frac{\partial}{\partial t}\Psi(\mathbf{r},t)+\Psi(\mathbf{r},t)\,\frac{\partial}{\partial t}\Psi^{*}(\mathbf{r},t)\,. \tag{3.179}$$

From (3.126),

$$\frac{\partial\Psi(\mathbf{r},t)}{\partial t}=\frac{1}{i\hbar}\left(-\frac{\hbar^{2}}{2m}\nabla^{2}+V(\mathbf{r},t)\right)\Psi(\mathbf{r},t)\,, \tag{3.180}$$

$$\frac{\partial\Psi^{*}(\mathbf{r},t)}{\partial t}=\frac{1}{-i\hbar}\left(-\frac{\hbar^{2}}{2m}\nabla^{2}+V(\mathbf{r},t)\right)\Psi^{*}(\mathbf{r},t)\,, \tag{3.181}$$

assuming the potential V is real valued. Substituting into (3.179), we have

$$\frac{\partial}{\partial t}\left(\Psi^{*}(\mathbf{r},t)\,\Psi(\mathbf{r},t)\right)=i\frac{\hbar}{2m}\left(\Psi^{*}(\mathbf{r},t)\,\nabla^{2}\Psi(\mathbf{r},t)-\Psi(\mathbf{r},t)\,\nabla^{2}\Psi^{*}(\mathbf{r},t)\right). \tag{3.182}$$

Using the vector identities

$$\nabla\cdot(\alpha\mathbf{A})=\mathbf{A}\cdot\nabla\alpha+\alpha\nabla\cdot\mathbf{A}, \tag{3.183}$$

$$\nabla\cdot\nabla\alpha=\nabla^{2}\alpha,$$

where α is a scalar and $\mathbf{A}$ is a vector,

$$\frac{\partial}{\partial t}\left(\Psi^{*}(\mathbf{r},t)\,\Psi(\mathbf{r},t)\right)=\nabla\cdot\frac{i\hbar}{2m}\left(\Psi^{*}(\mathbf{r},t)\,\nabla\Psi(\mathbf{r},t)-\Psi(\mathbf{r},t)\,\nabla\Psi^{*}(\mathbf{r},t)\right). \tag{3.184}$$

Comparing (3.184) with (3.176), and recalling that

$$\rho(\mathbf{r},t)=\Psi^{*}(\mathbf{r},t)\,\Psi(\mathbf{r},t) \tag{3.185}$$

is a probability density, we can recognize (3.184) as a conservation equation,

$$\frac{\partial\rho(\mathbf{r},t)}{\partial t}=-\nabla\cdot\mathbf{J}(\mathbf{r},t)\,, \tag{3.186}$$

where

$$\mathbf{J}(\mathbf{r},t)=\frac{-i\hbar}{2m}\left(\Psi^{*}(\mathbf{r},t)\,\nabla\Psi(\mathbf{r},t)-\Psi(\mathbf{r},t)\,\nabla\Psi^{*}(\mathbf{r},t)\right) \tag{3.187}$$

is called the *probability current density*. Multiplying by the charge q, we interpret $q\rho$ and $q\mathbf{J}$ as the probabilistic charge density and probabilistic current density, respectively, associated with the charge[†] q. The unit of probability current density is probability/m^2s.

[†]The normalization condition (3.8) leads to

$$\int_{\text{all space}} q\psi^{*}(\mathbf{r},t)\,\psi(\mathbf{r},t)\,d^{3}r=q,$$

in agreement with $q\psi^{*}\psi$ being a charge density, the integral of which provides the total charge.

From (3.187), we can appreciate that Ψ is often complex valued—if Ψ is real valued, there is no net current. In contradistinction, in classical time-harmonic (sinusoidal steady state) electromagnetics,

$$\mathbf{J}_e(\mathbf{r}, t) = \text{Re}\left\{\mathbf{J}_e(\mathbf{r}) e^{i\omega t}\right\}, \tag{3.188}$$

resulting in a real-valued quantity. In classical electromagnetics, as in sinusoidal steady-state circuit analysis, the use of complex phasors and the complex time-variation $e^{i\omega t}$ is merely for convenience. In quantum mechanics, the use of complex quantities is a necessity.

To aid in the interpretation of (3.187), it is useful to refer again to the electromagnetic case. Consider the flow of classical electrons in the x direction, crossing the y–z plane, where we assume that the current density is independent of x, $\mathbf{J}_e = \mathbf{J}_e(y, z, t)$. Then, the current at time t crossing the y–z plane is

$$I(t) = \int_y \int_z \mathbf{J}_e(y, z, t) \cdot \mathbf{a}_x \, dydz. \tag{3.189}$$

In a similar manner, in the quantum mechanical case

$$I(t) = \int_y \int_z q\mathbf{J}(y, z, t) \cdot \mathbf{a}_x \, dydz \tag{3.190}$$

is the probability current at time t crossing the y–z plane. This expression will be used later in describing electron transport through materials. (See, e.g., Section 10.2.3.)

Example

Assume a plane–wave wavefunction

$$\Psi(x, t) = Ae^{i(kx - Et/\hbar)}. \tag{3.191}$$

The probability current density is

$$\mathbf{J}(\mathbf{r}, t) = \frac{-i\hbar}{2m}\left(\Psi^*(\mathbf{r}, t)\nabla\Psi(\mathbf{r}, t) - \Psi(\mathbf{r}, t)\nabla\Psi^*(\mathbf{r}, t)\right) \tag{3.192}$$

$$= \frac{-i\hbar}{2m}\mathbf{a}_x |A|^2 \left(e^{-i(kx-Et/\hbar)}\frac{\partial}{\partial x}\left(e^{i(kx-Et/\hbar)}\right) - e^{i(kx-Et/\hbar)}\frac{\partial}{\partial x}\left(e^{-i(kx-Et/\hbar)}\right)\right) \tag{3.193}$$

$$= \mathbf{a}_x |A|^2 \frac{\hbar k}{m} = |A|^2 \frac{\mathbf{a}_x p}{m} = |A|^2 \frac{\mathbf{p}}{m} = |A|^2 \mathbf{v}, \tag{3.194}$$

where (2.15) was used. Therefore, in this case, the quantum current density is merely the product of the particle's probability density and the particle velocity.

3.5 MULTIPLE PARTICLE SYSTEMS

For much of this book, we will need to consider only the single particle Schrödinger equation, as presented in previous sections. With typically on the order of 10^{22} electrons per cubic centimeter for many materials, this should seem like a rather coarse approximation! To understand why the Schrödinger equation for a single particle is often adequate, we need to first examine how to model a system containing multiple particles.

Consider the case of two particles, particle 1 and particle 2. Recall that the state function describes the state of the system, which in this case is comprised of both particles. Therefore, the state function must depend on both particles,

$$\Psi(\mathbf{r}_1, \mathbf{r}_2, t), \tag{3.195}$$

where $\mathbf{r}_1$ and $\mathbf{r}_2$ are the "positions" of particles 1 and 2, respectively, in the following sense,

$$|\Psi(\mathbf{r}_1, \mathbf{r}_2, t)|^2 d^3r_1 d^3r_2 \tag{3.196}$$

is the joint probability of finding particle 1 in the vicinity d^3r_1 of point $\mathbf{r}_1$, and simultaneously finding particle 2 in the vicinity d^3r_2 of point $\mathbf{r}_2$.

The state function obeys (3.121),

$$i\hbar\frac{\partial\Psi(\mathbf{r}_1, \mathbf{r}_2, t)}{\partial t} = H\Psi(\mathbf{r}_1, \mathbf{r}_2, t), \tag{3.197}$$

where H is the system Hamiltonian that has the form

$$H = H_1 + H_2 + H_{12}. \tag{3.198}$$

In (3.198), H_1 is the Hamiltonian for particle 1 by itself, H_2 is the Hamiltonian for particle 2 by itself, and H_{12} represents the interaction between the two particles.

As an example, consider two charged particles in an otherwise empty space. Then, from (3.122),

$$\begin{aligned} H_1 &= -\frac{\hbar^2}{2m_1}\nabla_1^2, \quad H_2 = -\frac{\hbar^2}{2m_2}\nabla_2^2, \\ H_{12} &= \frac{q_1 q_2}{4\pi\varepsilon_0}\frac{1}{|\mathbf{r}_1 - \mathbf{r}_2|}, \end{aligned} \tag{3.199}$$

where $\nabla_{1,2}^2$ indicates the Laplacian in terms of each particle's coordinates,

$$\nabla_i^2 = \frac{\partial^2}{\partial x_i^2} + \frac{\partial^2}{\partial y_i^2} + \frac{\partial^2}{\partial z_i^2} \tag{3.200}$$

for particle i, and where $\mathbf{r}_i = (x_i, y_i, z_i)$, $i = 1, 2$. In (3.199), H_{12} is merely the electrostatic Coulomb interaction between the charged particles, i.e., the classical term for particle interaction.

The two-particle wavefunction, $\Psi(\mathbf{r}_1, \mathbf{r}_2, t)$, is said to exist in a six-dimensional *configuration space.* That is, particle 1 is described by three coordinates, and likewise for particle 2 (plus a time coordinate). For N interacting particles, one must consider a $3N$ dimensional configuration space. It is easy to see that the complexity grows rapidly with increasing the number of particles.

However, a vast simplification occurs if N particles are present, but if the particles do not interact with each other. Returning to the two-particle system, in the limit of noninteracting particles ($H_{12} = 0$), Schrödinger's equation becomes

$$i\hbar \frac{\partial \Psi(\mathbf{r}_1, \mathbf{r}_2, t)}{\partial t} = (H_1 + H_2)\, \Psi(\mathbf{r}_1, \mathbf{r}_2, t). \tag{3.201}$$

By setting

$$\Psi(\mathbf{r}_1, \mathbf{r}_2, t) = \Psi_1(\mathbf{r}_1, t)\, \Psi_2(\mathbf{r}_2, t), \tag{3.202}$$

we obtain two independent single-particle equations,

$$i\hbar \frac{\partial \psi_1(\mathbf{r}_1, t)}{\partial t} = H_1 \Psi(\mathbf{r}_1, t), \tag{3.203}$$

$$i\hbar \frac{\partial \psi_2(\mathbf{r}_2, t)}{\partial t} = H_2 \Psi(\mathbf{r}_2, t), \tag{3.204}$$

both of which can be solved independently of the other. Thus, for a system of N noninteracting particles one need "only" solve N single-particle equations. However, often the particles in question are identical (say, the 10^{22} particles in each cubic centimeter are all electrons), and indistinguishable (as explained in the following discussion). In this case, all 10^{22} equations are identical, and we only need to solve one single-particle equation. It is for this reason that often we only need to consider the single-particle Schrödinger equation. The simplicity gained by a one-particle approach, compared to a multiparticle approach, cannot be overestimated.

Of course, the validity of the single-particle Schrödinger equation relies on some assumptions that may appear questionable. First, it was assumed that the electrons don't interact with each other. One reason why electron-electron interactions may be ignored, as a first approximation, could be that the electrons are initially very distant from one another. Then, although the "system" really consists of all of the electrons, it is reasonable to assume, since the electrons are very far apart, and since the Coulomb potential between electrons varies as the reciprocal of distance, that the electrons don't influence each other. Then the single-particle Schrödinger equation can be solved for each particle (N independent equations for N different particles, or merely one equation if all of the particles are identical), obtaining the wavefunction. Often, in this case, the individually obtained wavefunctions are called *orbitals*, to distinguish them from the exact quantum state of the system. Each orbital will be a wavepacket concentrated where the likelihood of finding the electron is high. If electrons are brought closer together, at some point the individual orbitals will begin to overlap significantly. At this point the electrons will interact, and we may need to account for this interaction.

However, what about solid materials, where the electron density is on the order of 10^{22} electrons per cubic centimeter? In this case, the assumption of electrons being distant from each other clearly will not hold. In solids, one would expect that, with so many electrons present, they would be frequently bumping into each other. Amazingly, in many materials the assumption of noninteracting electrons is quite good. In general, this is a consequence of what is known as the *screening effect*: the electrons screen, or shield themselves, from each other. This is, in turn, related to the Pauli exclusion principle. Each electron repels the other electrons due to (1) the classical electrostatic repulsion (Coulomb interaction) between like charges, and (2) the exclusion principle, which says that electrons with the same spins tend to avoid each other. Because no two electrons can be in the same state, electrons with the same spin will not have a high probability of being located near each other. Thus, electrons tend to move through a material somewhat independently of the other electrons.

If we are truly talking about a collection of electrons in an otherwise empty space, then the electron–electron noninteracting assumption is all that is needed. However, we are often interested in solids, and so you might ask, what about the other atomic particles (the protons, neutrons, etc.), that make up atoms, which, in turn, make up the solid? They are, after all, quantum particles themselves, and therefore the system in question contains nonidentical particles, and we seem to be back to the multiparticle equation. It turns out that in crystalline materials, the so-called background lattice, which is everything associated with the regular array of atoms in a crystal except certain (outer shell) electrons that have been ionized, *does* have an important effect on the electrons. To be more specific, if the lattice is perfectly periodic, waves can pass freely through the structure at certain energies, whereas waves at other energies will not propagate. This leads to the concept of band theory and effective mass, which will be discussed in detail in Chapter 5. However, we can state at this point that band theory allows us to account for the lattice in a simple manner, via an effective mass in Schrödinger's equation, with the end result that we can solve the single-particle Schrödinger's equation in this case without treating the quantum particles that make up the background lattice individually. Again, the simplification provided by this approach cannot be overestimated. However, any disruption of perfect periodicity in the lattice, due to imperfections, thermal vibrations, etc., tends to scatter electrons, and must be accounted for separately.

Pauli Exclusion Principle. The consideration of multiple indistinguishable particles also provides some insight into the Pauli exclusion principle. Note that classical particles that may be identical are, nevertheless, distinguishable. Consider, for instance, identical bowling balls. One could mark each ball with a number or paint spot, thus making the balls distinguishable, without altering their properties. Systems consisting of large numbers of distinguishable particles follow Boltzmann statistics (page 268).

However, electrons are both identical and indistinguishable. Systems consisting of large numbers of indistinguishable particles follow either Fermi–Dirac or Bose–Einstein statistics (pages 268–269). Large numbers of indistinguishable particles that obey the Pauli exclusion principle follow Fermi–Dirac statistics, whereas indistinguishable particles that aren't bound to the exclusion principle follow Bose–Einstein statistics. In the following discussion a somewhat casual development of the Pauli exclusion principle for two particles is provided.

If two particles are indistinguishable, although perhaps interacting, then

$$|\Psi(\mathbf{r}_1, \mathbf{r}_2, t)|^2 \, d^3r_1 d^3r_2 \tag{3.205}$$

is the probability of finding one particle in d^3r_1 centered at $\mathbf{r}_1$ and one particle in d^3r_2 centered at $\mathbf{r}_2$. Since the particles are indistinguishable, the coordinates $\mathbf{r}_1$ and $\mathbf{r}_2$ are used simply because there are two particles—it is impossible to say which particle is which. This means that we must have

$$|\Psi(\mathbf{r}_1, \mathbf{r}_2, t)|^2 = |\Psi(\mathbf{r}_2, \mathbf{r}_1, t)|^2 . \tag{3.206}$$

Wavefunctions that satisfy (3.206) include the symmetric case

$$\Psi(\mathbf{r}_1, \mathbf{r}_2, t) = \Psi(\mathbf{r}_2, \mathbf{r}_1, t), \tag{3.207}$$

and the antisymmetric case

$$\Psi(\mathbf{r}_1, \mathbf{r}_2, t) = -\Psi(\mathbf{r}_2, \mathbf{r}_1, t). \tag{3.208}$$

The antisymmetric case (3.208) admits an interesting interpretation relevant to electrons. Consider the special case $\mathbf{r}_1 = \mathbf{r}_2$. Then, to satisfy (3.208), we must have

$$\Psi(\mathbf{r}_1, \mathbf{r}_1, t) = 0, \tag{3.209}$$

such that the probability of finding two such particles within the same small volume d^3r_1 is zero. It turns out that electrons obey this constraint, such that only one electron can occupy a certain state. This rule, which applies to fermions, is the Pauli exclusion principle.

More generally, the Pauli exclusion principle states that particles with half-integral spin quantum numbers (fermions; electrons, protons, neutrons) must have antisymmetric state functions, in the sense that a pairwise interchange of particles merely changes the sign of the state function, as in (3.208). Particles with integral spin quantum numbers (bosons; photons) must have symmetric state functions, (3.207).

If two fermions are considered, each of which satisfies a single particle state ψ_1 and ψ_2, respectively, then the composite wavefunction

$$\Psi(\mathbf{r}_1, \mathbf{r}_2) = \frac{1}{\sqrt{2}}\left(\psi_1(\mathbf{r}_1)\,\psi_2(\mathbf{r}_2) - \psi_1(\mathbf{r}_2)\,\psi_2(\mathbf{r}_1)\right) \tag{3.210}$$

will satisfy the antisymmetric condition (3.208). In a similar manner, for bosons, the wavefunction

$$\Psi(\mathbf{r}_1, \mathbf{r}_2) = \frac{1}{\sqrt{2}}\left(\psi_1(\mathbf{r}_1)\,\psi_2(\mathbf{r}_2) + \psi_1(\mathbf{r}_2)\,\psi_2(\mathbf{r}_1)\right) \tag{3.211}$$

will satisfy the symmetric condition (3.207).

3.6 SPIN AND ANGULAR MOMENTUM

As described previously, quantum particles have an intrinsic property called spin that has no classical counterpart. The name actually derives from early interpretations of experiments where electrons acted as though they were, in addition to orbiting the nucleus in an atom, spinning about their own axis. However, it was quickly understood that this was not the case. Spin is a purely quantum phenomenon that cannot be understood by appealing to everyday experiences—it should be regarded as just another property of quantum particles.

The Stern–Gerlach experiment in the early 1920s demonstrated the existence of spin, although spin was not known at the time and the experiment was performed for other reasons. Recall from classical electromagnetic theory that the force on a particle having charge q moving with velocity $\mathbf{v}$ in the presence of an electric field $\mathbf{E}$ and magnetic field $\mathbf{B}$ is given by the Lorentz force law,

$$\mathbf{F}_L = q\left(\mathbf{E} + \mathbf{v} \times \mathbf{B}\right). \tag{3.212}$$

Now assume that the electric field is not present, and consider passing a particle through a magnetic field. If the particle in question is charge neutral ($q = 0$), such as a nonionized atom, then it should suffer no Lorentz force deflection. However, another force may influence the particle. If the particle has a net magnetic moment[†] μ, then in passing through a magnetic field $\mathbf{B}$, it will experience a force

$$\mathbf{F}_\mu = \nabla\left(\mu \cdot \mathbf{B}\right). \tag{3.213}$$

Assume that the particles are moving along the y coordinate, the magnetic moment is constant, $\mu = \pm\mathbf{a}_z\mu_z$, and that the magnetic field is $\mathbf{B} = \mathbf{a}_z B_z(y)$. Then,

$$\mathbf{F}_\mu = \pm\mathbf{a}_y\mu_z\frac{\partial B_z(y)}{\partial y}, \tag{3.214}$$

so that the particles will be deflected in the $\pm y$ direction if the magnetic field is a function of y (i.e., not constant along y). The sign is chosen according to the direction of the magnetic moment vector.

In the Stern–Gerlach experiment, the atoms (i.e., charge-neutral quantum particles) are in the lowest energy state (as discussed further in Section 4.6), such that the electrons do not have any net angular momentum associated with electrons orbiting the nucleus. Thus, the magnetic moment of the atom should be zero. Since neutral atoms in the lowest energy state are used, $\mathbf{F}_L = \mathbf{F}_\mu = 0$, and, thus, the trajectory of the atoms should not be changed in passing through an inhomogeneous magnetic field. However, in the Stern–Gerlach experiment, it is found that the original beam of atoms splits into two components. One part of the beam is deflected in a certain direction (say, $+\mathbf{a}_y$), and the other part of the beam is deflected in the opposite direction ($-\mathbf{a}_y$). Thus, the atoms act in a manner that would be consistent with having a magnetic moment. Thus, the idea arose that electrons have an intrinsic magnetic moment, positive or negative but equal in magnitude, called spin. It can

[†]Magnetic moment shouldn't be confused with magnetic permeability, also denoted by μ. The magnetic moment of a current loop, i.e., a circulating charge, is $\mu = \widehat{\alpha} I A$, where I is the current in the loop, A is the cross-sectional area of the loop, and $\widehat{\alpha}$ is a unit vector normal to the loop's cross section, chosen to be in the direction given by the right-hand rule. The total magnetic moment of an atom is due to angular momentum (orbiting electrons forming small current loops) and spin.

be shown that the nuclei also have spin, which results in fine structure in the experimental results, but that the main effect is due to the electrons. Also, although the results could seemingly be explained classically if the electrons were spinning about their own axes, it can be shown that this is not the case.

Spin is quantized, taking either integral or half-integral values of $\hbar$. As mentioned in Section 2.4.3, particles with integral (in units of $\hbar$) spin are called bosons, and particles with half-integral (in units of $\hbar$) spin are called fermions.

Although not related to spin per se, recall that classical particles can possess angular momentum,

$$\mathbf{L} = \mathbf{r} \times \mathbf{p}, \tag{3.215}$$

where $\mathbf{r}$ is the particle's position vector and $\mathbf{p}$ is the particle's linear momentum. A good example is a particle in a circular orbit, such that

$$\mathbf{L} = \mathbf{a}_z mvr, \tag{3.216}$$

where $\mathbf{a}_z$ is a unit vector normal to the plane formed by $\mathbf{r}$ and $\mathbf{p}$ (following the right-hand rule), m is the mass of the particle, v is the particle's linear velocity, and r is the radius of the orbit. Recall that, in the absence of external forces, angular momentum is conserved, which explains why an ice skater spins faster as the skater's arms are lowered: r is reduced so that v must increase to maintain a constant angular momentum.

Whereas classical objects do not have any intrinsic spin, they do have angular momentum, although, perhaps not unexpectedly, in the quantum theory, angular momentum is quantized in integral units of $\hbar$, including 0 (i.e., $0, \hbar, 2\hbar, 3\hbar, \ldots$). Therefore, a quantum particle in general has both quantized spin and quantized angular momentum. Electronics based on the transport of spin, rather than the transport of charge, will be discussed in Section 10.4. Spin is also important in magnetic resonance imaging (MRI) technologies.

3.7 MAIN POINTS

In this chapter, the fundamental principles of quantum mechanics have been presented in simplified form. After studying this chapter, you should understand the four postulates of quantum mechanics, and related ideas. In particular, you should know

- the meaning of the state function;
- how to calculate the probability of finding a particle in a given region of space;
- how to determine the probability of measuring a certain observable, λ_n;
- the concepts of operators, eigenvalues, and eigenfunctions, and how to solve eigenvalue problems based on differential operators;
- the important quantum mechanical operators associated with momentum and energy;
- how to determine the expectation value (mean) of an observable;

- the general time-dependent and time-independent Schrödinger equations, and how to solve these equations in simple regions of space;
- the concept of the probability current density;
- the analysis of multiple particle systems, screening, the Pauli exclusion principle, and the concepts of spin and angular momentum.

3.8 PROBLEMS

1. For the matrix operator $L = \begin{bmatrix} -5 & 0 \\ 1 & 2 \end{bmatrix}$, show that eigenvalues and eigenvectors are

$$\lambda = 2, \qquad x = \begin{bmatrix} 0 \\ \alpha \end{bmatrix},$$

$$\lambda = -5, \qquad x = \begin{bmatrix} -7\beta \\ \beta \end{bmatrix},$$

where $\alpha, \beta \neq 0$. That is, show that the preceding quantities satisfy the eigenvalue problem $Lx = \lambda x$.

2. Consider the set of functions $\left\{ \frac{1}{\sqrt{2\pi}} e^{inx}, \ n = 0, \pm 1, \pm 2, \ldots \right\}$.

(a) Show that this is an orthonormal set on the interval $(-\pi, \pi)$.

(b) On the interval $(-\pi/2, \pi/2)$, is the set an orthogonal set, an orthonormal set, or neither?

3. Consider the set of functions $\left\{ \sqrt{\frac{2}{\pi}} \sin(nx), \ n = 1, 2, \ldots \right\}$ on the interval $(0, \pi)$.

(a) Show that this is an orthonormal set.

(b) Determine an operator (including boundary conditions) for which the preceding set are eigenfunctions. What are the eigenvalues?

4. For the differential operator $L = -d^2/dx^2$, $u(0) = u(a) = 0$, determine eigenvalues λ and eigenfunctions u. That is, solve

$$Lu = \lambda u,$$

where $u(x)$ is a nonzero function subject to the given boundary conditions. Normalize the eigenfunctions, and show that the eigenfunctions are orthonormal.

5. Repeat problem 3.4, but for boundary conditions $u'(0) = u'(a) = 0$, where $u' = du/dx$.

6. Assume that some observable of a certain system is measured and found to be λ_n for some integer n. By Postulate 2, we know that immediately after the measurement, the system is in state ψ_n, which is an eigenstate of the measurement operator $\widehat{o}$ (i.e., where $\widehat{o}\psi_n = \lambda_n \psi_n$).

(a) What can we conclude about the system's state immediately before the measurement?

(b) Assume that the identical measurement is then performed on 100,000 identical systems, and each time the measurement result is the same, λ_n. What can we infer about the system's state immediately before the measurement?

7. Assume that an electronic state has a lifetime of 10^{-8} s. What is the minimum uncertainty in the energy of an electron in this state?

8. In the example of solving the one-dimensional Schrödinger equation on page 65, we obtained the state functions

$$\Psi(x,t) = \psi(x)\, e^{-iE_n t/\hbar}$$

where

$$\psi(x) = \left(\frac{2}{L}\right)^{1/2} \sin\left(\frac{n\pi}{L}x\right), \quad n \text{ even}, \tag{3.217}$$
$$= \left(\frac{2}{L}\right)^{1/2} \cos\left(\frac{n\pi}{L}x\right), \quad n \text{ odd},$$

are eigenfunctions of the second derivative operator d^2/dx^2, and where energy eigenvalues were found to be

$$E_n = \frac{\hbar^2}{2m}\left(\frac{n\pi}{L}\right)^2. \tag{3.218}$$

(a) Show that the odd eigenfunction (sine) can be written as

$$\psi(x) = \frac{1}{\sqrt{2}}\left[\frac{1}{i\sqrt{L}} e^{i\frac{p_n}{\hbar}x} - \frac{1}{i\sqrt{L}} e^{-i\frac{p_n}{\hbar}x}\right] \tag{3.219}$$
$$= \psi_+ - \psi_-,$$

and determine a similar relation expression for the even eigenfunction. The term ψ_+ (ψ_-) represents a wave propagating with positive (negative) momentum. Thus, any state described by sine and cosine can be thought of as representing a superposition of positive and negative momentum states.

(b) Although the decomposition of a standing wave into two counterpropagating waves, as in part (a), is useful, it can be misinterpreted. Since the probability density $\psi(x,t)\,\psi^*(x,t)$ is independent of time, the expectation value of position, $\langle x \rangle$, is independent of time, and so, really, we should not think of the particle as "bouncing" back and forth in the confined space (otherwise, $\langle x \rangle$ would be a function of t). Determine the expectation value of momentum, using either (3.217) or (3.219), and discuss your answer in light of the preceding comment.

(c) Assume that the particle is in a state composed of the first two eigenfunctions,

$$\psi(x,t) = \frac{1}{\sqrt{2}}\left(\left(\frac{2}{L}\right)^{1/2} \cos\left(\frac{\pi}{L}x\right) e^{-i\frac{E_1 t}{\hbar}} + \left(\frac{2}{L}\right)^{1/2} \sin\left(\frac{2\pi}{L}x\right) e^{-i\frac{E_2 t}{\hbar}}\right).$$

Show that the expectation value of position as a function of time is

$$\langle x \rangle = \frac{16}{9}\frac{L}{\pi^2}\cos\left(\frac{3}{2}\frac{\hbar}{m}\frac{\pi^2}{L^2}t\right).$$

Interpret this solution, compared with the expectation value of position for a single stationary state ψ_n, which is time independent.

9. Since Schrödinger's equation is a homogeneous equation, the most general solution for the state function is a sum of homogeneous solutions (3.142),

$$\Psi(\mathbf{r}, t) = \sum_n a_n \psi_n(\mathbf{r})\, e^{-iE_n t/\hbar}. \tag{3.220}$$

Show that if $\Psi(\mathbf{r}, 0)$ is known, then an expression for the weighting amplitudes a_n can be determined. Assume that the eigenfunction ψ_n forms an orthonormal set. Hint: Multiply

$$\Psi(\mathbf{r}, 0) = \sum_n a_n \psi_n(\mathbf{r}) \tag{3.221}$$

by $\psi_m^*(\mathbf{r})$ and integrate. What is the interpretation of $|a_n|^2$?

10. Consider a particle with time-independent potential energy, and assume that the initial state of the particle is

$$\Psi(\mathbf{r}, t) = a_1\psi_1(\mathbf{r}, t) + a_2\psi_2(\mathbf{r}, t),$$

such that $P(\lambda_1) = |a_1|^2 = P_1$, $P(\lambda_2) = |a_2|^2 = P_2$, and $|a_1|^2 + |a_2|^2 = 1$. Show that

$$\langle E \rangle = P_1 \langle E_1 \rangle + P_2 \langle E_2 \rangle.$$

11. For the example of solving the one-dimensional Schrödinger's equation on page 65, determine the probability of observing the particle very near the boundary wall, $x = \pm L/2$. If the particle is in the $n = 2$ state, where is the particle most likely to be found?

12. For the example of solving the one-dimensional Schrödinger's equation on page 65, assume that the particle is in the $n = 2$ state. What is the probability that a measurement of energy will yield

$$E_2 = \frac{\hbar^2}{2m}\left(\frac{2\pi}{L}\right)^2?$$

What is the probability that a measurement of energy will yield

$$E_3 = \frac{\hbar^2}{2m}\left(\frac{3\pi}{L}\right)^2?$$

13. Consider a quantum encryption scheme using photons. Assume that a photon can only exist in either state 1, ψ_1, having energy E_1, or state 2, ψ_2, having energy E_2, or in a superposition of the two states, $\Psi = a\psi_1 + b\psi_2$. Assume that the states are orthonormal.

(a) If a photon exists in the superposition state $\Psi = a\psi_1 + b\psi_2$, what is the relationship between a and b?

(b) If a photon exists in the superposition state $\Psi = a\psi_1 + b\psi_2$, determine the probability of measuring energy E_2. Show all work and/or explain your answer.

(c) If the photon in a superposition state is sent over a network, explain how undetected eavesdropping would be impossible.

14. In Chapter 6, the reflection and transmission of a particle across a potential barrier will be considered. For now, assume that a potential energy discontinuity is present at $x = a$, and that to the left of the discontinuity, the wavefunction is given by

$$\Psi(x,t) = \left(e^{ikx} + Re^{-ikx}\right) e^{-iEt/\hbar},$$

and to the right of the discontinuity

$$\Psi(x,t) = Te^{iqx} e^{-iEt/\hbar},$$

where R and T are reflection and transmission coefficients, respectively, which will depend on the properties of the different regions and on the discontinuity in potential at $x = a$. Determine the probability current density on either side of the discontinuity.

15. In the example of solving the one-dimensional Schrödinger's equation on page 65, we obtained the state functions

$$\Psi(x,t) = \psi(x) e^{-iE_n t/\hbar}$$

where

$$\psi(x) = \left(\frac{2}{L}\right)^{1/2} \sin\left(\frac{n\pi}{L}x\right), \qquad n \text{ even},$$

$$= \left(\frac{2}{L}\right)^{1/2} \cos\left(\frac{n\pi}{L}x\right), \qquad n \text{ odd},$$

and where

$$E_n = \frac{\hbar^2}{2m}\left(\frac{n\pi}{L}\right)^2.$$

Determine the probability current density. Discuss your result.

16. Assume that the wave function

$$\psi(z,t) = 200e^{i(kz-\omega t)}$$

describes a beam of 2 eV electrons having only kinetic energy. Determine numerical values for k and ω, and find the associated current density in A/m.

Chapter

4

FREE AND CONFINED ELECTRONS

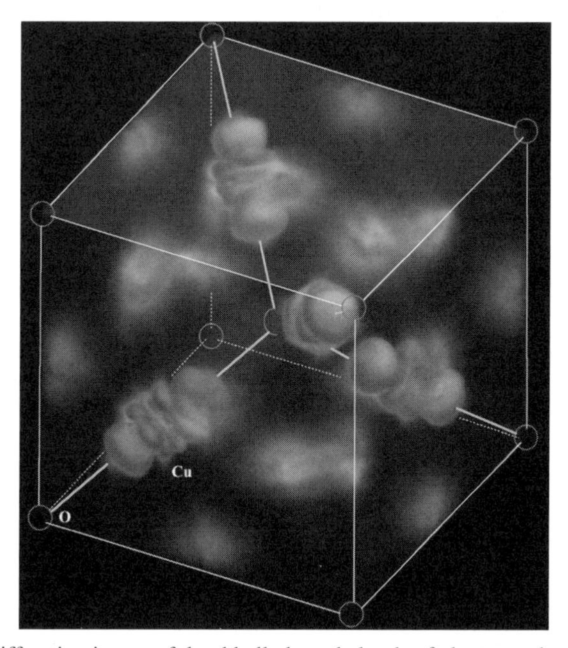

Electron and X-ray diffraction image of dumbbell-shaped clouds of electrons shared by covalent bonds between copper and oxygen atoms in cuprite, C_2O. The nuclei of the copper atoms (not shown) are at the center of the dumbbells, and those of the oxygen atoms (also not shown) are at the center and corners of the superimposed cube. The fuzzy clouds are less defined electron clouds representing covalent bonds between the copper atoms. (From J.M. Zuo, M. Kim, M. O'Keeffe, and J.C.H. Spence, *Nature* 401, 49–52 (1999). Used by permission.)

In the last chapter, Schrödinger's equation was introduced, along with the basic ideas of quantum mechanics. In this chapter, some examples of solving Schrödinger's equation are presented relevant to understanding physical phenomena associated with electrons confined to nanoscale regions of space, including quantum dots, wires, and wells. The chapter begins, however, with a discussion of free (unconfined) electrons, in part to provide a comparison to the bounded space result.

4.1 FREE ELECTRONS

As a first application of solving Schrödinger's equation, consider a free electron in an infinite space. By "free electron," we mean that there is no potential energy variation to influence the particle, i.e., $V(\mathbf{r}) = V_0$ (where V_0 can be zero; the important thing is that V is constant). For solid materials, the most common source of a potential is the atomic lattice, where the potential energy between an electron with charge q_e and an ionized atom of charge $-q_e$ is

$$V(r) = \frac{1}{4\pi\varepsilon_0}\frac{(q_e)(-q_e)}{r} = -A\frac{1}{r}, \tag{4.1}$$

where r is the distance between the electron and the ion. In one dimension,

$$V(x) = \frac{1}{4\pi\varepsilon_0}\frac{(q_e)(-q_e)}{|x|} = -A\frac{1}{|x|}, \tag{4.2}$$

as shown in Fig. 4.1.

Other sources of potential could be, for example, other electrons. Here it will be assumed that there is a single electron and no other particles. Therefore, there is obviously no boundary on the space since a material boundary would involve other electrons. Thus, the electron is totally unencumbered by interactions, and is considered free. It is worth noting that, although we will primarily discuss electrons since they are our main interest, most of the obtained results will apply generally to particles with mass.

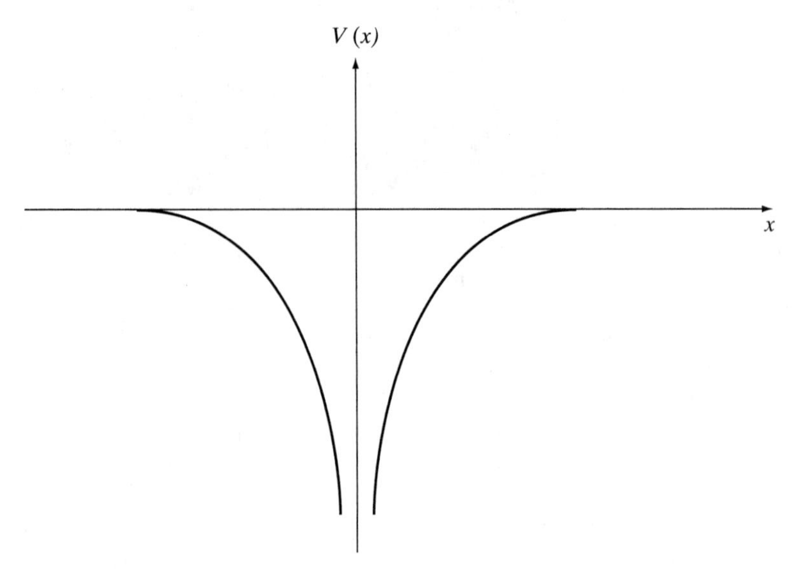

Figure 4.1 Potential $V(x)$ given by (4.2) versus position x, as a one-dimensional model of the potential due to an atom.

4.1.1 One-Dimensional Space

Considering first a one-dimensional problem, Schrödinger's equation (3.140),

$$\left(-\frac{\hbar^2}{2m_e}\frac{d^2}{dx^2} + V_0\right)\psi(x) = E\psi(x), \tag{4.3}$$

where m_e is the mass of the electron and E is the electron's energy, has the solution

$$\psi(x) = Ae^{ikx} + Be^{-ikx}. \tag{4.4}$$

In (4.4),

$$k^2 = \frac{2m_e(E - V_0)}{\hbar^2}, \tag{4.5}$$

where k is called the *wavevector* (in general), or wavenumber (for the scalar case), and where A and B are constants. The parabolic relationship between wavevector k and energy E, (4.5), is shown in Fig. 4.2; later this type of diagram will have great importance. Putting in the time variation (3.141), we have

$$\Psi(x,t) = \left(Ae^{ikx} + Be^{-ikx}\right)e^{-iEt/\hbar}, \tag{4.6}$$

from which we can see that the terms associated with A and B represent forward and backward traveling waves, respectively. A solution such as (4.6) is called a *plane wave solution* since the surfaces of constant amplitude and phase are plane surfaces.

Recall that two concepts of velocity for a wave were developed in Section 2.5, the phase velocity (2.30), $v_p = \omega/k$, and the group velocity (2.48), $v_g = \partial\omega/\partial k$. Although they

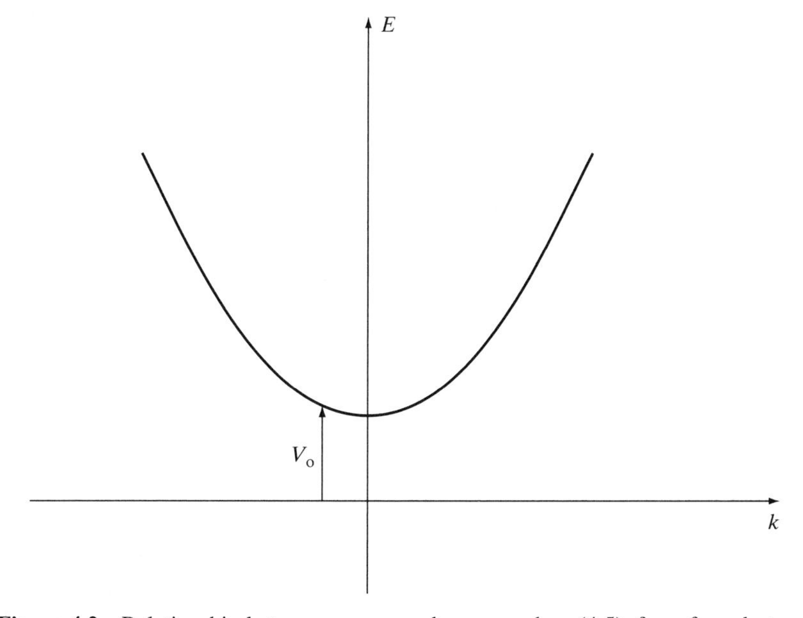

Figure 4.2 Relationship between energy and wavenumber, (4.5), for a free electron.

were derived from a consideration of wavepackets, they can be taken as possible definitions of wave velocities. For the solutions of Schrödinger's equation presented previously, setting $V_0 = 0$ for convenience and using $E = \hbar\omega$, we obtain

$$v_p = \frac{\omega}{k} = \frac{\hbar k}{2m_e} = \frac{p}{2m_e}. \tag{4.7}$$

Recalling that solutions of Schrödinger's equation should agree with classical physics in the classical limit, in order to see if the phase velocity agrees with our classical notion of velocity, we equate $p = m_e v$ for a classical electron to obtain

$$v_p = \frac{v}{2}. \tag{4.8}$$

Therefore, the phase velocity does not yield a reasonable value for the electron's velocity. However, the group velocity is

$$v_g = \frac{\partial \omega}{\partial k} = \frac{\hbar k}{m_e} = \frac{p}{m_e}, \tag{4.9}$$

which, with $p = m_e v$ for a classical electron, leads to

$$v_g = v, \tag{4.10}$$

the classical velocity. Therefore, as concepts of velocity, the group velocity is the more physically meaningful. For a classical electromagnetic plane wave in free space, where $k = \omega/c$, $v_p = v_g = c$.

Because the electron does not interact with anything, it can be assumed that it will be moving either along the positive or negative x coordinate. Let's assume that the electron is moving along the positive x coordinate ($B = 0$). Then, the probability of finding the electron in a range Δx centered at some x is

$$|\Psi(x,t)|^2 \Delta x = |A|^2 \Delta x, \tag{4.11}$$

which is independent of x. Therefore, the electron has equal probability to be found at any point in space. This is consistent with the uncertainty principle (2.39); since k is known exactly (i.e., it is assumed that we know the electron's energy), and, therefore, momentum is known by (2.15), the uncertainty in the electron's position must be infinite.

The probabilistic current density associated with the wave is, from (3.187),

$$\begin{aligned}\mathbf{J}(\mathbf{r},t) &= \frac{-i\hbar}{2m_e}\left(\Psi^*(\mathbf{r},t)\nabla\Psi(\mathbf{r},t) - \Psi(\mathbf{r},t)\nabla\Psi^*(\mathbf{r},t)\right) \\ &= \mathbf{a}_x |A|^2 \frac{\hbar k}{m_e}.\end{aligned} \tag{4.12}$$

The probabilistic charge density is

$$\rho = \Psi^*(\mathbf{r},t)\Psi(\mathbf{r},t) = |A|^2, \tag{4.13}$$

and, with $\hbar k = p = m_e v$,

$$\mathbf{J}(\mathbf{r}, t) = \mathbf{a}_x |A|^2 v = \mathbf{a}_x \rho v, \tag{4.14}$$

which has the usual interpretation of current density being the product of charge density and velocity.

The amplitude A is usually determined by normalization (3.8), although in this case the normalization cannot be performed since we have a plane wave existing everywhere in an infinite space. A resolution of this difficultly, and a fact that should always be kept in mind, is that Schrödinger's equation is a homogeneous linear differential equation. Its solution merely details the possible allowed states of the system, not which states will actually be filled.† (You can think of a filled state as a state "containing" an electron. Filling of the states will be considered in Section 8.2.) Furthermore, a sum of homogeneous solutions to a linear differential equation is itself a homogeneous solution, and so an electron may not be in a single state, but in a superposition of states having the form (3.53). As considered in Section 2.5, a superposition of plane waves is a wavepacket, which is suitably localized in space and energy (and time). A more realistic wavepacket solution can be normalized by (3.8), although this topic will not be pursued here.

4.1.2 Three-Dimensional Space

Now, moving to a three-dimensional space, Schrödinger's equation (3.139) is

$$\left(-\frac{\hbar^2}{2m_e}\left(\frac{\partial^2}{\partial x^2} + \frac{\partial^2}{\partial y^2} + \frac{\partial^2}{\partial z^2}\right) + V_0 - E\right)\psi(x, y, z) = 0, \tag{4.15}$$

where the Laplacian term is explicitly written in rectangular coordinates. The preceding equation can be easily solved using separation of variables, which was used in Section 3.2 to separate the time and space dependence of Schrödinger's equation. Assume a product solution of the form

$$\psi(x, y, z) = \psi_x(x)\,\psi_y(y)\,\psi_z(z). \tag{4.16}$$

Inserting (4.16) into (4.15), we have

$$\left(\frac{1}{\psi_x}\frac{\partial^2}{\partial x^2}\psi_x + \frac{1}{\psi_y}\frac{\partial^2}{\partial y^2}\psi_y + \frac{1}{\psi_z}\frac{\partial^2}{\partial z^2}\psi_z + \frac{2m_e}{\hbar^2}(E - V_0)\right) = 0. \tag{4.17}$$

The first term can be, at most, a function of x (but not y or z), the second term can be, at most, a function of y (but not x or z), and third term can be, at most, a function of z (but not x or y). These three terms then add to a constant to yield zero. The only way this can happen, for all possible values of x, y, and z, is if each of the three terms is actually equal to a constant.

†This is analogous to solving the source-free Newton's equations for a mass-spring system (or for a guitar string, bridge span, etc.) The solutions are the possible resonances of the system, although without knowing the input force, one does not know which resonances will be implicated in a force-driven problem.

Let the constant associated with the first term be $-k_x^2$, and similarly for the other two terms. Then

$$\frac{1}{\psi_x}\frac{\partial^2}{\partial x^2}\psi_x = -k_x^2 \rightarrow \psi_x(x) = Ae^{ik_x x} + Be^{-ik_x x}, \tag{4.18}$$

$$\frac{1}{\psi_y}\frac{\partial^2}{\partial y^2}\psi_y = -k_y^2 \rightarrow \psi_y(y) = Ce^{ik_y y} + De^{-ik_y y},$$

$$\frac{1}{\psi_z}\frac{\partial^2}{\partial z^2}\psi_z = -k_z^2 \rightarrow \psi_z(z) = Fe^{ik_z z} + Ge^{-ik_z z},$$

where we must have, from (4.17),

$$k_x^2 + k_y^2 + k_z^2 = \frac{2m_e}{\hbar^2}(E - V_0). \tag{4.19}$$

Since the electron is unencumbered by electrical or mechanical influences, we can assume that the electron is moving in a certain direction in the three-dimensional space. Therefore, one constant in each pair of constants (A or B, C or D, F or G) will be zero. For example, if the wave movement is along the positive x and y coordinates, but the negative z coordinate, then $B = D = F = 0$. The product solution can be written as

$$\psi(\mathbf{r}) = \left(A_0 e^{ik_x x} e^{ik_y y} e^{ik_z z}\right), \tag{4.20}$$

where the wavenumbers k_x, k_y, and k_z can still be chosen as positive or negative to fix the electron's direction, and where A_0 is the product of the three constants remaining in (4.18). We can write (4.20) compactly as

$$\psi(\mathbf{r}) = A_0 e^{i\mathbf{k}\cdot\mathbf{r}}, \tag{4.21}$$

where

$$\mathbf{k} = \mathbf{a}_x k_x + \mathbf{a}_y k_y + \mathbf{a}_z k_z, \quad k = |\mathbf{k}|, \tag{4.22}$$

$$k^2 = k_x^2 + k_y^2 + k_z^2 = \frac{2m_e(E - V_0)}{\hbar^2}$$

(hence the name wave*vector* for **k**). Putting in the time variation using (3.135), we have

$$\Psi(\mathbf{r}, t) = A_0 e^{i\mathbf{k}\cdot\mathbf{r}} e^{-iEt/\hbar}, \tag{4.23}$$

which is a plane wave moving in three-dimensional space. A more physically realistic solution would be a three-dimensional wavepacket, consisting of a superposition of waves having the form (4.23).

4.2 THE FREE ELECTRON GAS THEORY OF METALS

Assume now that we have, instead of a single electron, N electrons in an otherwise empty space (no lattice, etc.), where N is typically a very large number. For example, a copper atom has 29 electrons, 28 of which are strongly bound to the atom. However, the 29th electron is bound very loosely to the atom, and can essentially freely move about the material. Thus, copper has a free electron density of N electrons per cubic meter, where N is also the number of atoms per cubic meter (for copper, $N \sim 8.45 \times 10^{22}$ atoms/cm^3).

As discussed in Section 3.5, for a system of N electrons, the Schrödinger equation is much more difficult to solve compared with the case of a single electron. Part of this difficulty lies in the fact that the interaction among electrons will result in a very complicated potential term $V(\mathbf{r}_1, \mathbf{r}_2, \ldots, \mathbf{r}_N)$. However, if the interaction between electrons can be neglected, which is often a particularly appropriate approximation due in part to the Pauli exclusion principle, then each electron obeys the one-particle Schrödinger equation (3.139), where V will be a constant (since an otherwise empty space is considered).

The rationale behind ignoring electron–electron and electron–lattice interactions in a material was discussed in Section 3.5. The fact that in a metal the electron density is large actually implied being able to ignore electron–electron and electron–ion interactions, due to Coulomb screening and the exclusion principle. Therefore, to a first approximation, in many metals the N electrons tend to act freely, and, thus, these metals can be modeled as a collection of N non-interacting free electrons, often called a *free electron gas*. For many metals, this simple model is quite good at describing and predicting physically observed phenomena. The free electron gas model is discussed further in Chapter 10.

4.3 ELECTRONS CONFINED TO A BOUNDED REGION OF SPACE AND QUANTUM NUMBERS

The simplest solution of Schrödinger's equation has now been considered, i.e., the model of an electron, or some other particle, in an infinite space with no spatial energy variation. Of course, this is an unrealistic scenario, although it models a situation when an electron is relatively free. It also provides a comparison with the slightly more complicated, but more realistic case of an electron confined to a finite region of space. This confinement could be due to, for instance, an electron being bound to an atom, or to an electron being confined to a small nanoscopic material region such as a quantum dot. It turns out that this *particle-in-a-box* model is of the utmost importance in understanding nanoelectronic devices, since it is the simplest model that leads to the important topic of energy discretization.

Assume that the potential energy profile is constant (and, in fact, assume that $V_0 = 0$), but that the region of space is now finite in size. In particular, we will consider a rectangular box having dimensions $L_x \times L_y \times L_z$, although first we'll look at the one-dimensional problem of a finite line segment having length L_x.

4.3.1 One-Dimensional Space

Consider an electron confined to an interval $(0, L_x)$. In one dimension, Schrödinger's equation (3.140) with $V = 0$ is

$$-\frac{\hbar^2}{2m_e}\frac{d^2}{dx^2}\psi(x) = E\psi(x), \tag{4.24}$$

which has the solution[†]

$$\psi(x) = A \sin kx + B \cos kx, \tag{4.25}$$

which are obviously energy eigenfunctions (i.e., an eigenfunction of $H = -\left(\hbar^2/2m_e\right) d^2/dx^2$) where

$$k^2 = \frac{2m_e E}{\hbar^2}. \tag{4.26}$$

Outside of the line segment, we'll assume that the probability of finding the electron is zero (i.e., the electron is confined to the line segment by some sort of barrier), and so outside of the line segment, $\psi = 0$. This structure is often called a *quantum well* or *quantum box*. Quantum wells are very important in nanoelectronics, since a nanoelectronic device would naturally confine electrons to nanoscopic regions of space.

Continuity of the wavefunction (3.144) requires that

$$\psi(0) = \psi(L_x) = 0, \tag{4.27}$$

and so we find

$$B = 0, \tag{4.28}$$

$$k = \frac{n\pi}{L_x}, \quad n = 1, 2, 3, \ldots.$$

The wavefunction is, therefore,

$$\psi_n(x) = A \sin \frac{n\pi}{L_x} x, \tag{4.29}$$

which is normalized according to (3.8),

$$\int_0^{L_x} |\psi_n(x)|^2 \, dx = 1, \tag{4.30}$$

resulting in

$$\psi_n(x) = \left(\frac{2}{L_x}\right)^{1/2} \sin \frac{n\pi}{L_x} x, \tag{4.31}$$

or[‡]

$$\Psi_n(x, t) = \left(\frac{2}{L_x}\right)^{1/2} \sin \left(\frac{n\pi}{L_x} x\right) e^{-iE_n t/\hbar}. \tag{4.34}$$

[†]An exponential solution of the form (4.4) could be used, alternatively, although the sine/cosine form is preferred here since we will obtain bound states.

[‡]The solutions (4.34) exhibit even symmetry about the center of the box for n odd, and odd symmetry about the center of the box for n even. It is sometimes more convenient to consider a symmetric box, extending

Since k is discrete,

$$E = \frac{\hbar^2 k^2}{2m_e} = \frac{\hbar^2}{2m_e}\left(\frac{n\pi}{L_x}\right)^2 = E_n, \tag{4.35}$$

which denote eigenvalues of the Hamiltonian H. From (3.113), the expectation value of the electron's position is

$$\langle x \rangle = \int_0^{L_x} \Psi_n^* (x, t)\, x \Psi_n (x, t)\, dx \tag{4.36}$$

$$= \frac{2}{L_x} \int_0^{L_x} x \sin^2 \left(\frac{n\pi}{L_x} x\right) dx \tag{4.37}$$

$$= \frac{L_x}{2}, \tag{4.38}$$

i.e., the average position of the electron is the center of the box. Furthermore, since the electron only has kinetic energy ($V = 0$), the de Broglie wavelength is given by

$$\lambda_e = \frac{h}{\sqrt{2m_e E}}. \tag{4.39}$$

Using (4.35), we obtain

$$L_x = \frac{n\lambda_e}{2}, \tag{4.40}$$

so that an integral number of half de Broglie wavelengths fit inside the segment L_x. In addition, if we assume that the electron is in a superposition state,

$$\Psi (x, t) = \sum_n a_n \Psi_n (x, t), \tag{4.41}$$

then the probability of measuring energy E_m is

$$P (E_n) = \left| \int_0^L \Psi (\mathbf{r}, t)\, \Psi_n^* (\mathbf{r})\, dx \right|^2 = |a_n|^2 . \tag{4.42}$$

from $x = -L_x/2$ to $x = L_x/2$, where we would obtain (see the example on page 65)

$$\Psi (x, t) = \left(\frac{2}{L_x}\right)^{1/2} \cos \left(\frac{n\pi}{L_x} x\right) e^{-iE_n t/\hbar}, \quad n \text{ odd}, \tag{4.32}$$

$$= \left(\frac{2}{L_x}\right)^{1/2} \sin \left(\frac{n\pi}{L_x} x\right) e^{-iE_n t/\hbar}, \quad n \text{ even}, \tag{4.33}$$

where $n = 1, 2, 3, \ldots$.

At this point, it is important to appreciate two things.

- The size of the box *dictates* the possible energy levels of an electron confined to the box, via (4.35). That is, an electron confined to a box can only have energies given by the discrete values E_n, which depend only on the size of the box and on the electron's mass.
- Although the electron's wavefunctions (4.34) and energies (4.35) are the only possible solutions, since Schrödinger's equation is a homogeneous differential equation, the electron can be either in one of the discrete states Ψ_n, or in a superposition of states given by (3.142) (i.e., (4.41)).

We do not know the actual state that an electron will occupy; this will typically depend on temperature and other energy inputs. We have only obtained the *allowed, possible* states that the electron can occupy, and we have not discussed which states would actually be filled. We will get to the filling of states subsequently, at which point we will need to consider temperature.

Note that by simply confining the electron to a finite region of space, the solution obtained radically differs from the totally free electron case. In particular, for the free electron, the solution (4.4) represents a traveling wave (since it can be assumed that one of A or B is zero), and for the confined electron, the solution (4.29) represents a standing wave. The first several standing wave patterns are shown in Fig. 4.3.

Of equal, if not greater, importance, is the observation that the energy obtained for the particle without boundaries is continuous, whereas in a bounded region of space the particle's energy is discrete. The energy versus wavevector curve is parabolic, in a discrete sense, and is shown in Fig. 4.4, which should be compared to Fig. 4.2 on page 87 for the free electron.

For the spatially bounded particle, n in (4.31)–(4.35) is called the *quantum number*. Since we are dealing with electrons, to each state there is another quantum number, m_s, corresponding to spin, which does not affect the energy. Together, these two quantum numbers completely describe the state of the particle. The lowest state is the ground state, with $n = 1$.

One can gain an appreciation of the discreteness of energy by considering a numerical example. Let an electron be confined to an atomic-sized region of space, $L_x = 2 \times 10^{-10}$ m. The first two energy levels, E_1 and E_2 from (4.35), are

$$E_1 = 1.5 \times 10^{-18} \text{ J} \Rightarrow \frac{1.5 \times 10^{-18} \text{ J}}{1.6021 \times 10^{-19} \text{ C}} = 9.4 \text{ eV}, \tag{4.43}$$

$$E_2 = 6.0 \times 10^{-18} \text{ J} = 37.6 \text{ eV}.$$

In particular, the spacing between any successive energy levels is

$$E_n - E_{n-1} = 9.4\,(2n - 1) \text{ eV}, \tag{4.44}$$

and, thus, the discrete nature of the allowed energy levels is clearly evident. (Energy levels on the order of electron volts are easily measured.) If, instead, we consider a macroscopic

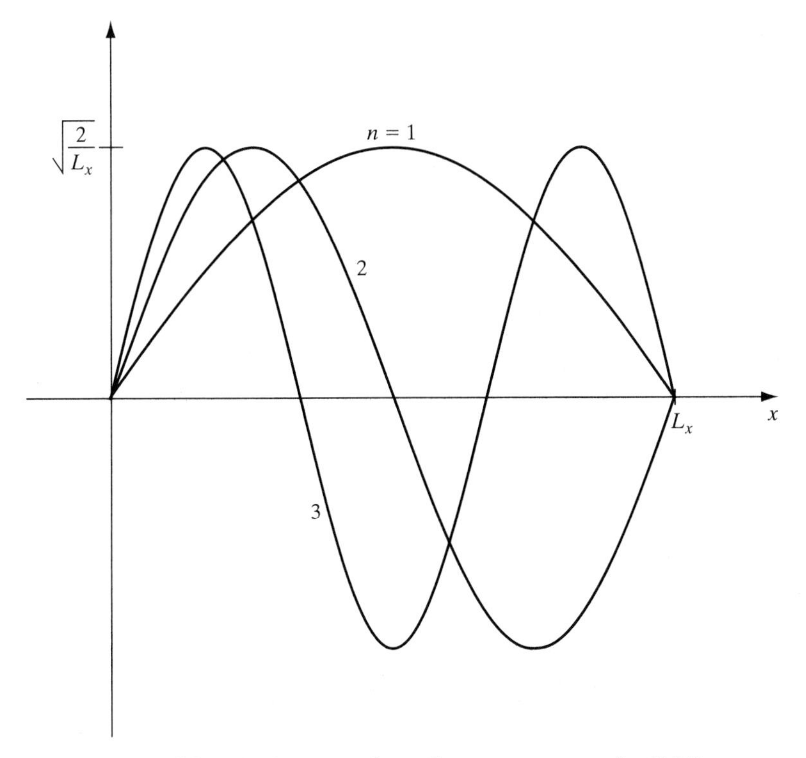

Figure 4.3 The first several standing wave patterns for (4.31).

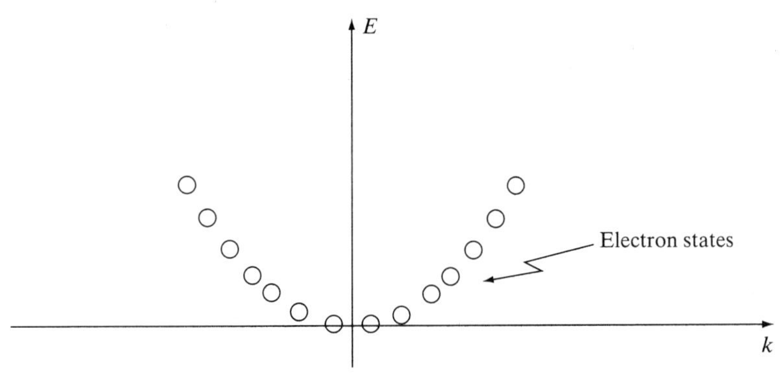

Figure 4.4 Energy versus wavenumber for discrete energy levels. Only values of energy and wavenumber represented by circles are allowed.

region of space, $L_x = 1$ cm, then (4.35) results in

$$E_1 = 6.0 \times 10^{-34}\ \text{J} = 3.8 \times 10^{-15}\ \text{eV}, \tag{4.45}$$

and the spacing between any two energy levels is

$$E_n - E_{n-1} = 3.8 \times 10^{-15}\,(2n-1)\ \text{eV}, \tag{4.46}$$

which is too small to be measurable. Thus, in a macroscopic region of space, the energy spacings are so close they appear to be continuous. This is yet another example of the "classical limit," where quantum mechanics provides the expected classical result known from experiments on macroscopic systems.

Viewing this from another perspective, let's assume that electron energies on the order of 10^{-3}–10^{3} electron volts are of most interest, commensurate with typical energy levels encountered in electronics and optics. Then, assuming that the electron has only kinetic energy, from (4.39), we see that the electron's de Broglie wavelength is in the range

$$\lambda_e = 10^{-8} - 10^{-11}\ \text{nm}. \tag{4.47}$$

In what will become a familiar theme throughout this text, note that when a particle's de Broglie wavelength is on the order of, or larger than, the size of a confining space, the particle "feels" the boundaries of the space and exhibits energy level discretization. When a particle's de Broglie wavelength is much smaller than the size of a confining space, conversely, the particle does not "feel" the finiteness of the space, and it behaves as if it were a free particle, which was the case examined in Section 4.1. Relating this to (4.40), we find that an electron with an energy level in the specified range actually inhabits one of the discrete levels E_n, although if $\lambda_e \ll L_x$, the electron is in one of the higher energy states (i.e., corresponding to a large value of n), and in this situation, the spacing between energy levels is so small we can model them as forming a continuum.

Note that E (the kinetic energy in this case, since $V = 0$) is not zero in the ground state. The lowest energy of a particle would be zero, classically, corresponding to the particle being at rest ($V = 0$, $E = 0$), which would occur at the temperature absolute zero ($T = 0$ K). Again, classically, at any nonzero temperature the electron would acquire thermal energy and start to vibrate.

Viewing this from the quantum mechanical picture, we can see a reason why the lowest energy, called the *ground state energy*, is nonzero. If the particle's kinetic energy were zero, then $p = 0$, and, via the uncertainty principle (2.39), the uncertainty in position would be infinity. Since we have bounded the particle's position to a finite range ($x = 0$ to $x = L_x$), then the particle's momentum can't be zero.[†] This ground state energy is also called *internal kinetic energy*, or *zero-point energy*.

Quantum Versus Classical Physics — What is Quantized? In classical physics, one has the concept of discrete resonant frequencies, but not the quantization of energy.

[†]We can view this phenomena as de Broglie waves reflected back and forth. For allowed E_n's, the interference is constructive; otherwise, the interference is destructive. However, this is a mathematical representation, and we shouldn't take this to mean that the particle is literally bouncing back and forth.

For example, in an electrical inductor–capacitor circuit, a discrete resonant frequency exists that depends on the inductance and capacitance values. However, the energy associated with the resonance is a continuous quantity, associated with the amplitude of the current or voltage in the circuit. Stored energy changes from electrical (energy stored in the capacitor) to magnetic (energy stored in the inductor) over the course of the oscillation. In a similar manner, the energy of a swinging pendulum converts between potential energy (the height of the bob) and kinetic energy (the velocity of the bob) as the bob oscillates at the discrete resonant frequency of the pendulum, although the energy of the bob is continuous (e.g., the height of the bob, which is the amplitude of the oscillation, is a continuous quantity). Guitar strings obviously lead to discrete resonance frequencies, although, again, energy is associated with the amplitude of the string displacement, and not to the discrete frequency of the resonance. Therefore, in classical physics, energy is associated with amplitude, which is not quantized, although frequency can be quantized. In the quantum picture, energy is often quantized, and energy and frequency quantization are connected by $E = \hbar\omega$.

4.3.2 Three-Dimensional Space

The example of an electron confined to a three-dimensional region of space, such as a quantum dot, follows easily from the previous result. Assume that the space has dimensions $L_x \times L_y \times L_z$, and that within the space $V = 0$. Schrödinger's equation (3.139) is

$$-\frac{\hbar^2}{2m_e}\nabla^2\psi(\mathbf{r}) = E\psi(\mathbf{r}), \tag{4.48}$$

and, assuming that the wavefunction can be written in product form,

$$\psi(x, y, z) = \psi_x(x)\,\psi_y(y)\,\psi_z(z), \tag{4.49}$$

the boundary conditions†

$$\begin{aligned} \psi_x(0) &= \psi_x(L_x) = 0, \\ \psi_y(0) &= \psi_y\left(L_y\right) = 0, \\ \psi_z(0) &= \psi_z(L_z) \doteq 0, \end{aligned} \tag{4.50}$$

lead to

$$\psi(x, y, z) = \left(\frac{8}{L_x L_y L_z}\right)^{1/2} \sin\frac{n_x\pi}{L_x}x\,\sin\frac{n_y\pi}{L_y}y\,\sin\frac{n_z\pi}{L_z}z. \tag{4.51}$$

We now have three quantum numbers,

$$\begin{aligned} n_x &= 1, 2, 3, \ldots, \\ n_y &= 1, 2, 3, \ldots, \\ n_z &= 1, 2, 3, \ldots, \end{aligned} \tag{4.52}$$

†This is sometimes called the hard-wall case, i.e., the electron is confined to the box by infinite hard walls.

and a fourth quantum number m_s that accounts for spin. Together, these four quantum numbers describe the state of the particle,† where different combinations of the quantum numbers refer to different states of the system.

The allowed discrete values of energy are given by

$$E_{n_x,n_y,n_z} = \frac{\hbar^2\pi^2}{2m_e}\left(\left(\frac{n_x}{L_x}\right)^2 + \left(\frac{n_y}{L_y}\right)^2 + \left(\frac{n_z}{L_z}\right)^2\right). \tag{4.53}$$

Of course, if $L_x = L_y = L_z = L$, then we obtain the simpler result

$$E = \frac{\hbar^2\pi^2}{2m_e L^2}\left(n_x^2 + n_y^2 + n_z^2\right) = E_n. \tag{4.54}$$

In this case, states with different quantum numbers but the same energy (e.g., $(n_x, n_y, n_z) = (1, 2, 3)$ and $(n_x, n_y, n_z) = (3, 1, 2)$) are called *degenerate*, and the number of states having the same energy is called the *degeneracy*.

As an aside, if we think of a cube of material of side L and we compress (squeeze) the material, then L decreases and, by (4.54), the energy levels increase. Thus, electrons in the material must increase their energy, and this energy gain comes from the work done by squeezing the material. The resulting pressure is called *Pauli pressure*, since the Pauli exclusion principle keeps multiple electrons (more than two) from occupying the same energy level. This pressure partially accounts for the resistance to squeezing of materials with high electron concentrations. (This also explains partially the avoidance of gravitional collapse.) Metals have a high electron concentration, although repulsive forces between nuclei, etc., also contribute to the pressure. If the exclusion principle did not hold and all electrons could be in the lowest state, their energy would still increase upon squeezing since E_1 would nevertheless increase as L decreased, although the effect would be much weaker.

Lastly, it is worth reiterating that we have solved the time-independent Schrödinger's equation to obtain the *possible* (i.e., allowed) electron states. What state an electron actually occupies will depend on other factors such as temperature, the presence of other electrons, and other energy sources.

4.3.3 Periodic Boundary Conditions

Rather than the boundary conditions (4.50), it is often convenient to consider periodic boundary conditions that result in traveling, rather that standing, wave solutions.‡ Periodic

†Readers familiar with electromagnetic theory will recognize the connection between the results in this section and the allowed resonances in a three-dimensional cavity (except, in that case, there is no spin integer).

‡If the region of space is large compared with atomic dimensions, then the influence of the boundary conditions on the motion of the particle will be small.

boundary conditions emulate an infinite solid, rather than a finite region, and are given by

$$\begin{aligned}\psi(x,y,z) &= \psi(x+L_x,y,z),\\ \psi(x,y,z) &= \psi(x,y+L_y,z),\\ \psi(x,y,z) &= \psi(x,y,z+L_z),\end{aligned} \tag{4.55}$$

leading to the solution of (4.48) as

$$\psi(\mathbf{r}) = \left(\frac{1}{L_x L_y L_z}\right)^{1/2} e^{i\mathbf{k}\cdot\mathbf{r}}, \tag{4.56}$$

where

$$\begin{aligned}\mathbf{k} &= \mathbf{a}_x k_x + \mathbf{a}_y k_y + \mathbf{a}_z k_z, \quad k = |\mathbf{k}|,\\ k^2 &= \left(k_x^2 + k_y^2 + k_z^2\right) = \frac{2m_e E}{\hbar^2},\\ k_x &= \frac{2n_x\pi}{L_x}, \quad k_y = \frac{2n_y\pi}{L_y}, \quad k_z = \frac{2n_z\pi}{L_z},\end{aligned} \tag{4.57}$$

$n_{x,y,z} = 0, \pm1, \pm2, \ldots$. Note that now we have the index $2n_{x,y,z}$, instead of $n_{x,y,z}$, and that both positive and negative values of k are allowed to account for waves moving in opposite directions. Allowed energy levels are given by

$$E_n = \frac{2\hbar^2\pi^2}{m_e}\left(\left(\frac{n_x}{L_x}\right)^2 + \left(\frac{n_y}{L_y}\right)^2 + \left(\frac{n_z}{L_z}\right)^2\right), \tag{4.58}$$

and, for a box having equal sides L,

$$E_n = \frac{2\hbar^2\pi^2}{m_e L^2}\left(n_x^2 + n_y^2 + n_z^2\right). \tag{4.59}$$

Note that the equation

$$E_n = \alpha\frac{\hbar^2\pi^2}{m_e L^2}\left(n_x^2 + n_y^2 + n_z^2\right) \tag{4.60}$$

represents either the hard-wall case for $\alpha = 1/2$ or, for $\alpha = 2$, the periodic boundary condition case.

4.4 FERMI LEVEL AND CHEMICAL POTENTIAL

An important concept that will be used throughout this text is the *Fermi level.* A related concept, the *chemical potential*, which has the symbol μ (not to be confused with permeability from electromagnetic theory, or magnetic moment), will first be introduced.

The precise definition of the chemical potential is complicated, and beyond what we need for the treatment described here (it involves the change in the Gibbs free energy with respect to a change in the number of particles in a system). However, we can think of the chemical potential as (approximately) the energy needed to add the Nth electron to a system of $N - 1$ electrons. For noninteracting electrons, this is simply the energy of the highest occupied state. To agree with the statistical definition utilized in Chapter 8 relating to Fermi level, we will define the chemical potential of a N-electron system by

$$[E_t(N+1) - E_t(N-1)]/2, \tag{4.61}$$

where $E_t(N)$ is the total energy of an N particle system.

Example

Consider a one-dimensional box of length L, as considered in Section 4.3.1, and assume that five noninteracting electrons are contained within the box. If we assume that electrons are in their lowest energy configuration, since two electrons can occupy each state (due to the Pauli exclusion principle), then the first two states are completely filled, and the third state is partially filled. The total energy of the $N + 1$-electron system is

$$E_t\,(6) = 2E_1 + 2E_2 + 2E_3,$$

where the energy levels E_n are defined by (4.35). The total energy of a four-electron system would be

$$E_t\,(4) = 2E_1 + 2E_2,$$

and, therefore, the chemical potential is

$$\mu\,(5) = [E_t\,(6) - E_t\,(4)]/2 = E_3 = \frac{\hbar^2}{2m_e}\left(\frac{3\pi}{L}\right)^2.$$

The *Fermi level* is defined as the value of the chemical potential at $T = 0$, i.e., $E_F = \mu\,(T = 0)$, and so the Fermi level (Fermi energy) is (nearly) the maximum energy of a particle at $T = 0$. It is common in semiconductor physics to use the term Fermi level rather than chemical potential, regardless of temperature. The main idea is that at absolute zero, a collection of N electrons will arrange themselves into the lowest available energy states, and form what is called a *Fermi sea* of electrons. The Fermi level is the surface of that sea at absolute zero. The concept of the Fermi energy is of considerable importance in nanoelectronics, and will be discussed further in Sections 5.4 and 8.2.

The chemical potential is a weak function of temperature, given approximately as

$$\mu\,(T) = E_F\left(1 - \frac{\pi^2}{12}\left(\frac{k_B T}{E_F}\right)^2 + \ldots\right) \tag{4.62}$$

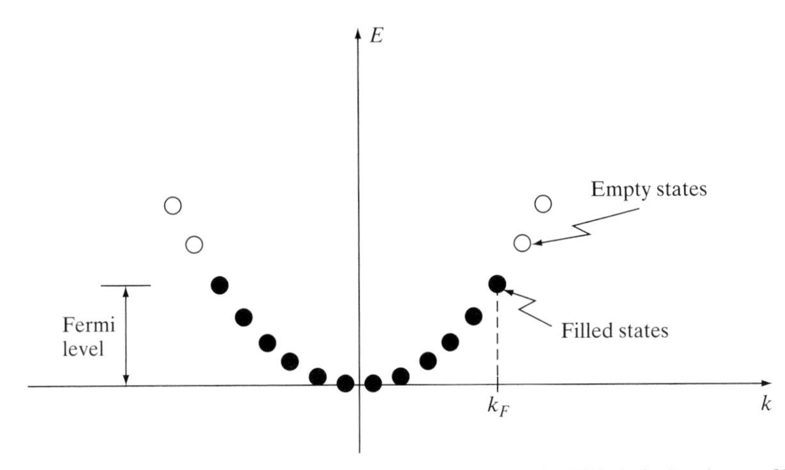

Figure 4.5 Energy versus wavenumber for discrete energy levels. Filled circles denote filled states, and empty circles denote empty states.

for a noninteracting Fermi gas (i.e., a collection of noninteracting fermions, which models many metals at low temperatures). In (4.62), k_B is *Boltzmann's constant*,

$$\begin{aligned} k_B &= 1.38 \times 10^{-23} \text{ J/K} \\ &= 8.62 \times 10^{-5} \text{ eV/K}, \end{aligned}$$

which can be thought of as a conversion factor between temperature and energy. As mentioned previously, with a slight abuse of notation, we usually denote $\mu(T)$ as simply the Fermi level, and use E_F to denote the value of the chemical potential at a given temperature.

With the concept of the Fermi level established, Fig. 4.4 on page 95 can be modified to account for the fact that at absolute zero, states below the Fermi level will be filled, and those above the Fermi level will be empty, as depicted in Fig. 4.5. Of course, spin must be accounted for, such that two electrons can occupy each state.

4.5 PARTIALLY CONFINED ELECTRONS—FINITE POTENTIAL WELLS

In Section 4.3, when an electron confined to a finite region of space (the particle-in-a-box model) was considered, the electron had no possibility of escape. Although the potential energy profile V was not specifically indicated, the assumption that the electron must be inside the box ($\psi = 0$ outside the box) is equivalent to the potential profile (in one dimension)

$$\begin{aligned} V &= 0, \quad 0 \leq x \leq L, \\ V &= \infty, \quad x < 0, \ x > L. \end{aligned} \tag{4.63}$$

The box can be thought of as an infinite quantum well, as shown in Fig. 4.6.

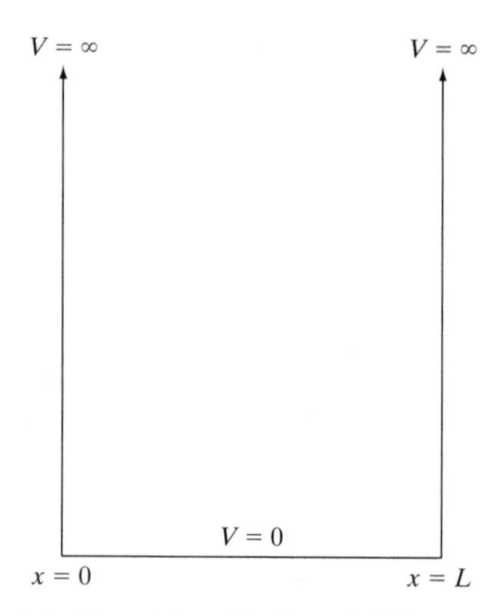

Figure 4.6 Potential profile for an infinite square well.

In the case of an infinite well, the particle has no chance of escaping the well. However, no jail is perfect, and we often want to allow for the possibility of the particle getting out of the well. Thus, we often need to model the walls as having finite height (finite potential), which is a more realistic model of what a particle such as an electron will "see" due to being confined in some way (say, by the potential of an atom).

4.5.1 Finite Rectangular Well

In this section, the symmetric potential well having finite, rather than infinite, walls (potential barriers) will be considered,

$$\begin{aligned} V &= 0, \quad -L \leq x \leq L, \\ V &= V_0, \quad x < -L, \ x > L, \end{aligned} \tag{4.64}$$

as shown in Fig. 4.7.

Comparing this figure with Fig. 4.1 on page 86, which shows the potential energy between an electron and an ionized atom, we can see that the profile in Fig. 4.7. approximates the influence of an ionized atom on an electron.† As will be shown in Chapters 6 and 9, this will also model the confining potential of material interfaces.

We introduce an electron having total energy $E < V_0$ into the potential energy well. The electron would be stuck in the well, classically. This is because outside the well, the electron's total energy would still be E, since there is no source of energy for the electron

†Of course, the electron has the same influence on the atom, although, since the mass of the atom is so much greater than that of the electron, we can assume the atom to be stationary.

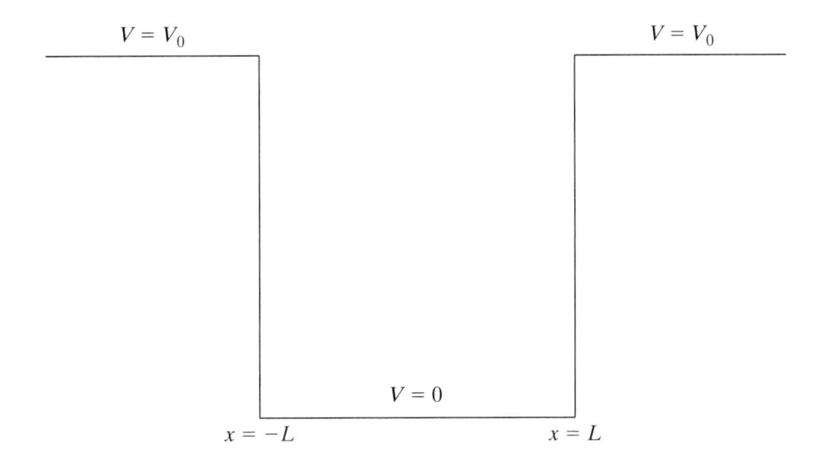

Figure 4.7 Finite square well potential.

(no incident radiation, nor collisions with other particles, etc.) Therefore, if the electron were outside the well, we would have

$$\begin{aligned} E &= E_{KE} + E_{PE} \\ &= E_{KE} + V_0 < V_0, \end{aligned} \tag{4.65}$$

where E_{KE} is kinetic energy and E_{PE} is potential energy. So

$$E_{KE} < 0,$$

which indicates that the electron's kinetic energy would be negative (i.e., $(1/2)\,mv^2 < 0$)! According to classical physics, this cannot occur, and therefore, classically, the electron must be in the well.

However, quantum mechanically, there is some (perhaps small) probability that the particle will be found outside the well. To see this, we start with Schrödinger's equation

$$\left(-\frac{\hbar^2}{2m_e}\frac{d^2}{dx^2} + V(x)\right)\psi(x) = E\psi(x), \tag{4.66}$$

where $V(x)$ corresponds to the profile given in (4.64). We solve this equation by considering separately the three regions (inside and outside of the well), solving Schrödinger's equation in each region, and applying boundary conditions at each interface between the regions to connect the three solutions together.

In the well region, which we'll call region II, we have

$$\left(\frac{\hbar^2}{2m_e}\frac{d^2}{dx^2} + E\right)\psi_2(x) = 0 \tag{4.67}$$

since $V = 0$ inside the well, leading to

$$\psi_2(x) = C \sin k_2 x + D \cos k_2 x, \tag{4.68}$$

with

$$k_2^2 = \frac{2m_e E}{\hbar^2}. \tag{4.69}$$

In the region $x < -L$, which we'll call region I, we have

$$\left(\frac{\hbar^2}{2m_e} \frac{d^2}{dx^2} - (V_0 - E) \right) \psi_1(x) = 0, \tag{4.70}$$

leading to

$$\psi_1(x) = Ae^{k_1 x} + Be^{-k_1 x}, \tag{4.71}$$

with

$$k_1^2 = \frac{2m_e (V_0 - E)}{\hbar^2}. \tag{4.72}$$

However, the wavefunction should be finite as $x \to -\infty$ and, assuming that $E < V_0$, then

$$B = 0. \tag{4.73}$$

In the region $x > L$, region III, we have

$$\left(\frac{\hbar}{2m_e} \frac{d^2}{dx^2} - (V_0 - E) \right) \psi_3(x) = 0 \tag{4.74}$$

leading to

$$\psi_3(x) = Fe^{k_3 x} + Ge^{-k_3 x}, \tag{4.75}$$

with

$$k_3^2 = \frac{2m_e (V_0 - E)}{\hbar^2} = k_1^2. \tag{4.76}$$

Finiteness of the wavefunction as $x \to +\infty$ leads to

$$F = 0. \tag{4.77}$$

In summary, we have

$$\begin{aligned} \psi_1(x) &= Ae^{k_1 x}, & x < -L, \\ \psi_2(x) &= C\sin k_2 x + D\cos k_2 x, & -L \leq x \leq L, \\ \psi_3(x) &= Ge^{-k_3 x}, & x > L, \end{aligned} \tag{4.78}$$

with

$$\begin{aligned} k_2^2 &= \frac{2m_e E}{\hbar^2}, \\ k_1^2 = k_3^2 &= \frac{2m_e(V_0 - E)}{\hbar^2}. \end{aligned} \tag{4.79}$$

It can be shown that symmetric potentials lead to either symmetric or antisymmetric solutions, and so either

$$\begin{aligned} C &= 0, \quad \text{symmetric solution, or} \\ D &= 0, \quad \text{antisymmetric solution.} \end{aligned} \tag{4.80}$$

Applying the four boundary conditions obtained from (3.143),

$$\begin{aligned} \psi_1(x = -L) &= \psi_2(x = -L), \\ \psi_2(x = L) &= \psi_3(x = L) \\ \psi_1'(x = -L) &= \psi_2'(x = -L), \\ \psi_2'(x = L) &= \psi_3'(x = L), \end{aligned} \tag{4.81}$$

to the symmetric case leads to the symmetric solution

$$\begin{aligned} \psi_1(x) &= D\cos(k_2 L)\, e^{k_1(x+L)}, & x < -L, \\ \psi_2(x) &= D\cos k_2 x, & -L \leq x \leq L, \\ \psi_3(x) &= D\cos(k_2 L)\, e^{-k_3(x-L)}, & x > L, \end{aligned} \tag{4.82}$$

where $k_1 = k_3$ and k_1, k_2 must satisfy the *eigenvalue equation*

$$k_2 \tan(k_2 L) = k_1. \tag{4.83}$$

One can appreciate that, although the finite well is usually a much more realistic model of a quantum well, the infinite well is often used to get a rough approximation of energy states within quantum structures. That is, the assumption of an infinite well leads to a simple formula for the energy states E_n (e.g., (4.35)), whereas the finite well does not yield a nice

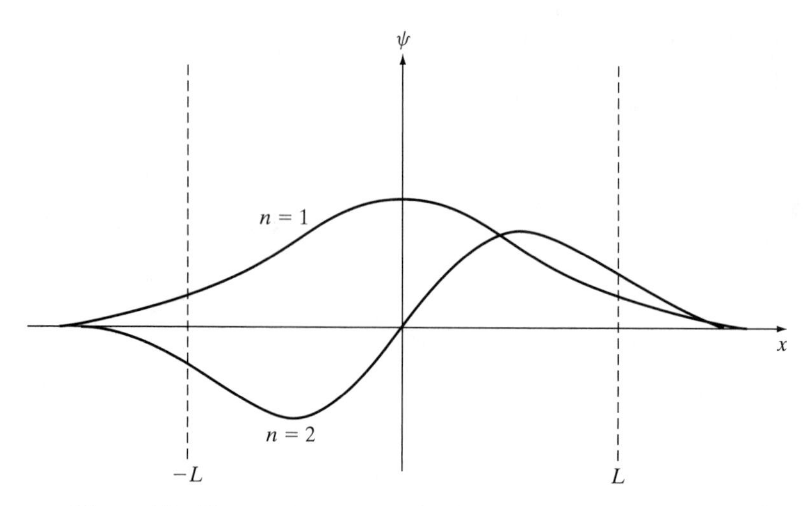

Figure 4.8 The two lowest states for the finite-height potential well.

formula, but only an eigenvalue equation that must be solved graphically or numerically for E (although discrete values E_n will still be obtained). Typical wavefunctions for the first two modes are shown in Fig. 4.8, from which it can be appreciated that the wavefunction is largest in the well region, but has nonzero "tails" extending outside the well. This is in contrast to the infinity-high wall case ($V_0 = \infty$), where the electron is absolutely bound to the well, shown in Fig. 4.3 on page 95. Therefore, for the finite potential energy barrier, electrons are quasi-confined, or localized, to the well, although they can also be found outside the well.

Applying the four boundary conditions (4.81) to the antisymmetric form of (4.78) (i.e., setting $D = 0$), we have

$$\begin{aligned} \psi_1(x) &= -C \sin(k_2 L)\, e^{k_1(x+L)}, & x &< -L, \\ \psi_2(x) &= C \sin k_2 x, & -L &\le x \le L, \\ \psi_3(x) &= C \sin(k_2 L)\, e^{-k_3(x-L)}, & x &> L, \end{aligned} \tag{4.84}$$

where $k_1 = k_3$ and k_1, k_2 must satisfy the eigenvalue equation

$$-k_2 \cot(k_2 L) = k_1, \tag{4.85}$$

which can be solved graphically or numerically. As in the case of the symmetric solution, in general, no simple formula for the energy states can be found.

Graphical Solution of the Eigenvalue Equation. Although it is usually desirable to solve (4.83) and (4.85) numerically, a simple graphical procedure can be applied that leads to considerable insight into the nature of the solutions (and, in fact, shows that discrete solutions do exist). In the following discussion, we will consider the eigenvalue equation for symmetric solutions, (4.83), although the same procedure works for the eigenvalue equation governing antisymmetric wavefunctions. In general, symmetric and antisymmetric solutions alternate (i.e., the lowest state is symmetric, and the next higher state is antisymmetric, etc.), just as was found for the infinite well case previously discussed.

First, note that

$$k_1^2 + k_2^2 = \frac{2m_e (V_0 - E)}{\hbar^2} - \frac{2m_e E}{\hbar^2} = \frac{2m_e V_0}{\hbar^2}, \tag{4.86}$$

such that

$$(k_1 L)^2 + (k_2 L)^2 = \frac{2m_e V_0 L^2}{\hbar^2}, \tag{4.87}$$

which is the equation of a circle in the k_1L–k_2L plane. We can also plot

$$k_1 L = k_2 L \tan(k_2 L) \tag{4.88}$$

in the k_1L–k_2L plane, and, obviously, the intersections will be the desired (discrete) solutions for energy E_n. A representative plot is shown in Fig. 4.9.

Only intersections in the upper-half plane are valid, since $k_1 < 0$ would cause the wavefunctions ψ_1 and ψ_3 to become infinitely large as $|x| \to \infty$ (far away from the well). We will label the N intersections in the upper-half plane by the index $n = 1, 3, 5, \ldots, N$ ($N = 1$ shown in Fig. 4.9, for two different V_0 values).

Several features of the solutions are evident from Fig. 4.9. From (4.87), the radius of the circle is

$$\sqrt{\frac{2m_e V_0 L^2}{\hbar^2}}, \tag{4.89}$$

so that for very small V_0 or L (small radius circle), there is only one solution. As V_0 or L increase, the radius of the circle increases, and so more discrete states will exist, although for any finite V_0 and L there will be a finite number (N) of solutions. In the limit $V_0 \to \infty$ or $L \to \infty$, a countable infinity of discrete solutions will exist, $N \to \infty$, in agreement with the infinite well discussed in Section 4.3.1.

In particular, we can determine at which point additional discrete states emerge. Increasing V_0 (or L) from low values, we see that when

$$\sqrt{\frac{2m_e V_0 L^2}{\hbar^2}} = m\pi, \tag{4.90}$$

and $m = 1, 2, 3, \ldots$, new intersections occur. For instance, solving for V_0 yields

$$V_0 = \frac{m}{2}\left(\frac{\hbar\pi}{L}\right)^2. \tag{4.91}$$

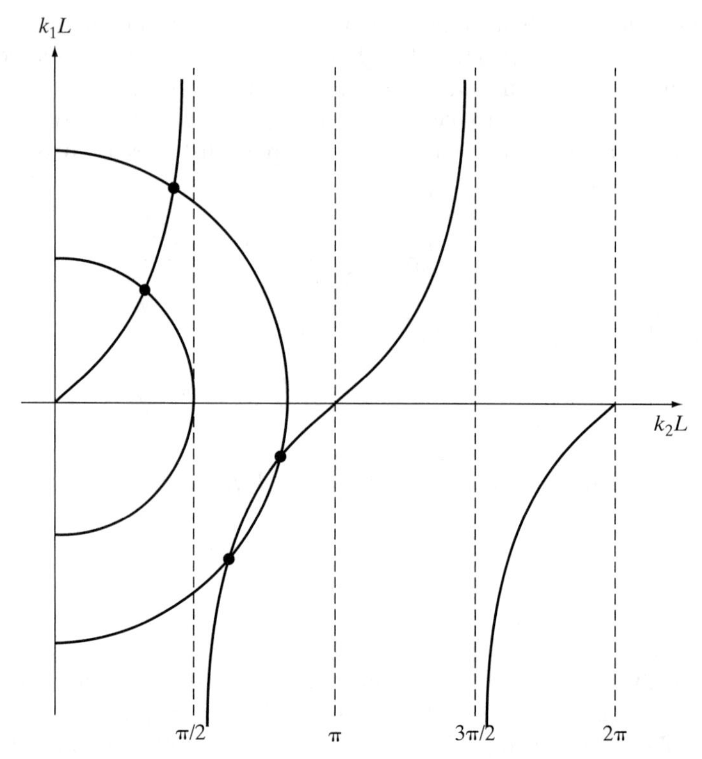

Figure 4.9 Graphical technique for determining the solutions of the eigenvalue equation (4.88). The tan function repeats indefinitely, with period π. The two circles represent two different values of confining potential V_0. Only intersections in the upper-half plane indicate the desired solutions.

As V_0 increases, the nth intersection moves along the upper parts of the tangent curve, approaching a value $k_2L = n\pi/2$, and the value $k_1L \to \infty$. Furthermore, in the limit $V_0 \to \infty$ the infinite well case should be recovered. This can be seen by using (4.79) and setting $k_2L = n\pi/2$, such that

$$k_2L = \sqrt{\frac{2m_eE}{\hbar^2}}L = \frac{n\pi}{2} \to E = \frac{\hbar^2}{2m_e}\left(\frac{n\pi}{2L}\right)^2,$$

in agreement with (4.35). (Here the well has length $2L$, and in (4.35), the length of the well is L.) Examining the eigenfunctions (4.82), we see that in this limit

$$\lim_{V_0\to\infty} \psi_1(x) = D\cos\left(\frac{n\pi}{2}\right)e^{\infty(x+L)} \to 0, \qquad x < -L, \tag{4.92}$$

$$\lim_{V_0\to\infty} \psi_2(x) = D\cos\left(\frac{n\pi}{2L}x\right), \qquad -L \leq x \leq L,$$

$$\lim_{V_0\to\infty} \psi_3(x) = D\cos\left(\frac{n\pi}{2}\right)e^{-\infty(x-L)} \to 0, \qquad x > L,$$

which becomes simply the even modes of the infinite well, (4.32), as expected. The point is that as the potential increases, the wavefunction becomes bound more tightly to the well, reflecting the increasing probability that the particle will be found in the well, and, as $V_0 \to \infty$, we recover the infinite well result.

In arriving at the bound solutions, it was assumed that $E < V_0$, and discrete allowed energy values were obtained. The electron is most likely to be found in the well, although it can also be found outside the well, with decreasing probability as we move away from the well. Other solutions exist for $E > V_0$, corresponding to a continuum of allowed energy values. In this case, the presence of the well merely perturbs the electron's wavefunction. Far from the well, we expect the wavefunction to correspond to a plane wave, since in these locations the electron is essentially free. The details will not be included here, although the idea is that for $E < V_0$, the electron is (at least somewhat) bound to the well, although it can be found outside the well with decreasing probability, and as $E \to V_0$, the electron becomes less bound to the well. For $E > V_0$, the electron can occupy a continuum of states, the well merely serving to perturb the electron's wavefunction. The same scenario occurs in classical electromagnetic theory for the modes of a finite-thickness dielectric layer, which tends to confine, though not perfectly, electromagnetic energy within the layer. It should be emphasized once again that we have found only possible allowed states for an electron. Whether or not an electron is in a certain state depends on other factors, as will be discussed in Chapters 5 and 8.

4.5.2 Parabolic Well—Harmonic Oscillator

Rather than the rectangular well considered in the previous section, here we briefly consider a parabolic potential profile, which has importance in modeling quantum heterostructures and split-gate devices. In this case, we assume the potential profile

$$V(x) = \frac{1}{2}Kx^2, \tag{4.93}$$

as shown in Fig. 4.10. which describes a classical harmonic oscillator. For example, a mass on a spring sees the force (4.93), where K is the spring constant and x is the displacement from equilibrium. This gives rise to harmonic motion $x(t) = A\cos\omega_0 t$, where the frequency of oscillation is $\omega_0 = \sqrt{K/m}$.

For the harmonic potential, Schrödinger's equation is

$$\left(-\frac{\hbar^2}{2m_e}\frac{d^2}{dx^2} + \frac{1}{2}\omega_0^2 m_e x^2\right)\psi(x) = E\psi(x). \tag{4.94}$$

Although the details will be omitted here, the solution of (4.94) is

$$\psi(x) = C_n H_n\left[\left(\frac{m_e\omega_0}{\hbar}\right)^{1/2} x\right] e^{-\frac{m_e\omega_0}{2\hbar}x^2}, \tag{4.95}$$

where C_n is a constant,

$$C_n = \left(\frac{1}{\sqrt{\pi}2^n n!}\right)^{1/2}\left(\frac{m_e\omega_0}{\hbar}\right)^{1/4},$$

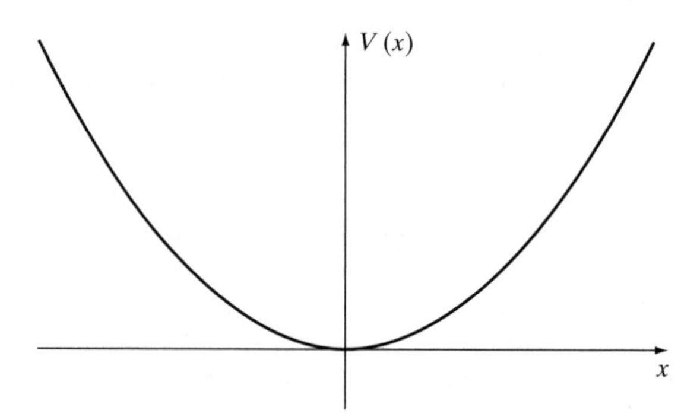

Figure 4.10 Harmonic oscillator (parabolic) potential.

and where H_n are Hermite polynomials, the first several of which are

$$H_0(x) = 1, \tag{4.96}$$

$$H_1(x) = 2x,$$

$$H_2(x) = 4x^2 - 2,$$

$$H_3(x) = 8x^3 - 12x, \ldots. \tag{4.97}$$

Despite the difference in the form of the solutions, the first several wavefunctions for the parabolic potential resemble those for the square well potential.

Energy levels are found to be

$$E_n = \left(n - \frac{1}{2}\right)\hbar\omega_0, \quad n = 1, 2, 3, \ldots, \tag{4.98}$$

and are equally spaced according to the index n.

4.5.3 Triangular Well

As a final example, we consider the case of the triangular well, shown in Fig. 4.11. As discussed in Section 6.3.1, triangular wells are often used to model the junction between two materials.

The potential profile is

$$V(x) = \begin{cases} \infty, & x < 0, \\ Cx, & x > 0, \end{cases} \tag{4.99}$$

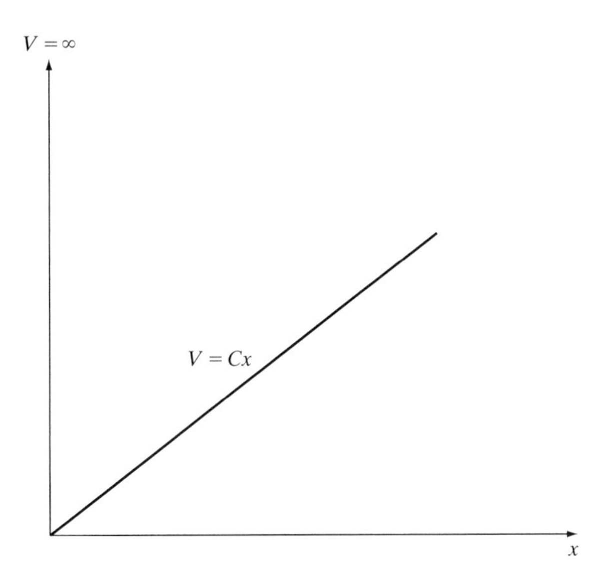

Figure 4.11 Triangular well potential.

where C is a constant. The solution of Schrödinger's equation with the triangular potential is fairly complicated, and the resulting wavefunctions are expressed in terms of *Airy functions*. The details will be omitted here, although the energy levels are given in a simple form,

$$E_n = \left(\frac{\hbar}{2m_e}\right)^{1/3} \left(\frac{3}{2}\pi C\right)^{2/3} \left(n - \frac{1}{4}\right)^{2/3}, \tag{4.100}$$

$n = 1, 2, 3, \ldots$.

It can be appreciated that for a rectangular well, $E_n \propto n^2$, for the parabolic well, $E_n \propto n$, and for the triangular well, $E_n \propto n^{2/3}$.

4.6 ELECTRONS CONFINED TO ATOMS—THE HYDROGEN ATOM AND THE PERIODIC TABLE

In Sections 4.3 and 4.5, we obtained discrete allowed energy states for an electron confined to a finite region of space, and we introduced the notion of quantum numbers. As a next step up in complexity, it is worthwhile to consider individual atoms, which confine electrons to atomic regions. Perhaps not surprisingly, it turns out that for atoms, electronic states are also quantized. By considering the simplest atom, hydrogen, we can introduce other quantum numbers that represent important physics not encountered in the simple quantum box previously considered. The hydrogen atom is one of the few realistic quantum mechanics problems that can be solved exactly, since problems with more electrons can usually only be solved numerically.

4.6.1 The Hydrogen Atom and Quantum Numbers

The hydrogen atom consists of one proton and one electron, interacting via the Coulomb force. Let the electron's position be denoted by (x_e, y_e, z_e) and the proton's position by (x_p, y_p, z_p). Thus, we have a two-particle Schrödinger's equation,

$$\left(-\frac{\hbar^2}{2m_e}\left(\frac{\partial^2}{\partial x_e^2}+\frac{\partial^2}{\partial y_e^2}+\frac{\partial^2}{\partial z_e^2}\right)-\frac{\hbar^2}{2m_p}\left(\frac{\partial^2}{\partial x_p^2}+\frac{\partial^2}{\partial y_p^2}+\frac{\partial^2}{\partial z_p^2}\right)+V\left(\mathbf{r}_e,\mathbf{r}_p\right)\right)\psi\left(\mathbf{r}_e,\mathbf{r}_p\right)$$
$$=E\psi\left(\mathbf{r}_e,\mathbf{r}_p\right), \tag{4.101}$$

where m_e and m_p are the electron and proton mass, respectively.

As is well known, the Coulomb potential energy depends only on the separation between the charges, $\mathbf{r}_e - \mathbf{r}_p$. It can be shown that in this case the problem can be separated into two, one-body problems,

$$\left(-\frac{\hbar^2}{2\mu_m}\left(\frac{\partial^2}{\partial x^2}+\frac{\partial^2}{\partial y^2}+\frac{\partial^2}{\partial z^2}\right)+V(\mathbf{r})\right)\psi(\mathbf{r})=E\psi(\mathbf{r}), \tag{4.102}$$

$$-\frac{\hbar^2}{2M}\left(\frac{\partial^2}{\partial X^2}+\frac{\partial^2}{\partial Y^2}+\frac{\partial^2}{\partial Z^2}\right)U(\mathbf{R})=E'U(\mathbf{R}), \tag{4.103}$$

with

$$\mu_m=\frac{m_e m_p}{m_e+m_p}, \quad M=m_e+m_p, \tag{4.104}$$

$$MX=m_e x_e+m_p x_p,$$

$$MY=m_e y_e+m_p y_p, \tag{4.105}$$

$$MZ=m_e z_e+m_p z_p, \tag{4.106}$$

where u_m is known as the *reduced mass*,† and where the total energy is

$$E_{tot}=E+E'. \tag{4.108}$$

†The reduced mass can also be written as

$$\frac{1}{\mu_m}=\frac{1}{m_e}+\frac{1}{m_p}. \tag{4.107}$$

Equation (4.103) is Schrödinger's equation for a particle of mass M in free space, and so, by the results of Section 4.1.2, we obtain plane wave solutions

$$U(\mathbf{R}) = A_0 e^{i\mathbf{k}\cdot\mathbf{R}}, \tag{4.109}$$

with

$$E' = \frac{\hbar^2 k^2}{2M}, \quad k = |\mathbf{k}|. \tag{4.110}$$

Turning to (4.102), since $m_e = 9.1095 \times 10^{-31}$ kg and $m_p = 1.6726 \times 10^{-27}$ kg, we can make the approximation

$$\mu_m = \frac{m_e m_p}{m_e + m_p} \simeq m_e. \tag{4.111}$$

The potential $V(r)$ is the well-known Coulomb potential,

$$V(r) = -\frac{q_e^2}{4\pi\varepsilon_0 r}, \tag{4.112}$$

and the Schrödinger equation (4.102) can be cast into spherical coordinates and solved via separation of variables. The details will be omitted here, although the result can be expressed as

$$\psi_{n,l,m_l}(r,\theta,\phi) = R_{n,l}(r)\, Y_l^{m_l}(\theta,\phi), \tag{4.113}$$

where

$n = 1, 2, 3, \ldots,$ is called the *principle quantum number*,

$l = 0, 1, 2, \ldots, n-1,$ is called the *angular momentum quantum number*, and

$m_l = 0, \pm 1, \pm 2, \ldots \pm l,$ is called the *magnetic quantum number*.

The principle quantum number n plays the role of the quantum number n found in the electron box (quantum well) examples considered in Sections 4.3 and 4.5. The angular momentum quantum number is so named because it relates to the (quantized) orbital angular momentum of the electron and the magnetic quantum number relates to the projection of the angular momentum vector. Note that the indices are interconnected. For example, for $n = 2$, we have

$$n = 2, \quad l = 0, \quad m_l = 0, \tag{4.114}$$
$$n = 2, \quad l = 1, \quad m_l = \pm 1.$$

Although we won't consider in detail the wavefunctions (4.113), Table 4.1 summarizes some of the functions that make up the wavefunction for low orders, and Fig. 4.12 shows a few of the low-order wavefunctions $\psi_{n,l,0}$. Recall that the probability of finding an electron is high where $\left|\psi_{n,l,m_l}\right|^2$ is relatively large, and that it is less likely to find an electron where $\left|\psi_{n,l,m_l}\right|$ is small.

TABLE 4.1 COMPONENTS OF THE HYDROGEN WAVEFUNCTIONS FOR LOW-PRINCIPLE QUANTUM NUMBERS.

n	l	m	$R_{n,l}(r)$	$Y_l^{m_l}(\theta, \phi)$
1	0	0	$\frac{2}{a_0^{3/2}} e^{-\frac{r}{a_0}}$	$\frac{1}{\sqrt{4\pi}}$
2	0	0	$\frac{1}{(2a_0)^{3/2}}\left(2 - \frac{r}{a_0}\right) e^{-\frac{r}{2a_0}}$	$\frac{1}{\sqrt{4\pi}}$
2	1	0	$\frac{1}{\sqrt{3}\,(2a_0)^{3/2}} \frac{r}{a_0} e^{-\frac{r}{2a_0}}$	$\sqrt{\frac{3}{4\pi}} \cos\theta$
2	1	± 1	$\frac{1}{\sqrt{3}\,(2a_0)^{3/2}} \frac{r}{a_0} e^{-\frac{r}{2a_0}}$	$\frac{1}{2}\sqrt{\frac{3}{2\pi}} \sin\theta\, e^{\pm i\phi}$

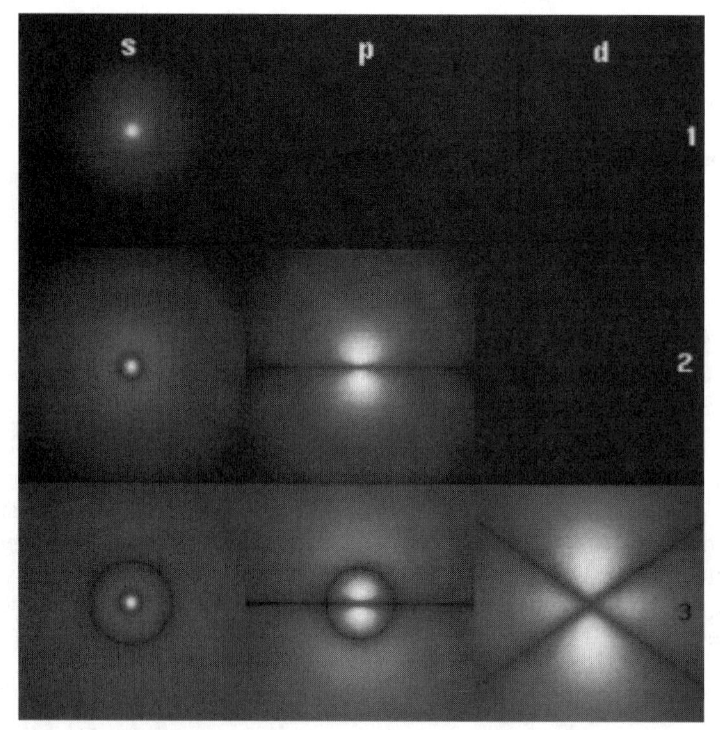

Figure 4.12 Several low-order hydrogen atom orbitals (plots of the probability density). The designations s,p,d refer to the quantum numbers $l = 0,1,2$, and the right-side designation is the principal quantum number n. In all cases m = 0. (Courtesy Wikipedia, The Free Encyclopedia.)

In Table 4.1, the constant a_0 is called the *Bohr radius,* and is given by

$$a_0 = \frac{4\pi\varepsilon_0\hbar^2}{m_e q_e^2} = 0.53 \text{ Å}. \tag{4.115}$$

This quantity, which comes about as a constant in the process of solving for the eigenfunctions ψ, does have important physical meaning. The Bohr radius is taken as the approximate "size" of the hydrogen atom, since for $r > a_0$, the wavefunction decreases rapidly, as can be seen from the function $R_{1,0}$ in Table 4.1. Somewhat more precisely, since the Coulomb potential varies as r^{-1}, the expectation value of r^{-1} gives the reciprocal of the approximate "size" of the hydrogen atom,

$$\begin{aligned}\left\langle \frac{1}{r} \right\rangle &= \int \psi_{100}^*(r)\, \frac{1}{r}\, \psi_{100}(r)\, d^3r \qquad (4.116)\\ &= \int_0^{2\pi}\int_0^{\pi}\int_0^{\infty} \frac{2}{a_0^{3/2}} e^{-\frac{r}{a_0}} \frac{1}{\sqrt{4\pi}} \left(\frac{1}{r}\right) \frac{2}{a_0^{3/2}} e^{-\frac{r}{a_0}} \frac{1}{\sqrt{4\pi}} r^2 \sin\theta\, dr\, d\theta\, d\phi\\ &= \left(\frac{2}{a_0^{3/2}}\right)^2 \int_0^{\infty} e^{-\frac{2r}{a_0}}\, r\, dr\\ &= \frac{1}{a_0}.\end{aligned}$$

It can be shown that energy is only dependent on the principle quantum number n, and its value is

$$E_n = -\frac{\mu_m q_e^4}{8\varepsilon_0^2 h^2 n^2}. \tag{4.117}$$

Therefore, for $n = 1$ (called the *ground state*) the corresponding energy is, replacing μ_m by m_e,

$$E_1 = -\frac{m_e q_e^4}{8\varepsilon_0^2 h^2} = -2.18 \times 10^{-18} \text{ J} = -13.6 \text{ eV}, \tag{4.118}$$

and, thus,†

$$E_n = -13.6 \frac{1}{n^2} \text{ eV}. \tag{4.119}$$

In general, and not only for the hydrogen atom, the principle quantum number n is called the *shell* number, and the states

$$l = 0, 1, 2, 3, 4, 5, 6, 7, \ldots \tag{4.120}$$

are called the $s, p, d, f, g, h, i, k, \ldots$ states, i.e.,

$$\begin{aligned} l &= 0, 1, 2, 3, 4, 5, 6, 7 \qquad (4.121)\\ &= s, p, d, f, g, h, i, k.\end{aligned}$$

† $R_Y = m_e q_e^4 / \left(8\varepsilon_0^2 h^2\right) = -13.6$ eV is known as the *Rydberg energy*.

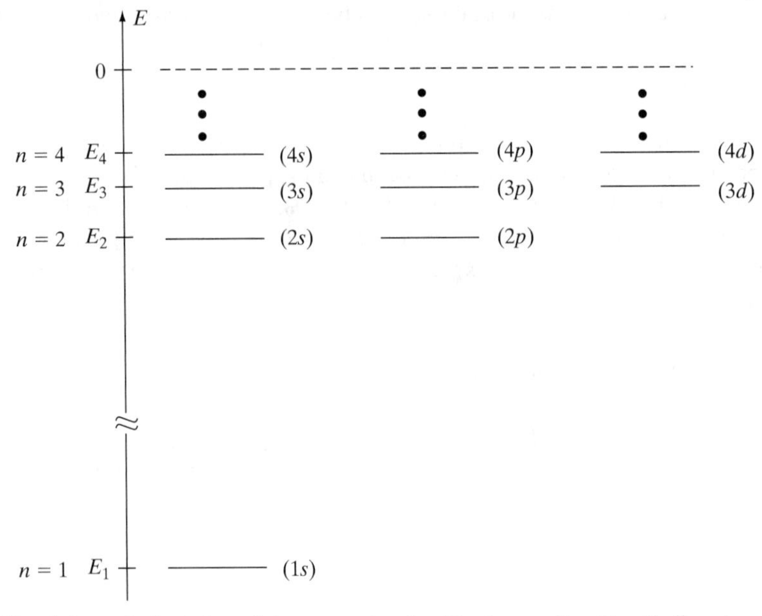

Figure 4.13 Schematic depiction of the energy levels of hydrogen. For $E > 0$, discrete energy levels cease to exist, although a continuum of states is present.

Thus, $n = 3, l = 1$ is the $3p$ state (third shell, second state ($l = 1$)). Energy levels for hydrogen are schematically depicted in Fig. 4.13.

The quantum numbers n, l, and m_l need to be augmented by a spin number, $s = \pm 1/2$. Together, the four quantum numbers (n, l, m_l, s) characterize the quantum state of the hydrogen atom.

Note that as an electron moves about a nucleus, it doesn't maintain constant radius or follow a set path. Unless some energy is supplied or extracted from the atom, the electron maintains constant total energy. Since its potential energy varies with distance away from the nucleus, an electron will change its kinetic energy (i.e., either speeding up or slowing down) to compensate for changes in radius as it moves about the nucleus.

The relatively simple two-particle hydrogen model is very important for modeling more complicated multiparticle configurations. In the next section, we will apply the hydrogen model to other atoms, leading to an approximate construction of the electronic configuration of many elements. In Chapter 5, the two-particle hydrogen model will be used to analyze semiconductor doping, and the creation of bound electron–hole pairs known as excitons.

4.6.2 Beyond Hydrogen—Multiple Electron Atoms and the Periodic Table

The hydrogen atom consists of one electron and one proton, and, for this system, Schrödinger's equation can be solved exactly. However, even for the next simplest atom—helium—which consists of two protons and two electrons, Schrödinger's equation cannot be solved in closed form. However, the usual procedure to gain some understanding of atoms beyond hydrogen is to invoke the Pauli exclusion principle. Recall that this principle dictates that two or more identical fermions (electrons are fermions) cannot occupy the same allowed state. We do the following for atoms consisting of multiple electrons:

1. Using the hydrogen model, start with the lowest energy level and count the number of states to be filled, dictated by the number of electrons associated with the atom in question.
2. Fill the states up one by one with electrons.
3. Once an energy level is full, proceed to the next higher energy level, and so on.

For example, for the first shell, $n = 1$, we have $l = m = 0$. There are two possible states, $(n, l, m_l, s) = (1, 0, 0, \pm 1/2)$, i.e.,

$$\text{state } 1: \; n = 1, \quad l = 0, \quad m_l = 0, \quad s = \frac{1}{2},$$

$$\text{state } 2: \; n = 1, \quad l = 0, \quad m_l = 0, \quad s = -\frac{1}{2}.$$

Therefore, hydrogen has one "1s" electron, and helium has two "1s" electrons, denoted as $1s^2$. Since helium has two electrons that fill the two $1s$ states and "close" the $n = 1$ shell, helium is chemically inert. In fact, the number of outer shell electrons is the single most important quantity in characterizing how an element will react with its environment. Elements with partially filled outer shells will tend to react with other elements, whereas elements with full (closed) outer shells will tend to be nonreactive. The number of outer shell electrons is so important that elements in the periodic table are grouped together (in columns) by the number of outer shell electrons.

Continuing in this fashion, we fill up the shells $n = 2, 3, 4, \ldots$ with available electrons. Note that, for example, the $n = 2$ shell has eight states,

$$\text{state } 1: n = 2, \quad l = 0, \quad m_l = 0, \quad s = \frac{1}{2},$$

$$\text{state } 2: n = 2, \quad l = 0, \quad m_l = 0, \quad s = -\frac{1}{2},$$

$$\text{state } 3: n = 2, \quad l = 1, \quad m_l = 0, \quad s = \frac{1}{2},$$

$$\text{state } 4: n = 2, \quad l = 1, \quad m_l = 0, \quad s = -\frac{1}{2},$$

$$\text{state } 5: n = 2, \quad l = 1, \quad m_l = 1, \quad s = \frac{1}{2},$$

$$\text{state } 6: n = 2, \quad l = 1, \quad m_l = 1, \quad s = -\frac{1}{2},$$

$$\text{state } 7: n = 2, \quad l = 1, \quad m_l = -1, \quad s = \frac{1}{2},$$

$$\text{state } 8: n = 2, \quad l = 1, \quad m_l = -1, \quad s = -\frac{1}{2},$$

so that, together, the $n = 1$ and $n = 2$ shells can accommodate 10 electrons. For example, neon has 10 electrons, thus filling the $n = 1$ and $n = 2$ shells, but not going into the $n = 3$ shell. Neon's electronic configuration is denoted by

$$1s^2 2s^2 2p^6,$$

since it has 2 "$1s$" electrons, 2 "$2s$" electrons, and 6 "$2p$" electrons. The next element, sodium, with 11 electrons, breaks into the $3s$ shell; its configuration is $1s^2 2s^2 2p^6 3s^1$.

Thus, it would seem that to determine the electronic configuration of any atom, one only has to know how many electrons it possesses. This procedure works surprisingly well in many cases; however, we need to remember that it is based on the simple hydrogen result. Most atoms are considerably more complicated than hydrogen, obviously. In particular, the hydrogen model applied to multi-electron atoms implicitly assumes that there is no interaction between electrons, which is not really true. Also, for elements other than hydrogen, it turns out that energy is not constant within a shell. Therefore, although the preceding procedure can be used as a starting point, it does not provide an exact description of the electronic configuration of other atoms. In particular, starting with potassium, some modifications to the described procedure must be made. In Table 4.2, the electronic configuration of the first 32 elements of the periodic table is provided, and, for convenience, the periodic table of the elements is shown in Fig. 4.14.

At this point, one comment should be made concerning spin and the filling of states. It turns out that spin interactions favor filling same-spin states first, before beginning to fill the other spin state. This is called *Hund's rule*. For example, carbon has the configuration $1s^2 2s^2 2p^2$. The $2p$ shell is partially filled, having two electrons, but in the ground state, both have the same spin. The same can be said of nitrogen, with three same-spin electrons. In the next element, oxygen, the addition of a fourth (out of six) $2p$ electron finally results in a spin down electron. This is shown in Table 4.3. The filling of spin states has particular importance in considering the magnetic properties of materials, since filled shells will have no net magnetic moment.

4.7 QUANTUM DOTS, WIRES, AND WELLS

So far in this chapter, we have studied electrons confined to finite regions of space via infinite potential energy barriers (Section 4.3), electrons localized to a region of space by finite potential energy barriers (Section 4.5), and electrons confined to (or associated with) atoms (Section 4.6). In all cases, we have found that the energy of the electron will be discrete, given by E_n. Furthermore, if we confine a particle to a region of space on the order of, or smaller than, the particle's de Broglie wavelength, the discreteness of possible energy states that the particle can occupy will be evident. Chapters previous to this one were intended to provide the necessary machinery to study electrons in various situations, and, in previous sections of this chapter, we examined the effects of confining electrons to finite regions of space. At this point, we can begin to consider the important topics of quantum dots, quantum wires, and quantum wells. Since these structures are often made from semiconductors, we will also revisit this topic in Chapter 9, after, among other things, considering some basic semiconductor physics in the next chapter.

TABLE 4.2 ELECTRON CONFIGURATION OF THE FIRST 32 ELEMENTS (OBTAINED BY SELF-CONSISTENT METHODS). Z IS THE ATOMIC NUMBER, I.E., THE NUMBER OF ELECTRONS.

Z	Element	1*s*	2*s*	2*p*	3*s*	3*p*	3*d*	4*s*	4*p*	4*d*
1	H, hydrogen	1								
2	He, helium	2								
3	Li, lithium	2	1							
4	Be, beryllium	2	2							
5	B, boron	2	2	1						
6	C, carbon	2	2	2						
7	N, nitrogen	2	2	3						
8	O, oxygen	2	2	4						
9	F, fluorine	2	2	5						
10	Ne, neon	2	2	6						
11	Na, sodium	2	2	6	1					
12	Mg, magnesium	2	2	6	2					
13	Al, aluminium	2	2	6	2	1				
14	Si, silicon	2	2	6	2	2				
15	P, phosphorus	2	2	6	2	3				
16	S, sulfur	2	2	6	2	4				
17	Cl, chlorine	2	2	6	2	5				
18	Ar, argon	2	2	6	2	6				
19	K, potassium	2	2	6	2	6		1		
20	Ca, calcium	2	2	6	2	6		2		
21	Sc, scandium	2	2	6	2	6	1	2		
22	Ti, titanium	2	2	6	2	6	2	2		
23	V, vanadium	2	2	6	2	6	3	2		
24	Cr, chromium	2	2	6	2	6	5	1		
25	Mn, manganese	2	2	6	2	6	5	2		
26	Fe, iron	2	2	6	2	6	6	2		
27	Co, cobalt	2	2	6	2	6	7	2		
28	Ni, nickel	2	2	6	2	6	8	2		
29	Cu, copper	2	2	6	2	6	10	1		
30	Zn, zinc	2	2	6	2	6	10	2		
31	Ga, gallium	2	2	6	2	6	10	2	1	
32	Ge, germanium	2	2	6	2	6	10	2	2	

We first need to specify what we mean by a quantum dot, a quantum wire, and a quantum well. Assume that electrons reside in a three-dimensional region of space, as shown in Fig. 4.15. From previous sections, we know that if $\lambda_e \ll L_x, L_y, L_z$, then the electrons will be free in all directions (i.e., they will act like free particles), and we have an effectively three-dimensional system. That is, the system in all directions is large compared with the size scale of the electrons. Even though the space is finite, and therefore the possible energy levels of the electron are discrete, because the space is relatively large, the discrete energy levels form, essentially, a continuum. For example, if we

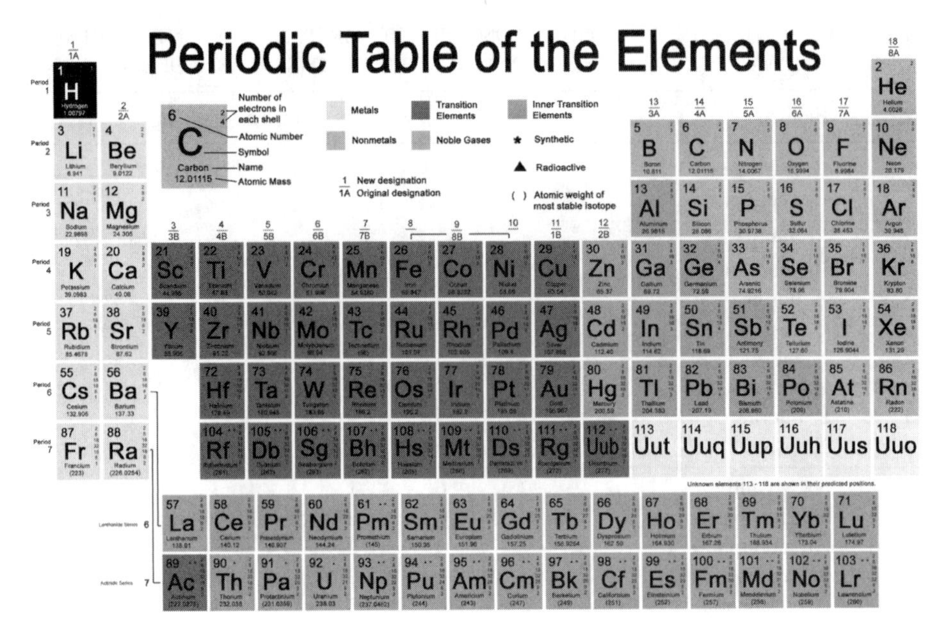

Figure 4.14 Periodic table of the elements. (Courtesy Carol and Mike Werner/Phototake NYC.)

TABLE 4.3 SPIN CONFIGURATION OF THE FIRST 10 ELEMENTS ACCORDING TO HUND'S RULE. NOTE THE DIFFERENCE BETWEEN CARBON, NITROGEN, AND OXYGEN.

Z	Element	1*s*	2*s*	2*p*
1	H, hydrogen	↑		
2	He, helium	↑↓		
3	Li, lithium	↑↓	↑	
4	Be, beryllium	↑↓	↑↓	
5	B, boron	↑↓	↑↓	↑
6	C, carbon	↑↓	↑↓	↑↑
7	N, nitrogen	↑↓	↑↓	↑↑↑
8	O, oxygen	↑↓	↑↓	↑↓↑↑
9	F, fluorine	↑↓	↑↓	↑↓↑↓↑
10	Ne, neon	↑↓	↑↓	↑↓↑↓↑↓

assume that the boundaries of the space have hard walls, then the energy levels are given by (4.53),

$$E = \frac{\hbar^2 \pi^2}{2m_e}\left(\left(\frac{n_x}{L_x}\right)^2 + \left(\frac{n_y}{L_y}\right)^2 + \left(\frac{n_z}{L_z}\right)^2\right). \tag{4.122}$$

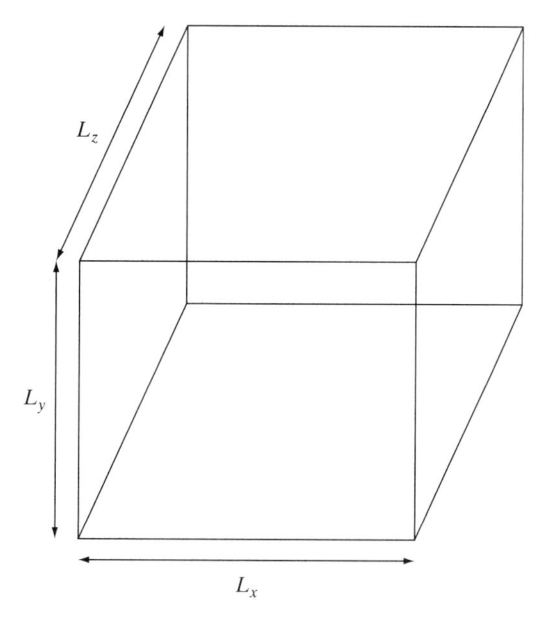

Figure 4.15 An effectively three-dimensional region of space; $L_{x,y,z} \gg \lambda_e$.

Since L_{xyz} are very large, and n_{xyz} are integers, the ratios n_x/L_x, n_y/L_y, and n_z/L_z vary almost continuously from very small values to very large values. Making the replacements

$$\frac{\pi n_x}{L_x} \rightarrow k_x, \quad \frac{\pi n_y}{L_y} \rightarrow k_y, \quad \frac{\pi n_z}{L_z} \rightarrow k_z, \tag{4.123}$$

where k_x, k_y, and k_z are continuous variables, we have

$$E = \frac{\hbar^2}{2m_e}\left(k_x^2 + k_y^2 + k_z^2\right) \tag{4.124}$$

$$= E_{\text{cont}}\left(k_x, k_y, k_z\right). \tag{4.125}$$

If L_{xyz} are large yet finite, $E_{\text{cont}}\left(k_x, k_y, k_z\right)$ is an approximately continuous energy profile, and, as $L_{xyz} \rightarrow \infty$, $E_{\text{cont}}\left(k_x, k_y, k_z\right)$ becomes the purely continuous energy for an electron in an infinite space, as derived in Section 4.1.2. (See, e.g., (4.19).)

In the preceding discussion, we used the de Broglie wavelength as the important "size" of the electron; that is, a region of space having length L is "large" if $L \gg \lambda_e$, and "small" if $L \leq \lambda_e$. While this is true, the de Broglie wavelength depends on the energy of the electron and, as discussed further in later chapters, the most important energy is the Fermi energy, introduced in Section 4.4. The de Broglie wavelength at the Fermi energy is called the *Fermi wavelength*, and is denoted by the symbol λ_F. Therefore, for a space to be sufficiently "large" so that the energy levels of the electron form an approximately continuous set, we usually require $L_x, L_y, L_z \gg \lambda_F$.

4.7.1 Quantum Wells

Now, assume that we make the space narrow in one direction, as depicted in Fig. 4.16.

By narrow in one direction, we mean, $L_x \leq \lambda_F \ll L_y, L_z$. In this case, we can write (4.122) as

$$E = \frac{\hbar^2\pi^2}{2m_e}\left(\frac{n_x}{L_x}\right)^2 + \frac{\hbar^2\pi^2}{2m_e}\left(\left(\frac{n_y}{L_y}\right)^2 + \left(\frac{n_z}{L_z}\right)^2\right) \tag{4.126}$$

$$= \frac{\hbar^2\pi^2}{2m_e}\left(\frac{n_x}{L_x}\right)^2 + \frac{\hbar^2}{2m_e}\left(k_y^2 + k_z^2\right) \tag{4.127}$$

$$= E_{n_x} + E_{\text{cont}}\left(k_y, k_z\right), \tag{4.128}$$

where, since L_x is relatively small, E_{n_x} represents discrete, one-dimensional energies. Since L_y and L_z are relatively large, two of the substitutions in (4.123) lead to $E_{\text{cont}}\left(k_y, k_z\right)$, which represents an approximately continuous energy profile. In this case, electron movement will be confined in the x-direction (i.e., electrons will "feel" the boundary in the x-direction), exhibiting energy quantization in that direction, and will be free in the two other directions. This makes for an effectively two-dimensional system, called a *two-dimensional electron gas*, also called a *quantum well*. The discrete energy levels given by n_x form what are called *subbands*. The idea is that an electron is in a certain discrete energy level n_x; that is, it is in a certain subband, but it is otherwise free in the

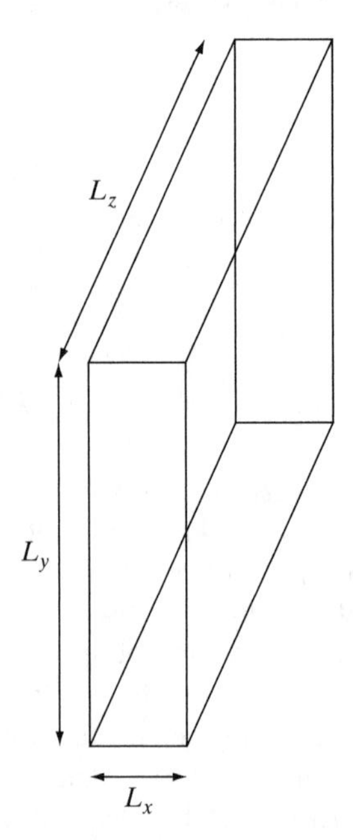

Figure 4.16 A quantum well, an effectively two-dimensional region of space, where $L_x \leq \lambda_F$, $L_{y,z} \gg \lambda_F$.

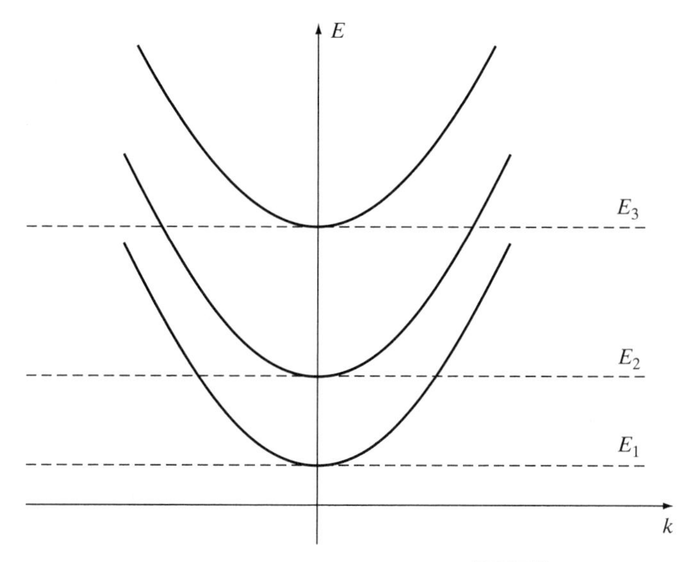

Figure 4.17 Energy versus continuous wavevector $k = \sqrt{k_x^2 + k_y^2}$, showing three subbands.

y–z plane. A depiction of the energy–wavenumber relationship is shown in Fig. 4.17. As previously mentioned, note that we have only found the allowed energy states; it remains to be seen how electrons "fill up" the various subbands, which will also be considered in Chapters 8 and 9.

To obtain the wavefunction we can consider the electron to be totally free in the y–z plane, but constrained in the x-direction by hard walls at $x = 0$, and $x = L_x$. Schrödinger's equation with $V = 0$ is

$$-\frac{\hbar^2}{2m_e}\nabla^2\psi(\mathbf{r}) = E\psi(\mathbf{r}), \tag{4.129}$$

and, given the longitudinal invariance of the problem and our experience with free electrons, we write the wavefunction in the product form

$$\psi(x, y, z) = \psi_x(x)\, e^{ik_y y} e^{ik_z z}, \tag{4.130}$$

where the plane-wave factor contains the continuous wavevectors k_y and k_z. The boundary conditions

$$\psi_x(0) = \psi_x(L_x) = 0$$

lead to

$$\psi(x, y, z) = \left(\frac{2}{L_x}\right)^{1/2} \sin\frac{n_x\pi}{L_x}x\, e^{ik_y y}\, e^{ik_z z}, \tag{4.131}$$

where $n_x = 1, 2, 3, \ldots$ and where the allowed energy is given by (4.127).

4.7.2 Quantum Wires

Continuing in the same fashion, now assume that we make the original space depicted in Fig. 4.15 narrow in two directions, $L_x, L_y \leq \lambda_F \ll L_z$, as depicted in Fig. 4.18. In this case, we can reorganize (4.122) as

$$E = \frac{\hbar^2 \pi^2}{2m_e}\left(\left(\frac{n_x}{L_x}\right)^2 + \left(\frac{n_y}{L_y}\right)^2\right) + \frac{\hbar^2 \pi^2}{2m_e}\left(\frac{n_z}{L_z}\right)^2 \tag{4.132}$$

$$= \frac{\hbar^2 \pi^2}{2m_e}\left(\left(\frac{n_x}{L_x}\right)^2 + \left(\frac{n_y}{L_y}\right)^2\right) + \frac{\hbar^2}{2m_e}k_z^2 \tag{4.133}$$

$$= E_{n_x,n_y} + E_{\text{cont}}(k_z), \tag{4.134}$$

where, since L_x and L_y are both relatively small, E_{n_x,n_y} represents discrete, two-dimensional subband energies, and, since L_z is relatively large, $E_{\text{cont}}(k_z)$ represents an approximately continuous energy profile. In this case, electron movement will be confined in the x–y plane (i.e., electrons will "feel" the boundaries in the x- and y-directions), exhibiting energy quantization in that plane, and will be free in the z-direction. This makes for an effectively one-dimensional system called a *quantum wire.*†

Regarding the electron as being totally free in one direction, but constrained by hard walls at $x = 0, L_x$ and $y = 0, L_y$, Schrödinger's equation with $V = 0$,

$$-\frac{\hbar^2}{2m_e}\nabla^2 \psi(\mathbf{r}) = E\psi(\mathbf{r}), \tag{4.135}$$

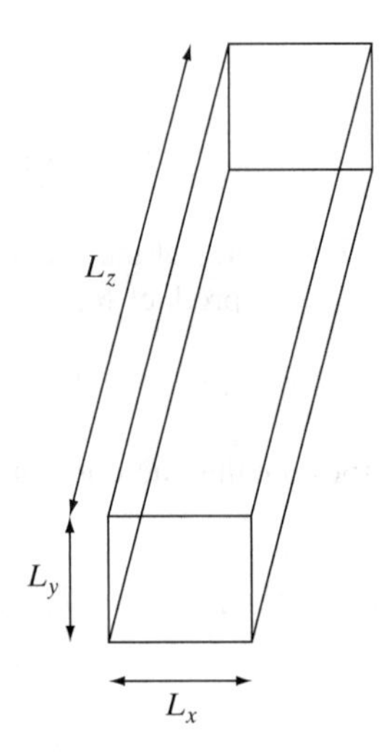

Figure 4.18 A quantum wire, an effectively one-dimensional space; $L_{x,y} \leq \lambda_F$, $L_z \gg \lambda_F$.

†Circular cross-section wires are also of interest, but won't be discussed here to avoid mathematical complications associated with the circular geometry.

will have solutions in the product form

$$\psi(x, y, z) = \psi_x(x)\,\psi_y(y)\,e^{ik_z z}. \tag{4.136}$$

Imposing the boundary conditions

$$\psi_x(0) = \psi_x(L_x) = 0, \tag{4.137}$$
$$\psi_y(0) = \psi_y(L_y) = 0,$$

we obtain

$$\psi(x, y, z) = \left(\frac{4}{L_x L_y}\right)^{1/2} \sin\frac{n_x \pi}{L_x} x \, \sin\frac{n_y \pi}{L_y} y \, e^{ik_z z}, \tag{4.138}$$

where $n_{x,y} = 1, 2, 3, \ldots$, and where energy is given by (4.133). An early quantum wire fabricated by electron beam lithography and etching is shown in Fig. 4.19.

4.7.3 Quantum Dots

Now assume that we make the space small in all three directions, $L_x, L_y, L_z \leq \lambda_F$, as depicted in Fig. 4.20. In this case, we can write (4.122) as

$$E = \frac{\hbar^2 \pi^2}{2m_e}\left(\left(\frac{n_x}{L_x}\right)^2 + \left(\frac{n_y}{L_y}\right)^2 + \left(\frac{n_z}{L_z}\right)^2\right) \tag{4.139}$$

$$= E_{n_x, n_y, n_z} \tag{4.140}$$

Figure 4.19 75-nm-wide etched quantum wire, (Reprinted with permission from Roukes, M. L. et al., "Quenching of the Hall Effect in a One-Dimensional Wire," *Phys. Rev. Lett.* 59 (1987): 3011. Copyright 1987, American Physical Society.)

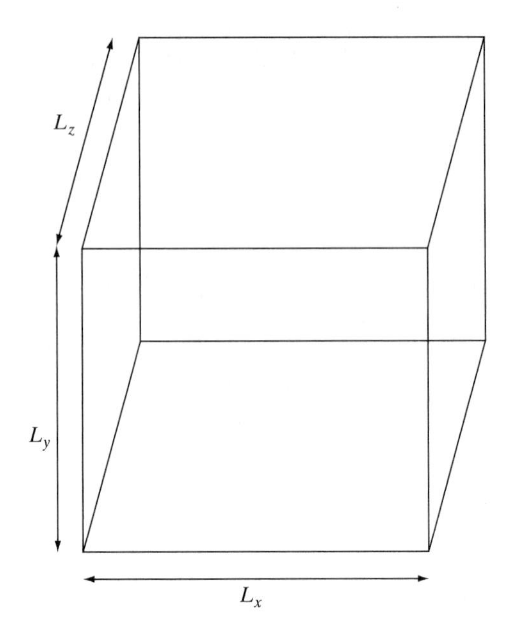

Figure 4.20 A quantum dot, an effectively zero-dimensional space; $L_{x,y,z} \leq \lambda_F$.

where, since L_x, L_y, and L_z are all relatively small, E_{n_x,n_y,n_z} represents discrete energies. Electron movement will be confined in all three directions (i.e., electrons will "feel" the boundaries in the x-, y-, and z-directions), exhibiting energy quantization in three dimensions, and will not be free in any direction. This makes for an effectively zero-dimensional system called a *quantum dot*.

Since the typical electron of interest has a Fermi wavelength on the order of a nanometer in metals, and many tens of nanometers in semiconductors (as further considered later in the text), quantum dots are nanoscale pieces of material, typically ranging in size from several nm to hundreds of nm. As such, quantum dots can contain from several hundred to several hundred thousand atoms. For example, as a rough estimate, assuming that the size of an atom is 0.1 nm (= 1 Å), then about 10 atoms can fit along a 1 nm line. A cube having sides of 1 nm will contain approximately $10^3 = 1,000$ atoms. A cube having sides of 10 nm will contain approximately 10^6 atoms. Thus, quantum dots are typically much larger than atoms, but are generally too small to act like a bulk solid. An electron in a quantum dot will act more like an electron in a molecule than an electron in a bulk solid, and for this reason, quantum dots are sometimes called *artificial molecules*. In general, one would tend to call nanoscopic material regions *quantum dots* if quantum confinement effects are important, and *nanoparticles* otherwise. Thus, a 10-nm-radius sphere of GaAs would be a quantum dot, whereas the same size sphere made from copper would be a nanoparticle.

Material Properties at the Nanoscale. Of particular importance for quantum dots is that material properties change dramatically at the nanoscale. One reason for this is the previously mentioned confinement effect of electrons in regions of space that are not large compared to the de Broglie or Fermi wavelength. Another reason for nanoscale materials to behave differently is that a large percentage of their atoms are located on the surface of the material, and the relatively large surface-to-volume ratio leads to material behavior different than the bulk specimen. As an example, consider again a 1 nm^3 cube, consisting of 1,000 atoms. Counting the six surfaces of the cube, we find that roughly $6 \times 10^2 = 600$ atoms, or more than half of the atoms, are on the surface of the cube, where they can easily interact with the outside environment (via unsatisfied chemical bonds, etc.) In general, the smaller the dot, the higher the percentage of atoms at the surface. Thus, one would then expect that quantum dots have higher chemical reactivity than bulk materials, which is generally found to be the case.

4.8 MAIN POINTS

This chapter presented the main ideas of what happens to particles, in particular, electrons, when they are confined to different-sized regions of space. After studying this chapter, you should know

- the difference between free and bound particles, and, in particular, the important idea that the possible energy states of a particle confined to a finite spatial region are quantized;
- the free electron gas model of conductors, and the assumptions behind this model;
- the role that the de Broglie and Fermi wavelength plays in determining what is a "small" region of space;
- the role of the Pauli exclusion principle in filling quantum states;
- the idea of the Fermi level and chemical potential, and how to compute these quantities for low electron density systems;
- the difference between an infinite (hard-wall) and finite confining potential, and, in the latter case, how to determine the allowed particle states;
- the general concept of where the periodic table comes from, and why elements are grouped as they are;
- the concepts of quantum wells, quantum wires, and quantum dots.

4.9 PROBLEMS

1. Write down the wavefunction $\psi(z, t)$ for a 3 eV electron in an infinite space, traveling along the positive z axis. Assume that the electron has only kinetic energy. Plug your answer into Schrödinger's time-dependent equation to verify that it is a solution.
2. Determine the wavefunction $\psi(z, t)$ for a 3 eV electron in an infinite space, traveling along the z axis at a velocity of 10^5 m/s. Determine the particle's potential energy, and plug your answer into Schrödinger's time-dependent equation to verify that it is a solution.
3. For classical light, the expression

$$c = \lambda f \tag{4.141}$$

holds, where c is the speed of light, f is the frequency, and λ is the wavelength. This expression also holds for quantized light (i.e., photons), where the classical wavelength λ is the same as the de Broglie wavelength. Does the expression

$$v = \lambda f, \tag{4.142}$$

where v is velocity, hold for electrons if λ is the de Broglie wavelength and $f = \omega/2\pi$ is the frequency, such that $E = \hbar\omega$? Assume that the electrons have only kinetic energy.

4. Consider an electron in a room of size $10 \times 10 \times 10$ m^3. Assume that within the room, potential energy is zero, and that the walls and ceilings of the room are perfect (so that the electron cannot escape from the room). If the electron's energy is approximately 5 eV, what is the state index $n^2 = n_x^2 + n_y^2 + n_z^2$? What is the approximate energy difference $E_{2,1,1} - E_{1,1,1}$?

5. Repeat problem 4.4 for an electron confined to a nanoscale space, $10^{-9} \times 10^{-9} \times 10^{-9}$ m^3.

6. Consider an electron having kinetic energy 2.5 eV. What size space does the electron need to be confined to in order to observe clear energy discretization? Repeat for a proton having the same kinetic energy.

7. Consider a 44 kg object (perhaps a desk) in a typical room with dimensions 10×10 m^2. Assume that within the room, potential energy is zero, and that the walls of the room are perfect (so that the room can be modeled as an infinitely deep potential well).

(a) If the desk is in the ground state, what is the velocity of the desk?

(b) If the desk is moving at 0.01 m/s, what is the desk's quantum state (i.e., what is the state index n)?

8. Consider the one-dimensional, infinite square well existing from $0 \leq x \leq L$. The wavefunction for a particle confined to the well is (4.34),

$$\psi(x,t) = \left(\frac{2}{L}\right)^{1/2} \sin\left(\frac{n\pi}{L}x\right) e^{-iE_n t/\hbar}. \tag{4.143}$$

If the particle is in the ground state, determine the probability, as a function of time, that the particle will be in the right half of the well (i.e., from $L/2 \leq x \leq L$).

9. Consider an electron in the first excited state ($n = 2$) of an infinitely high square well of length 2.3 nm. Assuming zero potential energy in the well, determine the electron's velocity.

10. Consider an electron confined to an infinite potential well having length 2 nm. What wavelength photons will be emitted from transitions between the lowest three energy levels?

11. Consider the one-dimensional, infinite square well existing from $0 \leq x \leq L$. Recall that the energy eigenfunctions for a particle confined to the well are (4.31),

$$\psi_n(x) = \left(\frac{2}{L}\right)^{1/2} \sin\left(\frac{n\pi}{L}x\right), \tag{4.144}$$

$n = 1, 2, 3, \ldots$. Now, assume that the particle state is

$$\Psi(x) = A(x(L - x)), \tag{4.145}$$

such that $\Psi(0) = \Psi(L) = 0$.

(a) Determine A such that the particle's state function is suitably normalized.
(b) Expand $\Psi(x)$ in the energy eigenfunctions, i.e., use orthogonality to find c_n such that

$$\Psi(x) = A(x(L - x)) = \sum_{n=1}^{\infty} c_n \psi_n(x). \tag{4.146}$$

(c) Determine the probability of measuring energy E_n, and use this result to determine the probability of measuring energy E_1 through E_6.

12. Repeat problem 4.11 if the particle state is

$$\Psi(x) = A\sqrt{(x(L - x))}. \tag{4.147}$$

13. Consider the one-dimensional, infinite square well existing from $0 \leq x \leq L$. If nine electrons are in the well, what is the ground state energy of the system?

14. Consider a one-dimensional quantum well of length $L = 10$ nm containing 11 non-interacting ground state electrons. Determine the chemical potential of the system.

15. Apply boundary conditions (4.81) to the symmetric form of (4.78) (i.e., set $C = 0$ in (4.78)) for the finite rectangular potential energy well considered in Section 4.5.1 to show that the wavefunction (4.82) results, where (4.83) must be satisfied.

16. Repeat problem 4.15 for the antisymmetric form of (4.78) (i.e., with $D = 0$), showing that (4.84) results, where the wavenumbers must obey (4.85).

17. Comparing energy levels in one-dimensional quantum wells, for an infinite-height well of width $2L$, energy states are given by (4.35) with L replaced by $2L$,

$$E_n = \frac{\hbar^2}{2m_e}\left(\frac{n\pi}{2L}\right)^2, \tag{4.148}$$

and for a finite-height well of width $2L$ the symmetric states are given by a numerical solution of (4.83),

$$k_2 \tan(k_2 L) = k_1. \tag{4.149}$$

Assume that $L = 2$ nm, and compute E_1 and E_3 (the second symmetric state) for the infinite-height well. Then compare it with the corresponding values obtained from the numerical solution of (4.149) for the finite-height well. (You will need to use a numerical root solver.) For the finite-height well, assume barrier heights $V_0 = 1{,}000, 100, 10, 1, 0.5$, and 0.2 eV. Make a table comparing the infinite-height and finite-height results (with percent errors), and comment on the appropriateness, at least for low energy states, of the much simpler infinite well model for reasonable barrier heights, such as $V_0 = 0.5$ eV.

18. For a $1s$ electron in the ground state of hydrogen, determine the expectation value of energy.
19. For an electron in the $(n, l, m) = (2, 0, 0)$ state of hydrogen, determine the expectation value of position.
20. Repeat problem 4.19 for an electron in the $(n, l, m) = (2, 1, 0)$ state of hydrogen.
21. For an electron in the $(n_x, n_y) = (1, 1)$ subband of a metallic quantum wire having $L_x = L_y = 1$ nm, if the total energy is 1 eV, what is the electron's longitudinal (i.e., z–directed) group velocity?
22. A 3 eV electron is to be confined in a square quantum dot of side L. What should L be in order for the electron's energy levels to be well quantized?

Chapter

5

Electrons Subject to a Periodic Potential — Band Theory of Solids

STM image of a nickel surface. (Image reproduced by permission of IBM Research, Almaden Research Center. Unauthorized use not permitted.)

In the previous chapter, we considered several different environments, including infinite and finite spatial regions, and we solved Schrödinger's equation to determine the possible allowed states for a particle in these environments. In this chapter we continue to use these ideas, but here we consider an electron in a crystalline material, leading to the important *band theory of solids*. It is hard to overestimate the importance of band theory. It turns out that band theory, in conjunction with the Pauli exclusion principle, can be used to explain the fundamental nature of insulators, conductors, and semiconductors. Band theory is also important in understanding semiconductor heterostructures, considered in Chapter 9, which are fundamental structures in nanoelectronics.

Another main point of this chapter is to introduce the concept of effective mass. One can imagine that it would be extremely difficult to account for all of the interactions between an electron and the various particles that make up a material. As described previously, in an exact model this interaction is taken into account via the potential energy term in Schrödinger's equation. However, as we will see, with some simplifying assumptions we can take into account the presence of a crystalline material by simply considering the electron to have a mass that is different from its empty space value. With this idea we can, for example, reconsider the quantum well, quantum wire, and quantum dot examples introduced in the last chapter, this time constructing these structures from a solid material, rather than being simply empty space. The most technically important materials for these applications are semiconductors.

In this chapter, attention is focused on bulk material properties; applications to nanoscale structures made from these materials will be considered in subsequent chapters.

5.1 CRYSTALLINE MATERIALS

The band theory of solids applies only to crystalline materials, strictly speaking, and so it is worthwhile to begin with a brief introduction to the different forms a solid may take.

Solid materials may be classified as *crystalline, polycrystalline, or amorphous.* A crystalline solid has a regular structure, consisting of a periodic array of atoms called the *lattice*. A polycrystalline solid has a well-defined structure in each of many small regions, but each region generally differs from its neighboring regions. In some ways, the opposite of a crystalline material is an amorphous solid, which does not exhibit any sort of regularity. The most common amorphous materials are glass and plastic.

Most materials of interest in electronics have been, traditionally, crystalline materials (e.g., semiconductors, such as silicon and gallium arsenide, and conductors, such as copper and gold). In this chapter, we consider the effect of such a periodic lattice on the behavior of electrons. Although in the past, naturally occurring crystalline materials were used for electronics applications, advances in materials processing technologies have begun to allow synthetic crystalline materials to be developed, engineered specifically to control electronic properties of the material.

Crystal Types. The fundamental property of a crystal is regularity in its atomic structure; the atoms in a crystal are arranged in a regular (periodic) array. To be fairly general, we need to develop two concepts: the idea of a lattice and of a basis.

A lattice is a set of points that form a periodic structure. The simplest lattice, called a *simple cubic* (sc) lattice, consists of points equally spaced at the corners of a three-dimensional cube, as shown in Fig. 5.1, although few materials have this structure.

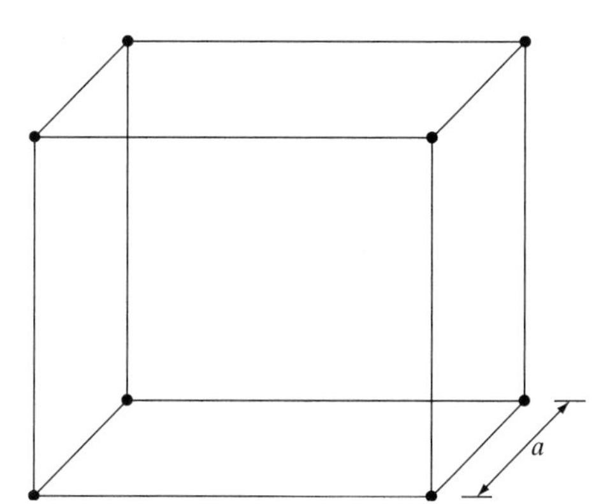

Figure 5.1 Simple cubic lattice. The lattice points are shown by a solid dot, located at the vertex of each corner. All sides have length a.

The lattice is defined by three fundamental vectors, $\mathbf{a}_1$, $\mathbf{a}_2$, and $\mathbf{a}_3$, known as *fundamental translation vectors*, such that the atomic arrangement looks identical when viewed from the point $\mathbf{r}$ and the point

$$\mathbf{r}' = \mathbf{r} + \mathbf{T}, \tag{5.1}$$

where

$$\mathbf{T} = u_1\mathbf{a}_1 + u_2\mathbf{a}_2 + u_3\mathbf{a}_3 \tag{5.2}$$

is called the *crystal translation vector* and where u_1–u_3 are integers. Many translation vectors are possible, and the set of three vectors form a parallelepiped. The parallelepiped with the smallest volume is called the *primitive cell*, constructed from *primitive translation vectors*. The primitive cell has only one lattice point (perhaps shared with other cells), and a crystal can be constructed from repetitions of the primitive cell. However, for envisioning the material to consist of repetitions of cells, the primitive cell is usually not the most convenient to work with, and here we will be primarily interested in other *unit cells*.

Whereas the lattice specifies the periodic arrangement of the crystal, it may not be the case that a single atom is located at each lattice point. A group of atoms called a *basis*, consisting of perhaps many atoms, is such that when the basis is placed at each lattice point the entire crystal is formed. That is, by definition of a lattice, the basis repeats in a periodic manner. The simplest example would be the case when the basis consists of one atom. An example where the basis consists of two different atoms is shown (in two dimensions) in Fig. 5.2.

Just slightly more complex than the sc lattice is the *body-centered cubic* (bcc) lattice, shown in Fig. 5.3. In particular, sodium and tungsten have this structure.

Of more importance for semiconductors is the *face-centered cubic* (fcc) lattice, shown in Fig. 5.4, which is constructed by adding to the simple cubic lattice additional points in the center of each square face. Materials exhibiting this type of lattice are copper (Cu), gold (Au), silver (Ag), nickel (Ni), and, importantly, silicon (Si), gallium arsenide

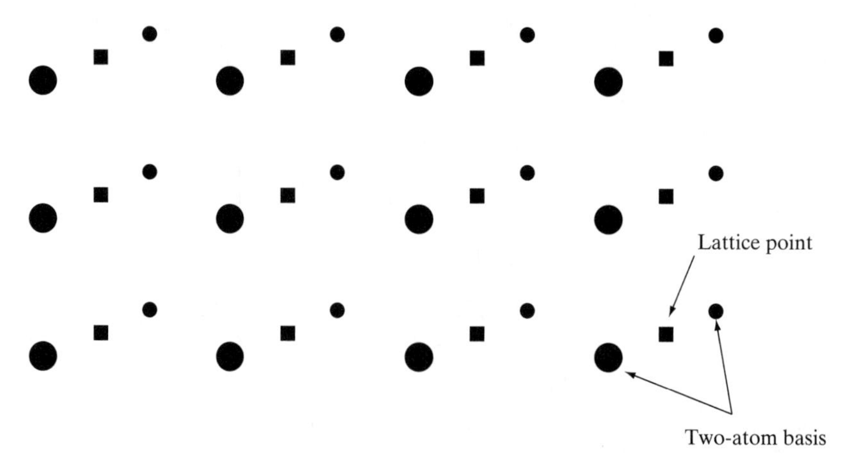

Figure 5.2 Crystal formed by a two-atom basis centered at each lattice point.

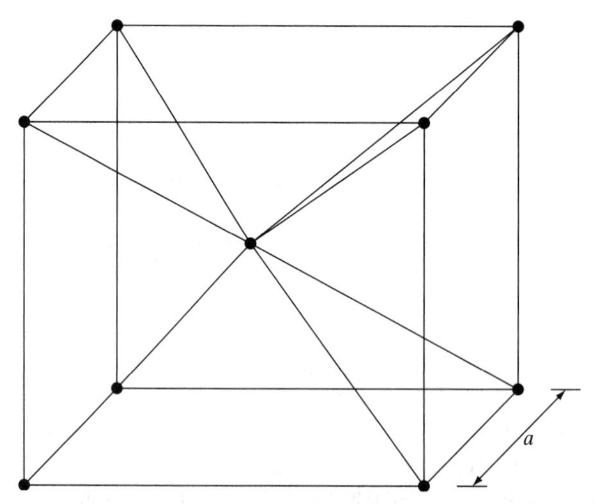

Figure 5.3 Body-centered cubic lattice. The lattice points are shown by a solid dot, located at the vertex of each corner, and in the center of the cube. All sides have length a.

(GaAs), and germanium (Ge). The lattice constant (a) for Si is 0.543 nm, and for Ge, $a = 0.566$ nm. The lattice constants of some other materials are listed in Table V in Appendix B.

To illustrate the idea of a primitive cell and unit cell, Figs. 5.1, 5.3, and 5.4 show what are called the *conventional cells*, which are the structurally simplest unit cells; the lattice constant a is the length of the side of the conventional cell. However, only for the simple cubic lattice is the conventional cell primitive (cell volume is a^3). The primitive cell for the important fcc lattice is shown in Fig. 5.5 (conventional cell volume is a^3, whereas the

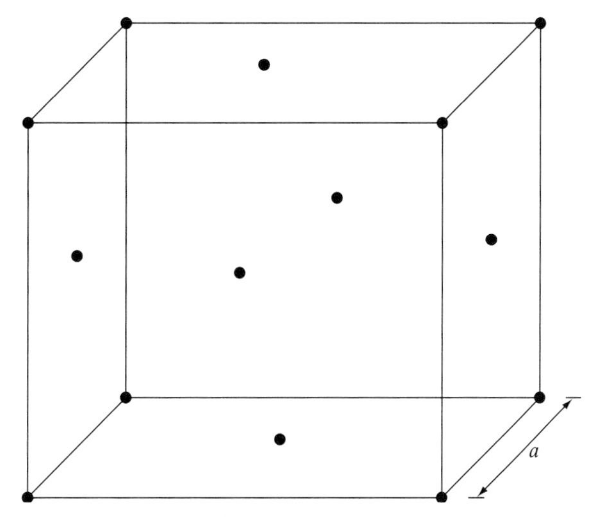

Figure 5.4 Face-centered cubic lattice. The lattice points are shown by a solid dot, located at the vertex of each corner, and in the center of each side of the cube. All sides have length a.

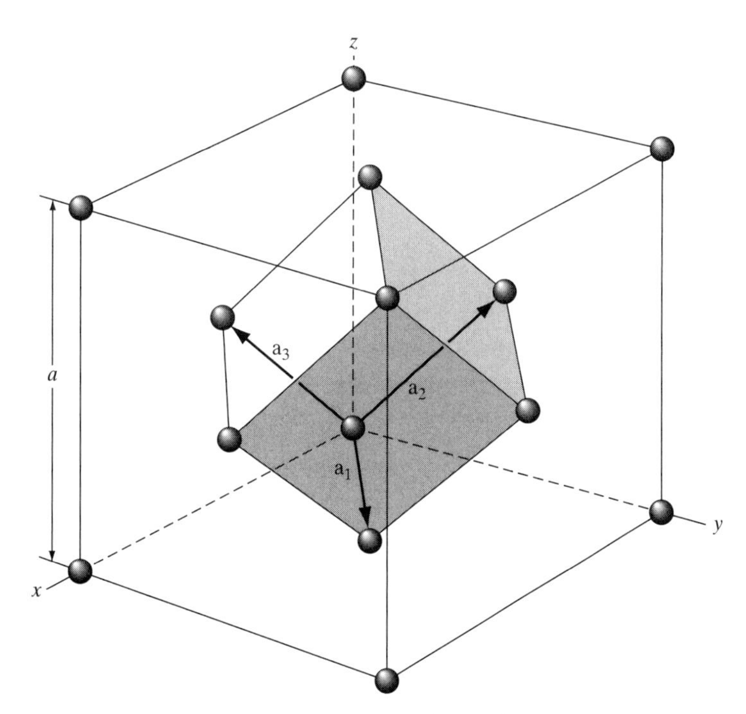

Figure 5.5 Primitive cell for the fcc lattice.

primitive cell volume is $a^3/4$), where the primitive translation vectors are

$$\mathbf{a}_1 = \frac{a}{2}\left(\mathbf{a}_x + \mathbf{a}_y\right), \quad \mathbf{a}_2 = \frac{a}{2}\left(\mathbf{a}_y + \mathbf{a}_z\right), \quad \mathbf{a}_3 = \frac{a}{2}\left(\mathbf{a}_z + \mathbf{a}_x\right). \tag{5.3}$$

Although important semiconductors exhibit the fcc lattice, the basis consists of two atoms. If the two atoms are identical, the material is said to have the *diamond structure*, which occurs for Si, Ge, and carbon (C). If the two atoms in the basis are different, the material is said to have a *zinc blende structure*. Semiconductors such as GaAs and AlAs have this structure, and are often called compound semiconductors (e.g., for GaAs, the two atoms in the basis are gallium and arsenic).

Having established the idea of periodic atomic structure, we consider next how electrons behave in a periodic potential.

5.2 ELECTRONS IN A PERIODIC POTENTIAL

When we want to examine the properties of an electron in a periodic lattice, we need to consider Schrödinger's equation such that the potential energy term $V(\mathbf{r})$ reflects the fact that the electron sees a periodic potential.

Consider a one-dimensional example, where the lattice points are spaced a distance a apart, and we assume an ionized atom (an ion) is located at each lattice point. A liberated electron moves about the material subject to the attractive Coulomb (electrostatic) force between the negatively charged electron and the positively charged ions. The Coulomb force between the electron and a single ion is

$$V(x) = \frac{1}{4\pi\varepsilon_0}\frac{(q_e)(-q_e)}{|x|} \tag{5.4}$$

from basic electrostatic theory, where x is the distance between the electron and the ion, and so the electron has the potential energy depicted in Fig. 5.6. The potential energy is obviously periodic, i.e., $V(x) = V(x+a)$.

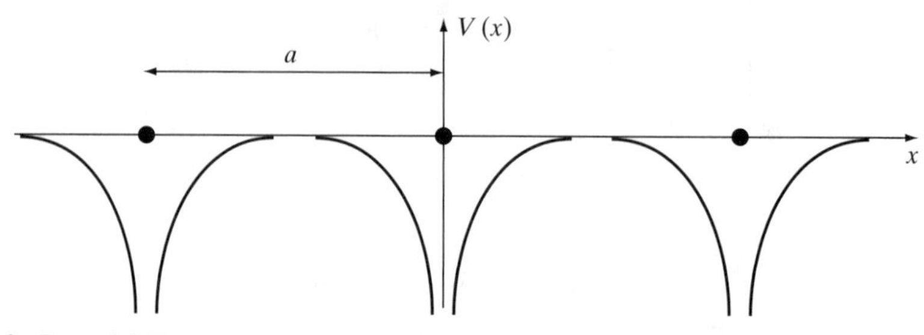

Figure 5.6 Potential $V(x)$ versus position x in a periodic lattice. Solid dots indicate the location of ions, and a is the period of the lattice.

Generalizing to three-dimensions, we find that Schrödinger's equation is

$$\left(-\frac{\hbar^2}{2m_e}\nabla^2 + V(\mathbf{r})\right)\psi(\mathbf{r}) = E\psi(\mathbf{r}), \tag{5.5}$$

where the potential energy term is periodic,

$$V(\mathbf{r}) = V(\mathbf{r} + \mathbf{T}), \tag{5.6}$$

and where $\mathbf{T}$ is the crystal translation vector. There is an important theorem called *Bloch's theorem* that applies to waves in periodic structures in general. (Recall that electrons have wave properties through de Broglie's relation.) For the case of Schrödinger's equation, it states that the solution of (5.5), when the potential is periodic, can be written as the product of a plane wave and a periodic function,

$$\psi(\mathbf{r}) = u(\mathbf{r})\, e^{i\mathbf{k}\cdot\mathbf{r}}, \tag{5.7}$$

where $\mathbf{k}$ is the wavevector to be determined (called the *Bloch wavevector*) and where u is periodic,

$$u(\mathbf{r}) = u(\mathbf{r} + \mathbf{T}). \tag{5.8}$$

Thus,

$$\psi(\mathbf{r} + \mathbf{T}) = u(\mathbf{r} + \mathbf{T})\, e^{i\mathbf{k}\cdot(\mathbf{r}+\mathbf{T})} = u(\mathbf{r})\, e^{i\mathbf{k}\cdot\mathbf{r}} e^{i\mathbf{k}\cdot\mathbf{T}} = \psi(\mathbf{r})\, e^{i\mathbf{k}\cdot\mathbf{T}}. \tag{5.9}$$

It is important to note that the Bloch theorem shows that electrons can propagate through a *perfect* periodic medium without scattering (i.e., without hitting the atoms). One could say, mathematically, that this is because the plane wave part of the solution, $e^{i\mathbf{k}\cdot\mathbf{r}}$, exists over the entire crystal, and therefore " sees" the whole crystal.

5.3 KRONIG–PENNEY MODEL OF BAND STRUCTURE

The potential depicted in Fig. 5.6 is fairly realistic in a one-dimensional sense, although Schrödinger's equation cannot be solved exactly when V has this form. As an approximate model, assume a one-dimensional crystal where

$$V(x) = \begin{cases} 0, & 0 \le x \le a_1, \\ V_0, & -a_2 \le x \le 0, \end{cases}$$

and where $a = a_1 + a_2$ is the period of the lattice, as shown in Fig. 5.7. This is known as the *Kronig–Penney model*.†

†This model provides an introduction to band theory. However, it is overly simplistic and leads to only a very rough approximation of carrier behavior in real materials.

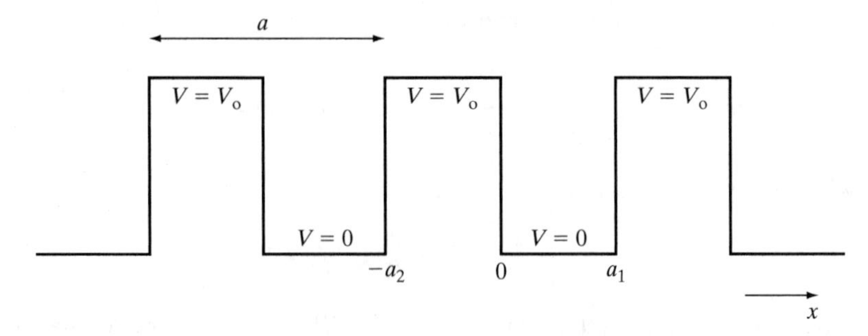

Figure 5.7 Kronig–Penney model of the potential due to a periodic lattice of period a.

In the region $-a_2 \leq x \leq 0$, the potential is $V = V_0$, and the solution of Schrödinger's equation is

$$\psi(x) = Ae^{i\beta x} + Be^{-i\beta x}, \tag{5.10}$$

where

$$\beta = \sqrt{\frac{2m_e(E - V_0)}{\hbar^2}}, \tag{5.11}$$

as shown in Section 4.1.1. In a similar manner, for $0 \leq x \leq a_1$ the potential is $V = 0$, and the solution of Schrödinger's equation is

$$\psi(x) = De^{i\alpha x} + Fe^{-i\alpha x}, \tag{5.12}$$

where

$$\alpha = \sqrt{\frac{2m_e E}{\hbar^2}}. \tag{5.13}$$

In one dimension, the Bloch form (5.7) is

$$\psi(x) = u(x)\,e^{ikx}, \tag{5.14}$$

in terms of the to-be-determined Bloch wavevector k, where

$$u(x) = u(x + a). \tag{5.15}$$

Therefore,

$$\begin{aligned}\psi(x \pm a) &= u(x \pm a)\,e^{ik(x \pm a)} = u(x)\,e^{ik(x \pm a)} \\ &= \psi(x)\,e^{\pm ika},\end{aligned} \tag{5.16}$$

and, using this relationship, we can write down the wavefunction in the following period, $a_1 \leq x \leq a_1 + a$, as

$$\begin{aligned}\psi(x) &= \left\{Ae^{i\beta(x-a)} + Be^{-i\beta(x-a)}\right\}e^{ika}, \qquad a_1 \leq x \leq a, \\ &= \left\{De^{i\alpha(x-a)} + Fe^{-i\alpha(x-a)}\right\}e^{ika}, \qquad a \leq x \leq a_1 + a.\end{aligned} \tag{5.17}$$

Enforcing continuity of ψ and ψ' at $x = 0$ and $x = a_1$ leads to the eigenvalue equation

$$\cos ka = \cos(\alpha a_1)\cosh(\delta a_2) - \frac{\alpha^2 - \delta^2}{2\alpha\delta}\sin(\alpha a_1)\sinh(\delta a_2), \tag{5.18}$$

if $0 < E < V_0$, and

$$\cos ka = \cos(\alpha a_1)\cos(\beta a_2) - \frac{\alpha^2 + \beta^2}{2\alpha\beta}\sin(\alpha a_1)\sin(\beta a_2), \tag{5.19}$$

if $E > V_0$, where

$$\delta = \sqrt{\frac{2m_e(V_0 - E)}{\hbar^2}}. \tag{5.20}$$

In the preceding equations, the energy E is the only unknown parameter. For a solution to exist, we must have

$$-1 \leq \cos ka \leq +1, \tag{5.21}$$

and so the right side of (5.18), denoted as $r(E)$, must obey this condition. A typical plot of $r(E)$ vs. E is shown in Fig. 5.8.

This figure makes clear the fact that there are certain allowed values of energy, called *allowed energy bands*, and certain unallowed values of energy, called *band gaps*. That is, if E is in an allowed energy band, Schrödinger's equation for the Kronig–Penney model has a solution, and if E is not in an allowed energy band, there is no solution (because we

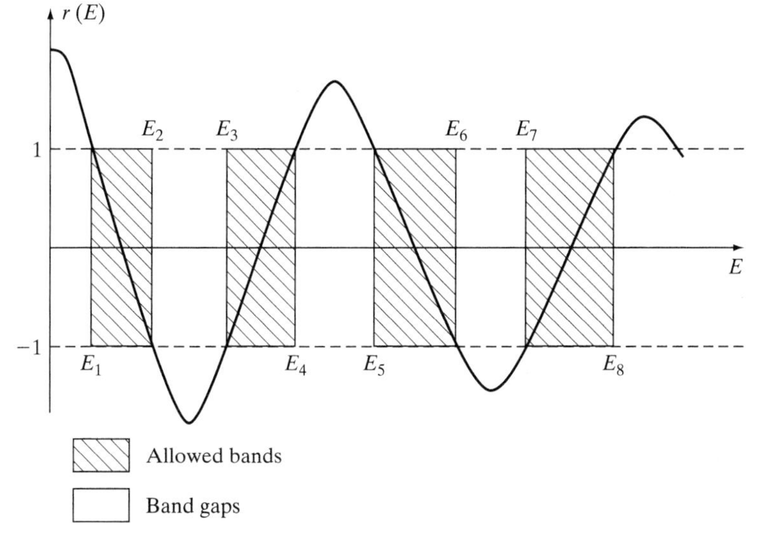

Figure 5.8 Plot of $r(E)$ versus E, showing allowed energy bands and band gaps.

must have $|\cos ka| \leq 1$ in (5.21)). Within an allowed band energy can take any value, i.e., it is not discretized. Note that as energy increases, the allowed energy bands increase in width, and so the forbidden bands decrease in width. As we will see later in this chapter, the concept of energy bands leads to the fundamental characteristics of conductors, insulators, and semiconductors.

Since

$$\cos ka = r(E), \tag{5.22}$$

we can generate an important figure called the *dispersion diagram*, which is a diagram of energy versus wavenumber (E vs. k). To generate the dispersion diagram, start at $E = 0$ and compute $r(E)$. If $|r(E)| = |\cos ka| > 1$, we are at a forbidden energy (i.e., in a bandgap), and we need to increase E a bit and try again. If $|r(E)| \leq 1$, we are at an allowed energy (i.e., in an energy band), and in this case the corresponding wavenumber is

$$k = \frac{1}{a}\cos^{-1}(r(E)). \tag{5.23}$$

Since cosine is an even function, $-k$ will also be a solution. By increasing E by a small amount and checking the value of $r(E)$, we can generate the plot of allowed and unallowed energy bands. One form of the result will look like Fig. 5.9. This depiction is known as the *extended zone scheme.*

The various sections of wavenumber space are divided into what are called *Brillouin zones*, with the range

$$-\frac{\pi}{a} \leq k \leq \frac{\pi}{a} \tag{5.24}$$

denoting the important *first Brillouin zone*. The second Brillouin zone is the range

$$-\frac{2\pi}{a} \leq k < -\frac{\pi}{a}, \quad \frac{\pi}{a} < k \leq \frac{2\pi}{a}, \tag{5.25}$$

and so on for higher zones.

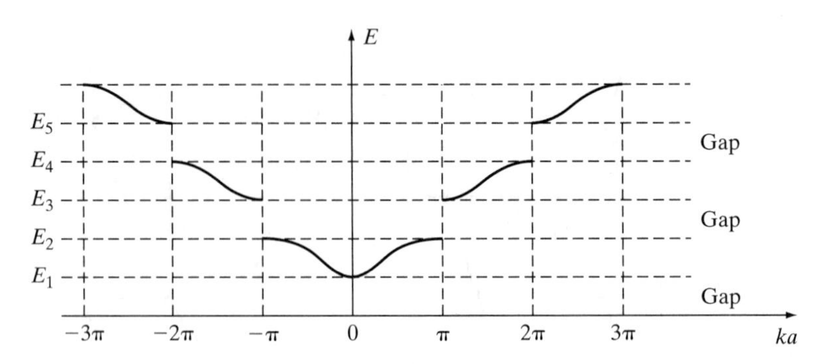

Figure 5.9 Energy band energy E versus Bloch wavenumber k in the extended zone scheme.

Due to the periodic nature of the crystal, there is no real difference between the wavevector k and the wavevector $k \pm n2\pi/a$, $n = 0, \pm 1, \pm 2, \ldots$ (that is, the structure is periodic in both "real space" and "wavevector space," also known as *reciprocal space*). This can be proved in general, although here one can simply note that (5.23) is multivalued due to the inverse cosine function. That is, assume that a given E value determines the right side of (5.18) (or (5.19)), leading to $\pm k$ from (5.23). Any other wavenumber value $\pm k + n2\pi/a$, $n = 0, \pm 1, \pm 2, \ldots$ will also satisfy (5.18) (or (5.19)), since $\cos(k + n2\pi/a)\,a = \cos ka$ for n an integer. Therefore, for a given allowed energy, there are an infinite number of k values, and the E–k plot can also be drawn with E oscillating in a continuous fashion within the band as k increases. This depiction, shown in Fig. 5.10, is known as the *repeated zone scheme*).

Lastly, another depiction arises from noting that the energy bands in the higher Brillouin zones can be all translated to the first Brillouin zone by shifts of $n2\pi/a$. This results in what is called the *reduced zone scheme*, depicted in Fig. 5.11. In the remainder of the text, band diagrams will be depicted in the reduced zone scheme, which is the most common format for describing band structure. However, it should be noted that all three depictions (extended, repeated, and reduced) convey the same information.

5.3.1 Effective Mass

In the previous chapter, various nanostructures, such as quantum wells, quantum wires, and quantum dots, were considered. However, the structures were basically empty boxes for electrons. At this point, it may seem difficult to imagine how one can incorporate the fact that a nanostructure is made from a real material, such as a semiconductor. It turns out that there is a relatively simple approximate method to do this, involving a concept known as *effective mass*.

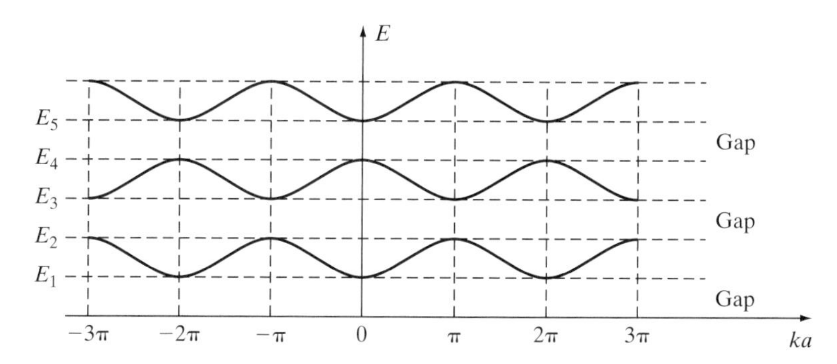

Figure 5.10 Energy band energy E versus Bloch wavenumber k in the repeated zone scheme.

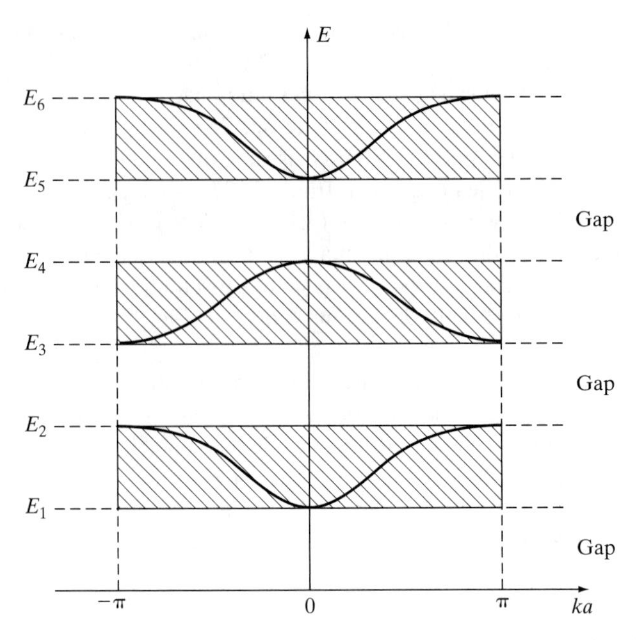

Figure 5.11 Energy band energy E versus Bloch wavenumber k in the reduced zone scheme.

In a nonrelativistic sense, an electron in empty space has a well-defined, constant mass. However, it is often useful to view mass as merely a proportionality factor between force and acceleration. (Recall Newton's second law, $\mathbf{F} = m\mathbf{a}$.) This view is particularly appropriate when studying electrons in crystals, since in this case it turns out that often electrons appear to act as if their mass is different from the free-space value. This simply means that in a crystal, electrons do not respond to external forces in the same way that free electrons do.

We view an electron quantum mechanically as represented by a wavepacket, with the electron's velocity being its group velocity (as discussed in Section 2.5),

$$v_g = \frac{\partial \omega}{\partial k} = \frac{1}{\hbar} \frac{\partial E}{\partial k}. \tag{5.26}$$

The influence of the electron's environment is contained in the energy relation $E(k)$. For example, for an electron in free space, we use (4.5),

$$E(k) = V_0 + \frac{\hbar^2 k^2}{2m_e}. \tag{5.27}$$

For an electron in a periodic potential, even assuming the simplistic Kronig–Penney model, $E(k)$ cannot be given by a simple formula, but it can be determined by the procedure developed on pages 140. Methods for determining the energy-wavenumber relationship for more realistic models of materials are beyond the scope of this book, although the band structure can be exceedingly complex. However, usually the complete band structure of a material does not need to be known, since often only electrons in certain regions of a band are of interest. For example, considering semiconductors, typically one is interested in the behavior of electrons near a band edge (say, in the center of the first Brillouin zone near $k = 0$), since these are the electrons that will be most important for conduction. In that case, only the local behavior of the E–k curve will be important.

The main idea is the following. We know how to solve Schrödinger's equation for an electron in an "empty" region of space, either confined or unbounded, where we obtain a parabolic E–k relationship. (See, e.g., (4.5) or (4.26).) The parabola has the form $E = V_0 + \alpha k^2$, where $\alpha = \hbar^2/2m_e$. Considering the importance of electrons near bandedges in semiconductors, and the occurrence of parabolic-like dispersion behavior near bandedges in real materials (the band structure of Ge, Si, and GaAs is shown in Fig. 5.12), it makes sense

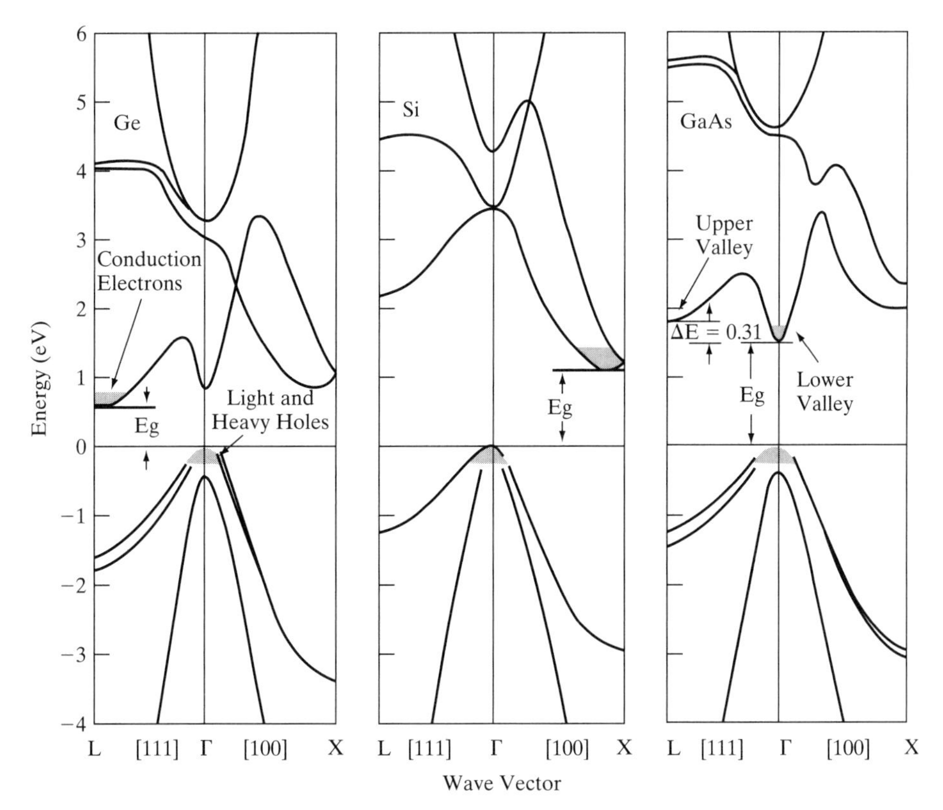

Figure 5.12 Bandstructure of Ge, Si, and GaAs. The Γ point is in the center of the first Brillouin zone ($k = (0, 0, 0)$), and the X and L points represent the zone boundaries along certain directions in three-dimensional k−space. In this case, $X = (1, 0, 0)\, 2\pi/a$, and $L = (1, 1, 1)\, \pi/a$, where a is the length of the cube edge in the fcc lattice. The curves do not look symmetric about the center point since moving from Γ to X and from Γ to L represents moving in different physical directions in the crystal. (Based on a figure from S. M. Sze, *Physics of Semiconductor Devices*, John Wiley & Sons, 1969. Used by permission.)

to locally model dispersion behavior using a parabolic relationship. This leads to the form

$$E(k) = E(k_0) + \beta (k - k_0)^2, \tag{5.28}$$

which is expected to be valid near bandedges centered at the point $k = k_0$, and where β will depend on the material in question (i.e., β governs the rate of expansion of the parabola). Since the band structure can be calculated or measured, β can be determined for a given material.† However, (5.28) can be made to look like the "empty" space result if we set $\beta = \hbar^2/2m^*$, where m^* is called the effective mass (which will, like β, depend on the material and band in question). Substituting this value of mass for m_e in Schrödinger's equation essentially incorporates the effect of the material on the electron, as long as one is interested in the behavior of the electron in the vicinity of a parabolic region of a band.

An explicit equation for m^* can be obtained by taking two derivatives of (5.28) with respect to k, leading to

$$m^* = \hbar^2 \left(\frac{\partial^2 E}{\partial k^2} \right)^{-1}. \tag{5.29}$$

Note that the effective mass is proportional to the reciprocal of the curvature of the E–k plot. From (5.26),

$$\frac{\partial v_g}{\partial k} = \frac{1}{\hbar} \frac{\partial^2 E}{\partial k^2}, \tag{5.30}$$

and so an equivalent expression for effective mass is

$$m^* = \hbar \left(\frac{\partial v_g}{\partial k} \right)^{-1}. \tag{5.31}$$

One can also develop a different definition of effective mass. Using $\hbar k$ as momentum,‡ with an electron represented by a wavepacket moving at the group velocity, one can define an effective mass such that

$$m^* v_g = \hbar k. \tag{5.32}$$

This is consistent with Newton's law (force equals the time rate of change of momentum),

$$F = \frac{dp}{dt}, \tag{5.33}$$

†As seen in Fig. 5.12, for any material there will be many different regions of parabolic E–k behavior, and, thus, many different values of β will exist, although typically only one or a few such parabolic regions will be of interest.

‡The quantity $\hbar k$ is called the *crystal momentum*. This is *not* the physical momentum of the electron. Rather, it is a quantity that describes the electron's Bloch state within a band, and is the correct quantity to use in Newton's law to obtain the electron's dynamics in a crystal.

leading to the equation of motion

$$q_e \mathcal{E} = \hbar \frac{dk}{dt}, \tag{5.34}$$

where $\mathcal{E}$ is the magnitude of the electric field for this scalar problem. Using (5.26) in (5.32), we obtain

$$m^* = \hbar^2 k \left(\frac{\partial E}{\partial k} \right)^{-1}. \tag{5.35}$$

As discussed in [8], this definition of effective mass can be used where the E–k curve is not parabolic. If the E–k relation is simply $E = V_0 + \beta k^2$, the two definitions lead to the same effective mass, $m^* = \hbar^2/2\beta$. The important point is that at most points of interest within an allowed energy band, the electron in a crystal moves as if it were free, except with an effective mass.

Effective mass can be positive or negative. A positive effective mass $m^*/m_e > 1$ means that the electron's velocity increase is less than what it would be for an electron in free space (i.e., the electron seems heavier), the difference indicating momentum transfer to the lattice. A positive effective mass $m^*/m_e < 1$, the usual case in semiconductors at the bandedge, means that the electron's velocity increase is more than it would be for an electron in free space (i.e., the electron seems lighter), the difference indicating momentum transfer from the lattice. For example, for GaAs, $m_e^* = 0.067 m_e$, resulting in a significant narrowing of the E–k curve, as depicted in Fig. 5.13. A negative effective mass (often

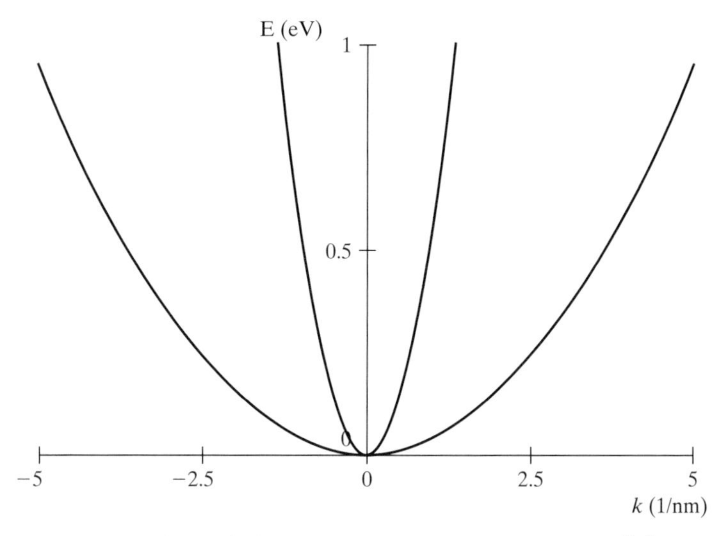

Figure 5.13 E (eV) versus k (nm^{-1}) for an electron in free space, $E(k) = \hbar^2 k^2/2m_e$ (wider curve), and for an electron at the bottom of the conduction band (set to reference level $E = 0$) in GaAs using the parabolic approximation, $E(k) = \hbar^2 k^2/2m_e^*$ (narrower curve). The effective mass $m_e^* = 0.067 m_e$ for GaAs results in significant narrowing of the E–k curve relative to the free-space environment.

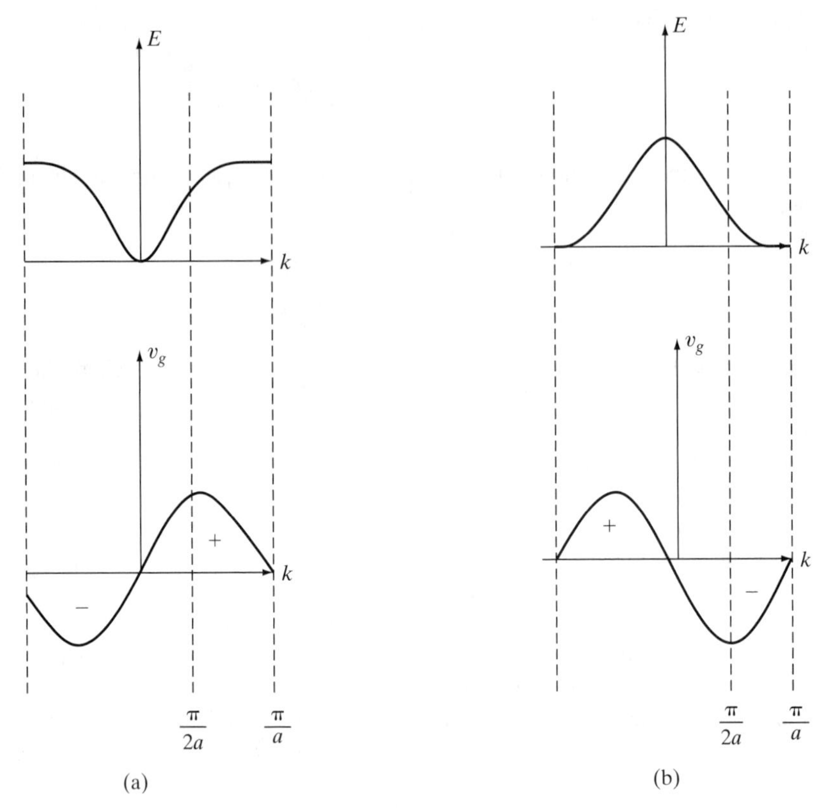

Figure 5.14 Energy and group velocity $v_g = (1/\hbar)(\partial E/\partial k)$ for two representative bands (left, a typical conduction band, and right, a typical valence band; bands will be discussed further in the next section). The period of the crystal is a.

found near the top of an energy band) means that, conversely, the particle is accelerated in a direction opposite to the direction of the applied force.†

Figure 5.14 depicts the relationship between energy and group velocity for a typical portion of an E–k curve.

If we consider what happens when an electric field is applied to a crystal, a clearer picture of the behavior of electrons in energy bands develops. When a constant electric field of magnitude $\mathcal{E}$ is applied to the crystal, an electron will experience a Lorentz force, $q_e\mathcal{E}$. This force will accelerate the electron, increasing its wavenumber. To be specific, the equation of motion (5.34) can easily be solved to yield

$$k(t) = k(0) + \frac{q_e\mathcal{E}}{\hbar}t, \tag{5.36}$$

†To see this, consider $\mathbf{F} = m^*\mathbf{a}$. Since $\mathbf{F} = q_e\mathbf{E}$ with $q_e < 0$, then $m^* < 0$ means that in response to an applied electric field, the electron moves in a direction opposite to that which would occur in free space.

and so k increases linearly with respect to time. (For simplicity, we can assume $k(0) = 0$.) As k increases towards the value $k = \pi/2a$, the electron's velocity increases, as expected, as does the effective mass.

At $k = \pi/2a$, the E–k curve has an inflection point (and so the parabolic band assumption is no longer valid), and the electron's velocity has reached a maximum. As k increases further, the electron decelerates, and finally, as k reaches the Brillouin zone boundary at π/a, the electron's velocity goes to zero, indicating that the electron wavefunction is represented by a standing, rather than traveling, wave.

As k increases further and moves away from the zone boundary into the second Brillouin zone,† a conduction band electron will initially have a negative group velocity, indicating that the electron is moving in the reverse direction. The electron continues to reverse direction periodically as k increases through zone boundaries.

Therefore, we have the following picture. We apply a static (d.c.) electric field to the crystal, and as a result of band structure, the electron oscillates back and forth. (These are called *Bloch oscillations*.) Can this really be so? In practice, this behavior is not seen. Defects and impurities will generally be present,‡ as well as lattice vibrations called *phonons*,§ and

†Despite what could be inferred from the extended zone scheme depicted in Fig. 5.9, as k increases through Brillouin zone boundaries, the energy of the electron does not "jump." $E(k)$ evolves smoothly, as depicted in the repeated zone scheme of Fig. 5.10. Sometimes the electron is said to be Bragg scattered at the zone boundaries, but this gives a rather unphysical picture, and so this idea is avoided here.

‡As an interesting aside, the following figure shows an STM image of electron standing waves extending outward from defects/impurities in a Cu surface.

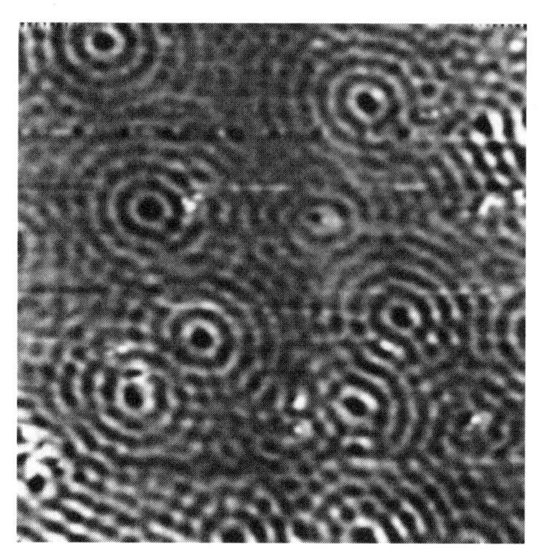

A 42 nm × 42 nm STM image of Cu ($T = 150$ K). Defects create standing electron waves emanating radially outward from the defect. (Reprinted with permission from Petersen, L., P. Laitenberger, E. Lægsgaard, and F. Besenbacher, "Screening Waves from Steps and Defects on Cu(111) and Au(111) Imagined with STM: Contribution from Bulk Electrons," *Phys. Rev. B*, 58 (1998): 7361. Copyright 1998, American Physical Society.)

§Phonons are lattice vibrations (vibrations of the atoms in a lattice) treated quantum mechanically. The energy of these vibrations (i.e., heat) is related to the kinetic energy of the atoms in the material. For a lattice vibration of frequency ω, energy is quantized as $\hbar\omega$, as with all quantum particles. Heat is carried through a

the electron will collide with these obstacles such that the wavevector will not be able to increase very much. That is, we can only solve (5.34) over the time period between collisions, and, therefore, the wavevector does not simply linearly increase. This is discussed further in Section 10.1. The wavevector only has a chance to increase a small amount in the first zone before a collision occurs, and, therefore, the k value will not reach the zone boundary $k = \pm\pi/a$.

For example, assume that the average time between collisions (called the *momentum relaxation time*) is $\tau = 2.47 \times 10^{-14}$ s, which is the value for copper at room temperature. If $\mathcal{E} = -100$ V/m, then $k\left(t = 2.47 \times 10^{-14}\right) = 3{,}753$ m^{-1}. The first Brillouin zone boundary occurs at $k = \pm\pi/a$, and, using $a = 3.61$ Å for the lattice constant for copper, we obtain $k = 8.7 \times 10^{9}$ m^{-1} at the zone boundary. Thus, for a typical collision time, the wave vector only gets about 10^{-5} percent of the way to the Brillouin zone boundary before its momentum is stopped by a collision! It is obvious that even very large applied fields or long values[†] of τ will not change the situation too much. However, Bloch oscillations are real, and have been experimentally observed in a number of periodic structures, most notably semiconductor superlattices. (See Section 6.3.4.)

Note that electron collisions are regarded as the *cause* of electrical resistance in the classical, free-electron gas model of conductivity, as discussed in Section 10.1. However, in the quantum mechanical picture, the collisions are actually *necessary* to allow d.c. current to flow! If no collisions occur, such as may be the case in an ultrapure sample at low temperatures, an a.c. current will result from a d.c. applied field!

Lastly, note that for free electrons, the E–k curve (4.5) is a parabola,

$$E = \frac{\hbar^2 k^2}{2m_e}, \tag{5.37}$$

and, therefore,

$$\frac{\partial^2 E}{\partial k^2} = \frac{\hbar^2}{m_e}, \tag{5.38}$$

such that

$$m^* = m_e, \tag{5.39}$$

i.e., the effective mass is the ordinary mass, as expected. It is because the dispersion relation for an electron in a material does not have the simple form (5.37) that the concept of effective mass arises.

In summary, via the effective mass, we represent the influence of a crystalline lattice on an electron. The specific nature of the periodic potential is contained in the $E\,(k)$ relationship,

material as a flow of phonons. Similar to photons, phonons transport energy and momentum, but not mass, and obey Bose–Einstein statistics. (See Section 8.2.)

There are two broad classes of phonons: relatively low-energy modes called acoustic phonons, which are related to sound propagation in a material, and higher energy modes called optical phonons, so named because optical energies can excite them.

[†] τ is temperature and material dependent, although one could consider practical values into the picosecond range.

which usually must be obtained numerically or experimentally. We can often greatly simplify solving problems involving an electron in a crystal by solving an equivalent problem of an electron in free space, except where the mass of the electron is given by m^*. For example, we can solve Schrödinger equation

$$\left(-\frac{\hbar^2}{2m}\nabla^2 + V(\mathbf{r})\right)\psi(\mathbf{r}) = E\psi(\mathbf{r}), \tag{5.40}$$

where $V(\mathbf{r})$ accounts for the crystal lattice (and is, therefore, horrendously complicated), or we can solve the much simpler equation

$$\left(-\frac{\hbar^2}{2m^*}\nabla^2 + V_0\right)\psi(\mathbf{r}) = E\psi(\mathbf{r}), \tag{5.41}$$

where m^* is the effective mass, accounting for the crystalline lattice, and where V_0 is a constant potential depending on the problem. The method of replacing (5.40) with (5.41) is known as the *effective mass approximation* of Schrödinger's equation.

The effective masses of electrons, m_e^*, and holes, m_h^*, in Si and GaAs are given in Table 5.1, and values for some other common semiconductors are given in Table IV in Appendix B. Holes will be discussed later in this chapter.

To further complicate the situation, in many crystals there is a different effective mass in each different direction; near bandedges this is given as

$$m_{x,y,z}^* = \hbar^2\left(\frac{\partial^2 E}{\partial k_{x,y,z}^2}\right)^{-1}. \tag{5.42}$$

However, for many metals, the effect of the lattice is generally screened by the large density of electrons (Section 3.5), such that often in a metal one can set $m^* = m_e$.

In an amorphous material such as SiO_2, the concept of energy bands is not quite appropriate. However, often electrons in SiO_2 can be modeled as having an effective mass of the order of $0.4m_e$–$0.9m_e$, and the effective mass Schrödinger equation can be used to model electrons in, for example, metal-oxide-semiconductor structures such MOS capacitors.

TABLE 5.1 EFFECTIVE MASS IN Si AND GaAs.

	Effective Mass	
Semiconductor	m_e^*/m_e	m_h^*/m_e
Si	0.26	0.50 0.24
GaAs	0.067	0.50 0.082

5.4 BAND THEORY OF SOLIDS

The material presented in the previous section leads to what is called the *band theory of solids*, the central idea of which is that, due to the periodic potential associated with the crystalline lattice, there are allowed and disallowed energy bands. Furthermore, we have only found the possible bands and the bandgaps, but have not considered if a certain band will be "filled." Whether or not a band has electrons in it depends on the number of electrons in the system, and the energy of the electrons. Several different situations arise:

1. If an allowed band is completely empty of electrons, obviously there are no electrons in the band to participate in electrical conduction. This can happen, for example, in a high-energy band, where the energies of the band are above the energies of any of the system's electrons.
2. Perhaps a bit more surprising is the fact that when an allowed band is completely filled with electrons, those electrons cannot contribute to electric conduction either. This is because electrons are fermions, and must obey the Pauli exclusion principle, so that no two electrons can be in the same state. Therefore, if an electron is given energy, due to, say, a voltage applied to the material, the electron must move to a (perhaps only slightly) higher energy state. However, if all such states are already filled, the electron has no empty state to move into, and, therefore, the electron cannot gain any energy and contribute to conduction.[†] This would be the case, for example, for a band of energies that are much less than the Fermi energy of the system's electrons.

To partially summarize, no conduction can take place in a material that has energy bands that are either completely filled or completely empty. An analogy is to a jar of marbles. An empty band is like an empty jar, and a filled band is like a jar filled so full that no marbles can move (thus, there are no empty spots for a marble to move into). Only when the jar is partially full can marbles move within the jar.

3. Thus, we are left to conclude that only electrons in a partially filled energy band can contribute to conduction.

In general, the lower bands in a material will be completely filled with electrons. (Recall that in Section 4.6 we discussed the hydrogen atom and the periodic table, and that states are typically filled one by one, starting at low energies.) The main question is whether or not the uppermost band that contains electrons is completely filled, or only partially filled. Materials that have partially filled uppermost bands are called conductors (mostly metals), such that when energy is supplied by an external source, electrons can move into a higher unoccupied state within the band.

To illustrate these concepts, a *real-space band diagram* for a typical metal is shown in Fig. 5.15. It is obvious that the highest band to contain electrons is partially full, and so electrical conduction can take place.

[†]Of course, if a very large amount of energy is supplied such that the electron can jump across the energy gap into a different, higher, empty, or partially filled band, that electron can contribute to conduction.

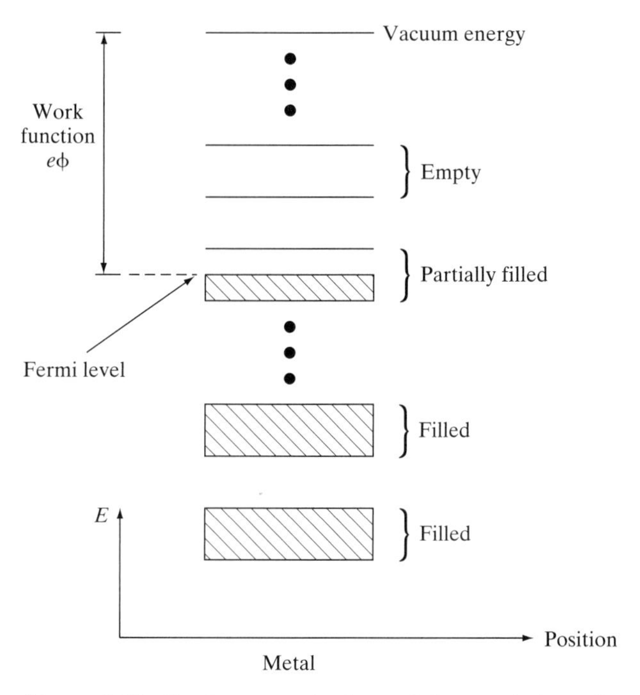

Figure 5.15 Band structure (real space) for a typical metal.

The *work function*, $e\phi$, is the energy difference between the vacuum level and the Fermi level (the uppermost energy of the electrons at $T = 0$ K). It is thus the energy needed to completely liberate an electron from the metal at $T = 0$ K, and, given the narrow tail of the Fermi–Dirac distribution, which will be described in Section 8.2, this energy approximately holds at other temperatures as well. Energy can be supplied, for example, thermally, or electromagnetically, as in the photoelectric effect. For example, considering shining light on a metal surface, *Einstein's relation* is

$$\hbar\omega = e\phi + E_{KE}, \tag{5.43}$$

where $\hbar\omega$ is the energy of the incident light, and E_{KE} is the kinetic energy of the emitted electrons. For copper, the work function is on the order of 4.75 eV (ranging from 4.5 to 5.0 eV depending on the crystal orientation), and for gold, $e\phi$ is in a similar range. Thus, as discussed in Section 2.2, the photoemitted electrons will have kinetic energy $E_{KE} = \hbar\omega - e\phi$, which only depends on frequency, and not on intensity.

Materials having completely filled lower bands and empty upper bands, such that conduction cannot take place, are either insulators or semiconductors at low temperatures, as shown in Fig. 5.16. The highest band that is filled at $T = 0$ K in semiconductors is called the *valence band*, and the unfilled bands above the valence band are called *conduction bands*.

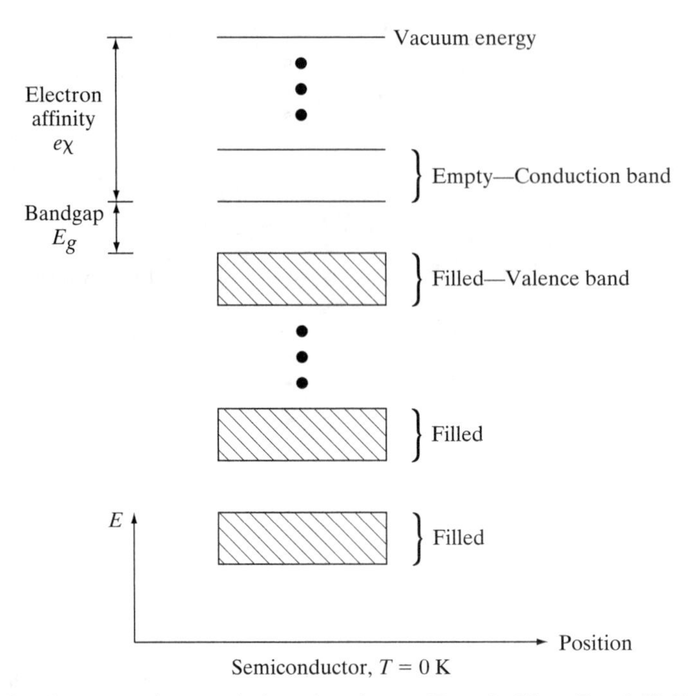

Figure 5.16 Band structure for a typical semiconductor ($E_g \sim 1$ eV) at $T = 0$ K. This picture also holds for a typical insulator ($E_g \sim 8$ eV) for a range of temperatures.

From the point of band theory, the main difference between a semiconductor and an insulator is the size of the bandgap. For semiconductors, the energy gap is not too large (for Si, $E_g \simeq 1.12$ eV, and for GaAs, $E_g \simeq 1.43$ eV), so that as temperature increases some thermally excited electrons have enough energy to jump across the gap and reach a previously empty band, thus being able to contribute to conduction. Other than thermal energy, an external excitation such as an applied voltage can also supply enough energy for electrons to jump across the gap and contribute to conduction, especially when dopant atoms are present in semiconductors (to be discussed in the next section). The band structure after some energy input (in this case, thermal) is depicted in Fig. 5.17. The energy difference between the vacuum level and the bottom of the conduction band is called the *electron affinity*, $e\chi$.

For insulators, the bandgap is large (perhaps 8–10 eV; SiO_2 has $E_g \simeq 8$ eV), so that thermally excited electrons, or those excited by an applied voltage, do not generally have enough energy to cross the gap. Of course, even for good insulators, the actual gap is finite. If enough energy is supplied, thermally or by an applied voltage, electrons will cross the gap and move into an unoccupied state in a higher band, thus being able to contribute to conduction.

For example, consider the gate oxide in a MOSFET assuming a $V_0 = 1.5$ volt power supply and SiO_2 as the gate oxide. The associated energy is $eV_0 = 1.5$ eV, which is much less than the bandgap energy of SiO_2. Therefore, in this case, valence electrons will not be able to cross the gap and contribute to conduction. However, this energy may be large

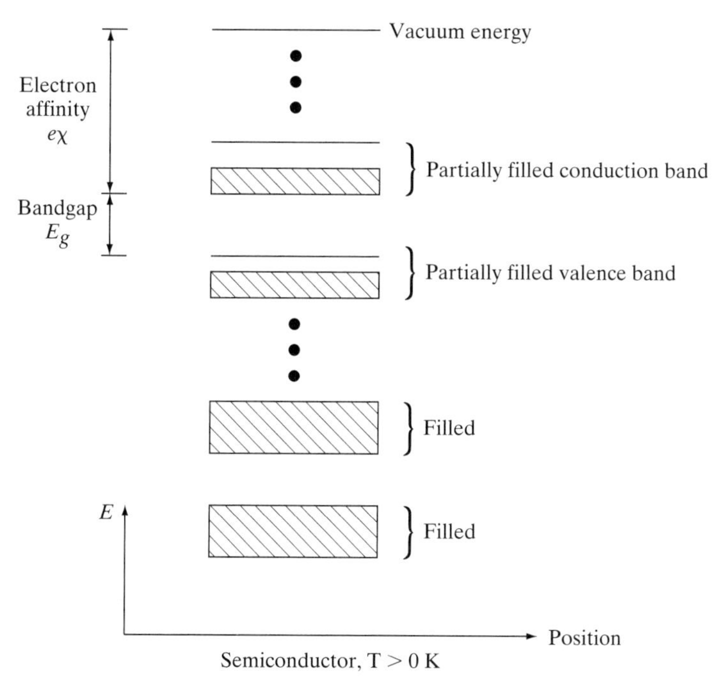

Figure 5.17 Band structure for a typical semiconductor at $T > 0$ K, showing that some of the electrons that were in the valence band at $T = 0$ K (Fig. 5.16) have moved, via thermal energy, into the first conduction band. Note that E_g is slightly temperature dependent.

enough to excite electrons associated with impurities, (which lie at energy levels in the band gap, as discussed in the next section), into the conduction band.†

Recalling the electronic configuration of the elements in the periodic table, we find that if the outermost shell, (which, for an individual atom, is analogous to the topmost conduction band containing electrons for a solid), is partially filled, due to an odd number of electrons, the material will generally be a conductor. If the outermost shell is completely filled, due to an even number of electrons, the material will be an insulator or a semiconductor, depending on the size of the bandgap. This is sort of a general rule of thumb, and doesn't always hold. For example, beryllium has four electrons, yet is a metal. (Beryllium belongs to the class of divalent metals in group IIA of the periodic table.)

Holes. When an electron is elevated from the valence band to a conduction band, it leaves an empty spot in the valence band called a *hole*. Holes are not implicated in an electron gas model of metals, nor generally in the characterization of metals at all, due to the very large numbers of "free" conduction band electrons. However, in a semiconductor, holes play a prominent role, and for each electron that contributes to conduction, there is an associated hole that can also contribute to conduction. It can be seen that holes are capable

†We reiterate that SiO_2 is an amorphous material, and, as such, the concept of a bandgap, which was developed for crystals, is an approximation.

of moving in the valence band by considering the following. Assume that electron e_1 moves to the conduction band, leaving behind a hole h_1 in the valence band. The valence band is no longer full, and, therefore, another valence band electron e_2 can move, and it will tend to move into h_1. This fills the original hole, but leaves another hole, h_2, in the position that e_2 previously occupied. A valence band electron e_3 moves into h_2, leaving behind hole h_3, and so on. In this way, holes (spots in the lattice where an electron is missing) can move in the valence band. Holes carry positive charge (i.e., the absence of negative charge), and move in the direction of an applied electric field. Physically, it should be kept in mind that in both bands it is actually electrons that are moving.

Furthermore, the hole effective mass in many semiconductors is separated into heavy and light hole masses, since the valence band is often divided into light and heavy bands. (See Fig. 5.12 on page 143.) Actually, the valence band often consists of several bands, and in each band the effective mass is different. Bands having large effective mass are called *heavy bands*, and bands having lower effective mass are called *light bands*. The heavy and light hole effective masses for Si and GaAs are listed in Table 5.1, and values for other common semiconductors are provided in Table IV in Appendix B.

5.4.1 Doping in Semiconductors

It is often necessary to know how many electrons can contribute to conduction. In metals, the conduction band is partially full, even at $T = 0$ K, and there are a lot of electrons to contribute to electrical conduction.† As described previously, in semiconductors at $T = 0$ K, the valence band is full and the conduction band is empty, so that no electrons can participate in conduction. As temperature increases, thermal energy of the crystal elevates some electrons into the conduction band, leaving a hole in the valence band. The resulting (*intrinsic*) carrier densities of electrons and holes are labeled n_i and p_i, respectively. It is obvious that, since each time an electron is elevated into the conduction band a hole is created in the valence band, $n_i = p_i$.

The intrinsic carrier concentration in a typical semiconductor is very low because thermal energy is usually small compared to the bandgap energy. As a result, in Si at room temperature, for example, $n_i \sim 1.5 \times 10^{10}$ cm^{-3}; for comparison, copper has on the order of 10^{22} cm^{-3} carriers. However, the carrier concentration in semiconductors can be modified by adding impurities called *dopants*. Undoped semiconductors are called *intrinsic semiconductors*, and doped semiconductors are called *extrinsic semiconductors*.

For illustration purposes, consider Si, a Group IV element (i.e., silicon has four outer shell electrons). If we introduce impurity elements from Group V (i.e., an element having five outer shell electrons, such as phosphorus), each impurity will bond (covalently) with four neighboring silicon atoms, using up four of its electrons in the process, as shown in Fig. 5.18. The fifth electron will still be bound to the nucleus of the impurity (i.e., the impurity atom will have a net positive charge), although much more weakly than it would be in the elemental state. A small amount of energy (typically, on the order of 0.05 eV) will

†That there are *some* electrons to contribute to conduction in a partially filled band is obvious. That there are, in fact, *a lot* of electrons that can contribute to conduction is established by considering the density of states, which will be discussed in Chapter 8.

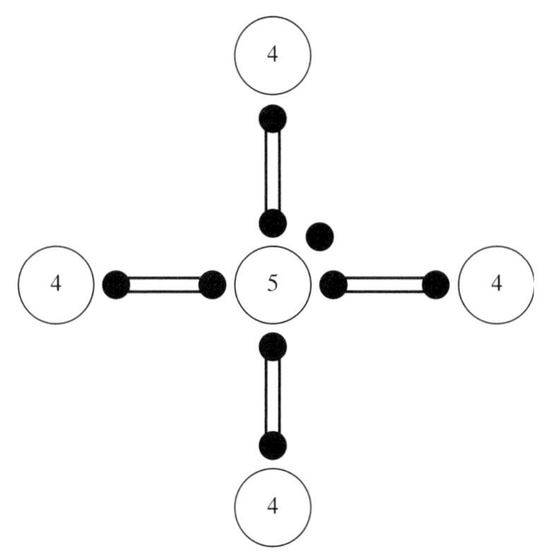

Figure 5.18 Part of a Group IV element lattice with the insertion of a Group V element. The double parallel lines indicate a covalent bond. One "extra" electron will be present, loosely bound to the Group V atom.

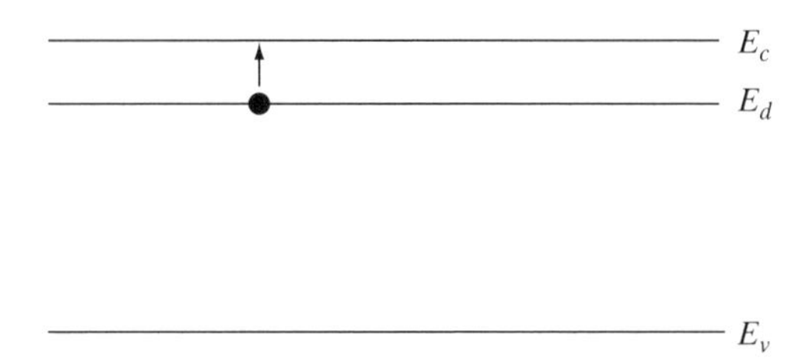

Figure 5.19 Real-space energy band diagram showing the level of a donor electron, and its transition to the conduction band upon adding a small amount of energy.

liberate the electron and move it to the conduction band. The impurity is called a *donor*, and we can view this on an energy band diagram as the electron being at a donor level, E_d, lying just below the conduction bandedge, as shown in Fig. 5.19. A small amount of thermal energy will be able to bridge the energy gap $E_c - E_d$, and the electron will contribute to conduction. For example, considering silicon with the donor being phosphorus, $E_c - E_d \simeq 0.044$ eV at room temperature. Thus, donor atoms are easily ionized, resulting in what is called an *n type* semiconductor.

As a rough approximation, since the excess electron is loosely held to the donor atom, the situation resembles the hydrogen atom in that one can consider a two-particle problem (the excess electron, and everything else). This hydrogen model can provide a rough approximation for these shallow dopants. The fact that the dopant atom is immersed in a semiconductor, rather than free space, is accounted for by using the appropriate effective mass and permittivity. Energy states of the hydrogen atom are given by (4.117), and substituting the effective mass in place of m_e, and the permittivity of the semiconductor in place of the vacuum permittivity, we have

$$E_n = -\frac{m_e^* q_e^4}{8\varepsilon_r^2 \varepsilon_0^2 h^2 n^2}. \tag{5.44}$$

If this dopant atom is to be ionized, there must be a transition from the $n = 1$ state to $n \to \infty$, leading to the donor ionization energy

$$\Delta E_d = \frac{m_e^* q_e^4}{8\varepsilon_r^2 \varepsilon_0^2 h^2}. \tag{5.45}$$

For Si, using $m_e^* = 0.26m_e$ and $\varepsilon_r = 11.7$, we obtain $\Delta E_d = 0.025$ eV, within an order of magnitude of measured values. Measured values of donor ionization energies for Si and Ge are given in Table 5.2.

In a similar manner, the impurity element can come from Group III of the periodic table, having three outer shell electrons (e.g., boron). In forming a covalent bond with its four neighboring silicon atoms, there will be one electron missing, forming a hole, as shown in Fig. 5.20. An electron with a small amount of energy in the valence band

TABLE 5.2 MEASURED VALUES OF DONOR IONIZATION ENERGIES FOR Si AND Ge (meV) ([6]).

	Ionization Energy (meV)		
Semiconductor	P	As	Sb
Ge	12.0	12.7	9.6
Si	45	49	39

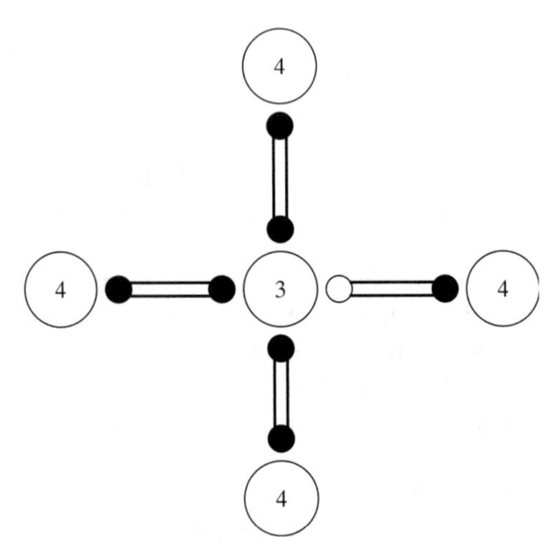

Figure 5.20 Part of a Group IV element lattice with the insertion of a Group III element. The double parallel lines indicate a covalent bond. One "extra" hole will be present, loosely bound to the Group III atom.

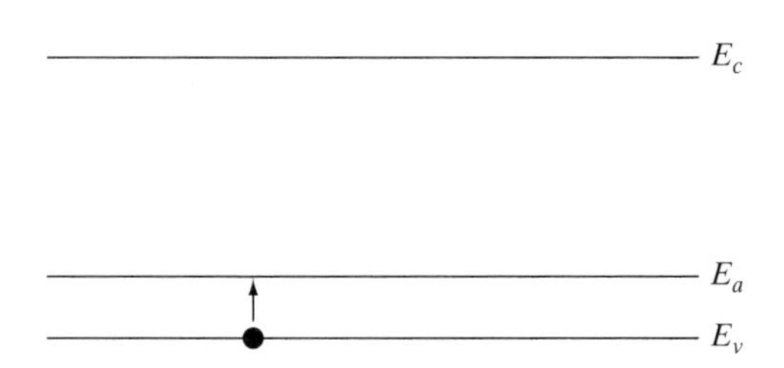

Figure 5.21 Real-space energy band diagram showing acceptor level.

can move into the hole, and so the impurity is called an *acceptor*. From a band diagram perspective, the acceptor impurity energy level lies just above the valence bandedge, as shown in Fig. 5.21. For a silicon lattice, typical values of $E_a - E_v$ are on the order of 0.05 eV (e.g., if the dopant is boron, $E_a - E_v \simeq 0.45$ eV). Acceptor ionization levels can be determined from (5.45) using the effective mass of holes. The resulting material is known as a *p type* semiconductor.

5.4.2 Interacting Systems Model

There is another way to view bandstructure that is often helpful, especially in understanding how two systems interact when brought together. It turns out that if a quantum system has energy levels $E_1, E_2, E_3, \ldots$, then if two such identical systems (perhaps two atoms, or two quantum wells) are brought together, it can be shown that each energy level will split into two levels,

$$E_n \to E_n^+, E_n^-, \tag{5.46}$$

where $E_n^{\pm}$ is an energy value slightly above/below the energy value E_n of the isolated system. This is depicted in Figs. 5.22 and 5.23.

In Fig. 5.22, two identical systems, each having energy levels E_1 and E_2, are spaced sufficiently far apart so that they don't interact. Then, if the systems are brought near to

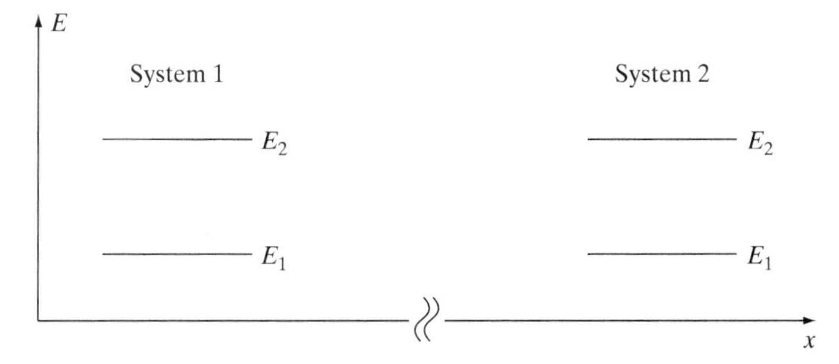

Figure 5.22 Two identical quantum systems, each having two energy levels, are spaced far apart. No interaction occurs.

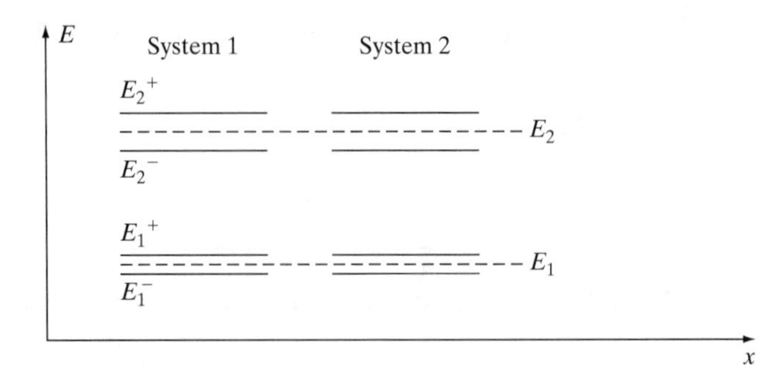

Figure 5.23 The two identical quantum systems from Fig. 5.22 are brought into close proximity, resulting in energy level splitting.

each other, as depicted in Fig. 5.23, the energy levels split as

$$E_1 \rightarrow E_1^+, E_1^-, \tag{5.47}$$

$$E_2 \rightarrow E_2^+, E_2^-. \tag{5.48}$$

The splitting is due to the overlap of each system's wavefunctions (really orbitals, as discussed in Section 3.5). For example, in the case of two atoms that come together to form a molecule, the atomic orbitals associated with each atom begin to overlap as the atoms are brought together. This can be seen by considering a simplified linear model of forming a lithium (Li) molecule. Lithium has the electronic configuration $1s^2 2s^1$, and in forming the molecule Li_2, the s shell atomic orbitals form *antibonding* and *bonding* molecular orbitals, as depicted in Fig. 5.24(a). In the ground configuration, the bonding molecular state is filled with the two $2s^1$ electrons (one from each atom), and the antibonding state is empty, as depicted in Fig. 5.24(b).

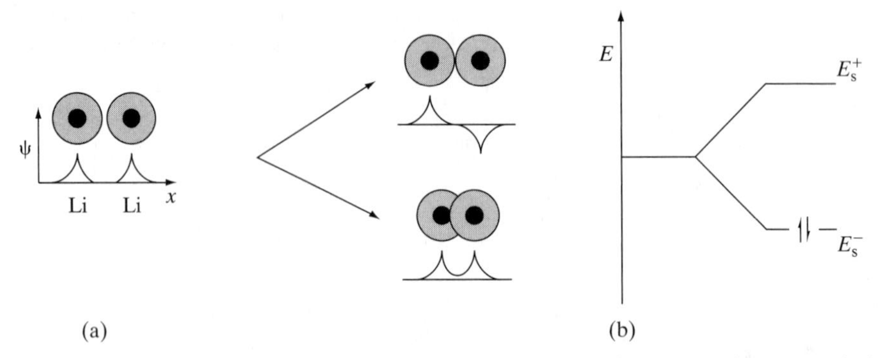

Figure 5.24 Depiction of combining atomic orbitals of lithium to form molecular states of Li_2. (a) Two Li atoms are brought into close proximity. The atomic s shells interact, forming antibonding (top) and bonding (bottom) molecular orbitals. The $2s^1$ electron from each atom will go into the lowest energy state, as depicted in (b).

If N identical atoms are brought together, each energy level of an isolated atom, $E_1, E_2, E_3, \ldots$, will split into N levels,

$$E_1 \to \begin{cases} E_{1,1} \\ E_{1,2} \\ \vdots \\ E_{1,N} \end{cases}, \qquad E_2 \to \begin{cases} E_{2,1} \\ E_{2,2} \\ \vdots \\ E_{2,N} \end{cases}, \qquad E_3 \to \begin{cases} E_{3,1} \\ E_{3,2} \\ \vdots \\ E_{3,N} \end{cases}, \qquad \ldots, \tag{5.49}$$

where each group is centered on the corresponding isolated atomic level, E_n. As $N \to \infty$, the N discrete levels centered on each E_n merge into a quasi continuum of allowed energy levels, forming an energy band centered on E_n. If two adjacent levels of the original, widely spaced system, say, E_n and E_{n+1}, are well separated in energy, then the resulting energy bands centered on E_n and on E_{n+1} will not overlap, creating a bandgap. This is depicted in Fig. 5.25, where the s and p states of an atomic shell in an atom (Section 4.6.2) split into separate s and p bands in a solid. The top half of either resulting band (s or p in Fig. 5.25) is the antibonding half, and the bottom half is the bonding half.

If, for example, the atomic s shell in Fig. 5.25 is partially filled (i.e., an ns^1 shell), then the resulting s band will be partially filled (the bonding half), resulting in a conductor. If the shells in question are full (e.g., ns^2), then the resulting bands will be full, usually resulting in an insulator or semiconductor.

However, the situation is often much more complicated. For instance, assume that the s shell is full, and that the next p shell is empty. The resulting s band will be full, and the p band will be empty. However, these bands may overlap, allowing conduction. This

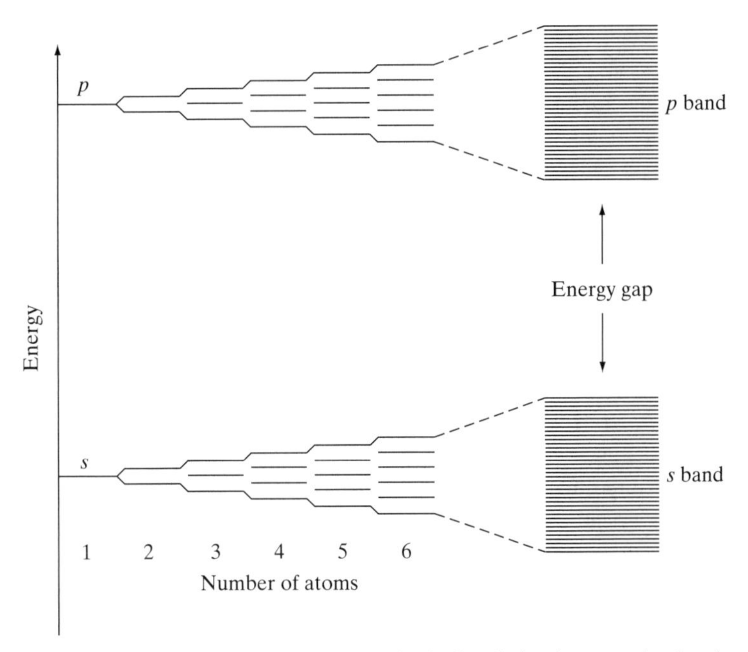

Figure 5.25 Depiction of s and p states in an atomic shell splitting into s and p bands as atoms are brought together to form a solid.

happens in, for example, Mg, which has the atomic configuration† $1s^2 2s^2 2p^6 3s^2 3p^0$ (i.e., full s shell, empty p shell) and yet is a metallic solid due to band overlap. Moreover, it is often energetically favorable for atomic states to become mixed (*hybridized*) in forming atomic bonds. This often occurs when the s and p shells are sufficiently close in energy such that the atomic orbitals can interact, resulting in sp molecular orbitals. This is extremely common, and occurs in most semiconductors.

The idea of combining atoms to form molecules and solids can be made quantitatively accurate by a method known as the linear combination of atomic orbitals (LCAO), although this is beyond the scope of the text. Nevertheless, it is an intuitively appealing model, since in the limit of large separation we expect that systems will not interact, but will begin to interact as they are brought closer together.

5.4.3 The Effect of an Electric Field on Energy Bands

In order to analyze electronic devices, we must be able to understand the effect of applying an electric field or a potential difference to a material. For simplicity, we will consider a one-dimensional example and a constant electric field $\mathbf{E} = \mathbf{a}_x \mathcal{E}_0$. From basic electrostatics, the work done (energy expended) in moving a charge q from 0 to x in the presence of the field is

$$W(x) = -\int_0^x \mathbf{F} \cdot d\mathbf{l} = -\int_0^x (\mathbf{a}_x q \mathcal{E}_0) \cdot \mathbf{a}_x dx = -q\mathcal{E}_0 x, \tag{5.50}$$

where $\mathbf{F} = q\mathbf{E}$ is the force on the charge q by the field‡ $\mathbf{E}$. For an electron, $q = -e < 0$, and energy $-q_e \mathcal{E}_0 x = e\mathcal{E}_0 x$ is required to move the electron against the field. The energy required to move the electron to the point x increases linearly as x increases. In moving in the opposite direction, the electron gives up energy to the field.

Figure 5.26(a) shows an electron in the conduction band of a material. To accommodate the idea of needing to linearly increase the electron's energy if it is to move in the positive x-direction, and decrease its energy in moving in the negative x-direction, we tilt the energy band diagram under the influence of the applied field, as shown in Fig. 5.26(b). We can think of the tilted bands as encouraging the electrons to "roll downhill," or requiring an energy input to climb uphill. From a voltage standpoint, positive voltage depresses (pushes down) the energy level.

5.4.4 Bandstructures of Some Semiconductors

Given the importance of semiconductors in electronics applications, we will very briefly examine the band structure of several types of semiconductors.

†$3p^0$ and higher shells are not usually listed, although it was included here to emphasize the point that the s shell is full and the next p shell is empty.

‡Since voltage is work per unit charge, $V(x) = W(x)/q$ (referenced to $V(0) = 0$); if we know the potential V rather than the electric field $\mathcal{E}$, energy is simply qV. Be aware that the symbol V can represent voltage or potential energy, although usually its meaning is clear from the context.

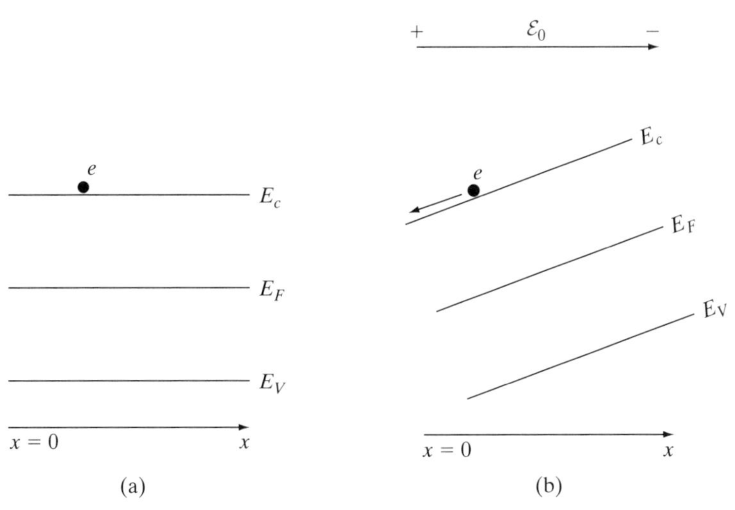

Figure 5.26 (a) Real-space band diagram in the absence of an applied electric field. (b) Real-space band diagram in the presence of an applied field $\mathcal{E}_0$.

For most semiconductors, the top of the valence band occurs at $k = 0$. However, the bottom of the conduction band may occur at some other value of k, and we can divide semiconductors into two general categories.

Direct bandgap semiconductors are such that the bottom of the conduction band occurs at $k = 0$, as shown in Fig. 5.27, where E_c and E_v are the conduction and valence bandedges, respectively, and where the bandgap is $E_g = E_c - E_v$. For a direct bandgap semiconductor, near $k = 0$ the energy bands can be modeled by

$$E(k) \simeq E_c + \frac{\hbar^2 k^2}{2m_e^*}, \qquad \text{conduction band} \tag{5.51}$$

$$E(k) \simeq E_v - \frac{\hbar^2 k^2}{2m_h^*}, \qquad \text{valence band.}$$

Materials such as GaAs and InP are direct bandgap semiconductors. (The bandstructure for GaAs is shown in Fig. 5.12 on p. 143.)

Indirect bandgap semiconductors are semiconductors for which the bottom of the conduction band does not occur at $k = 0$, but at some other wavenumber value, as shown in Fig. 5.28. Semiconductors such as Si, Ge, and AlAs are indirect bandgap semiconductors. (The bandstructure for Ge and Si is also shown in Fig. 5.12.) Actually, silicon has a complicated bandstructure, and the bottom of the conduction band occurs at six equivalent minima in k-space, resulting in six conduction band valleys.

Indirect semiconductors highlight the difference between crystal momentum and particle momentum, since at the bottom of the conduction band in an indirect semiconductor, the particle momentum is zero, although the crystal momentum $\hbar k$ is nonzero. Particle momentum notwithstanding, the electron dynamics are governed by the equation of motion

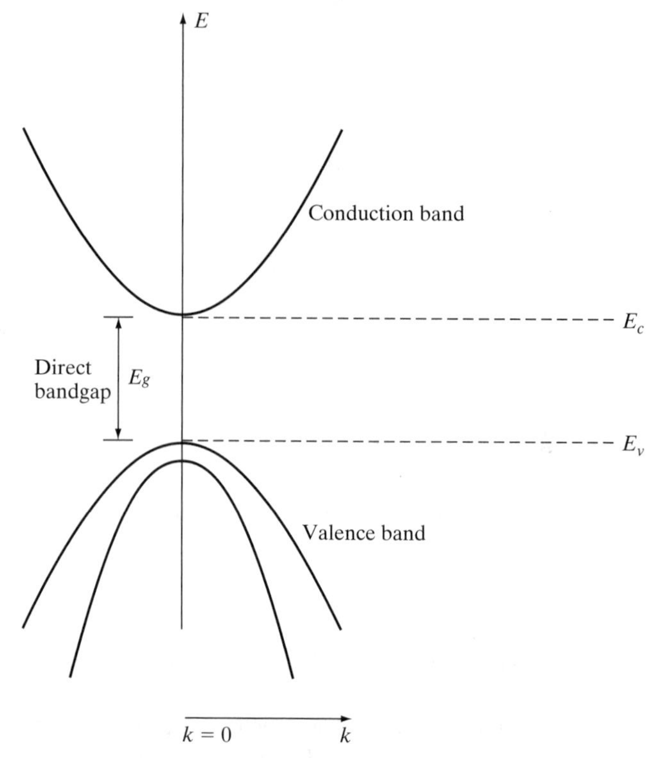

Figure 5.27 Energy bands (wavenumber space) for a typical direct bandgap material (light and heavy hole valance bands are shown).

involving the crystal momentum (5.34). The bandgap properties of several important semiconductors are given in Table 5.3.

Considering the conduction and valence bands, it should be noted that electron kinetic energy is given by $E - E_c > 0$, and hole kinetic energy by $E_v - E > 0$, as depicted in Fig. 5.29.

5.4.5 Electronic Band Transitions—Interaction of Electromagnetic Energy and Materials

So far, we have been considering primarily materials in thermal equilibrium, where electrons and holes have, on average, thermal energy corresponding to the temperature of the material. When an electron is thermally excited from a lower state (the valence band, or an impurity level) to the conduction band, the electron is available to contribute to conduction. In general, the electron will exist in the conduction band for some period of time before falling to a lower state, a process called *recombination* (as in an electron recombining with a hole). For a material in thermal equilibrium, the generation rate of carriers is equal to the recombination rate, so that the carrier concentration remains constant.

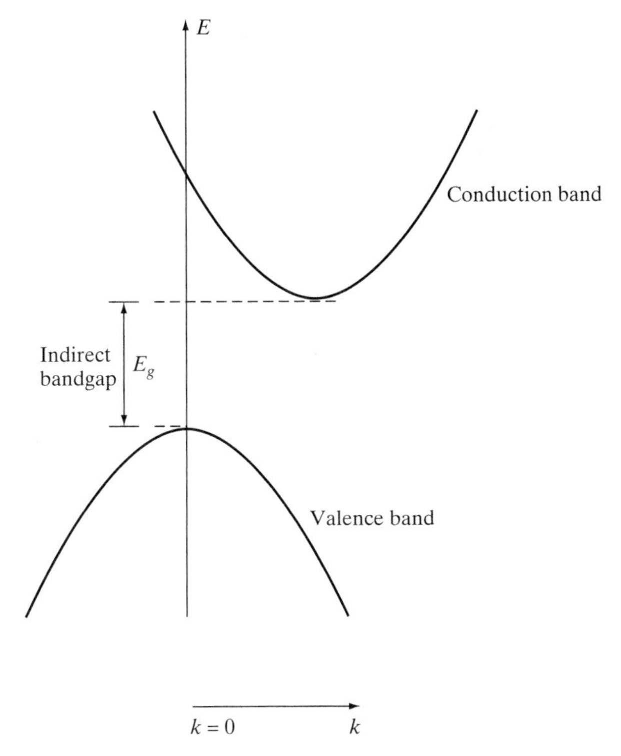

Figure 5.28 Energy bands (wavenumber space) for a typical indirect bandgap material.

TABLE 5.3 BANDGAP PROPERTIES OF SEVERAL IMPORTANT SEMICONDUCTORS. FOR GAP TYPE, I INDICATES AN INDIRECT BANDGAP SEMICONDUCTOR, AND D A DIRECT GAP SEMICONDUCTOR.

Crystal	Gap type	E_g (eV) @ 0 K	E_g (eV) @ 300 K
Si	I	1.17	1.11
Ge	I	0.74	0.66
GaAs	D	1.52	1.43
InP	D	1.42	1.27
AlAs	I	2.23	2.16
CdS	D	2.58	2.42
CdSe	D	1.84	1.74
CdTe	D	1.61	1.44

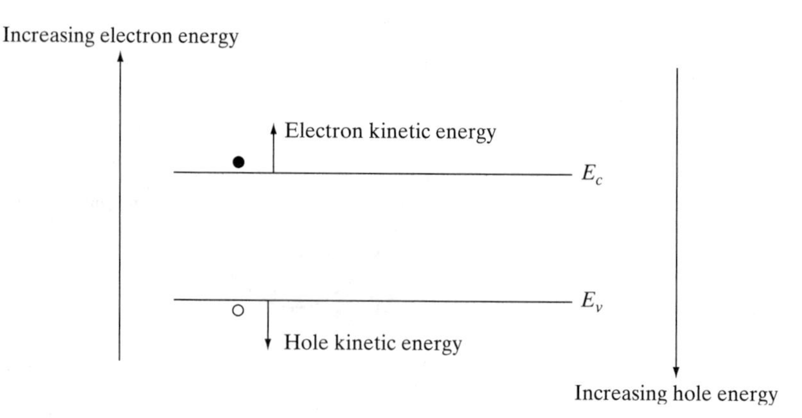

Figure 5.29 Depiction of electron and hole energy.

Nonequilibrium conditions refer to the presence of charge carriers by other means, typically by either direct electrical injection, or by electromagnetic processes (e.g., photons incident on the material). Since the interaction of electromagnetic energy and materials is obviously very important in electrical applications, in this section, a brief survey of several important interaction mechanisms will be provided.

Direct Bandgap Semiconductors. Figure 5.30 shows typical energy band diagrams for a direct bandgap material, where (a) depicts the energy bands in real space and

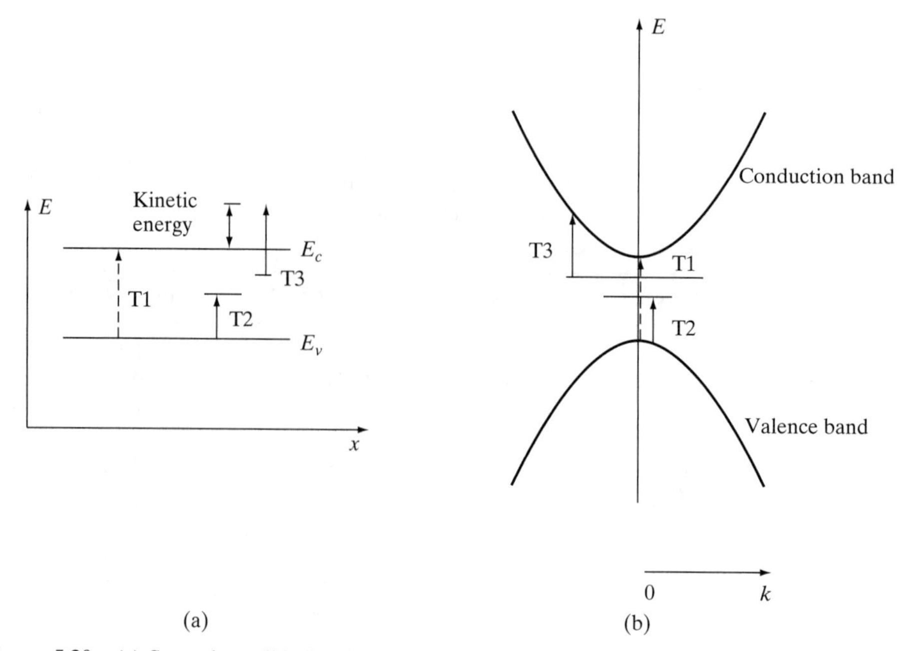

Figure 5.30 (a) Several possible band transitions as shown on the real-space band diagram. (b) The same transitions in wavenumber space.

(b) shows the energy bands in wavenumber space. Three possible transitions arising from the interaction of a photon and the material are depicted, denoted† as T1–T3. In all transitions, we must have conservation of energy and momentum.

T1 involves a transition from the valence bandedge (or below) to the conduction bandedge (or above). It is a *direct transition*, involving absorption of a photon. To elevate an electron from the valence bandedge to the conduction bandedge requires an energy input of $E_g = E_c - E_v$. Energy conservation requires that

$$\hbar\omega = E_g,$$

for a bandedge-to-bandedge transition (at $k = 0$), where $\hbar\omega$ is the energy of the incident photon. Photons having larger energies can cause larger transitions (i.e., perhaps from deeper in the valence band, or to deeper in the conduction band, or both). This can result in particles gaining kinetic energy (the excess energy measured from the bandedge), or energy going into lattice vibrations.

For the direct vertical transition T1 shown, conservation of momentum, which, via de Broglie's relation (2.15) corresponds to conservation of wavevector, states that

$$\Delta k = k_{pt},$$

where Δk is the change in the electron's wavevector, and k_{pt} is the wavevector of the photon. For optical photons, where $\lambda_{pt} \sim 400$–700 nm, the photon wavevector is in the range

$$k_{pt} = \frac{2\pi}{\lambda} \sim 1.5 \times 10^7 - 9.0 \times 10^6 \ m^{-1}.$$

However, electron wavevectors tend to be much larger. For example, for a crystal with lattice constant $a = 0.5$ nm, at the Brillouin zone boundary, the electron's wavevector is

$$k_e = \frac{\pi}{a} = 6.28 \times 10^9 \ m^{-1}.$$

So k_{pt} is rather small on the scale of the first Brillouin zone, and so, approximately, $\Delta k = 0$. Thus, we get the vertical transition shown as T1 in Fig. 5.30(b). This type of transition can be used to measure the bandgap of a direct bandgap semiconductor, by determining the smallest energy of photons that are absorbed by the semiconductor.

Transitions T2 and T3 represent possible transitions associated with impurity states. These impurities can be defects in the lattice, donor or accepter atoms, or other elements that may be present. This leads to the creation of *impurity states* in the bandgap, localized in positions that physically correspond to the location of the impurity. Therefore, the uncertainty in position of the impurity (say, Δx) is small, and, via the uncertainty principle, the uncertainty in Δk will be large. This is why the impurity level is depicted as being spread

†The same types of transitions can occur thermally, as previously discussed. However, considering that room temperature thermal energy is approximately 25 meV, only relatively small thermal transitions are likely.

over a relatively wide range of wavenumbers in Fig. 5.30(b), but over a small range of positions in Fig. 5.30(a).

Of course, although we have considered absorption of a photon† and elevation of an electron to a higher electronic state, the reverse process can also occur. That is, an electron can fall from an elevated state to a lower energy state by emitting a photon. This is the idea behind, for example, laser operation and material fluorescence. In fact, often an electron is raised to a point higher in the conduction band by absorption of a photon, and rapidly falls down to the bottom of the band by emitting quantized lattice vibrations (phonons), a process that is known as *nonradiative relaxation*. The electron then may fall from the bottom of the conduction bandedge to the valence band, by, for example, emitting a photon. Other types of transitions are also possible.

Indirect Bandgap Semiconductors. When one considers the interaction of energy and an indirect bandgap semiconductor, such a Si or Ge, the model changes somewhat. As depicted in Fig. 5.31, the transition from the top of the valence band to the bottom of the conduction band (i.e., the minimum energy transition) cannot be a vertical transition in wavenumber space. Because of the small wavevector associated with photons, photons enable essentially vertical transitions, providing energy but not enough momentum (wavenumber) for the necessary indirect transition. Some other interaction will be necessary for the transition to occur, and the answer is provided by phonons, which provide the necessary change in wavevector. Phonons have relatively large wavevectors, although their energy is typically low, and so phonons result in, essentially, horizontal transitions. Therefore, we have a three "particle" interaction: the photon, the phonon, and the electron. In a classical model, we would view the transition as arising from the interaction of the electromagnetic field, the electron, and the vibrating lattice. In the quantum model, we quantize the electromagnetic field as a photon, we quantize the vibrating lattice as a phonon, and the mutual interaction of these with the electron results in the transition, as depicted in Fig. 5.31. Conservation of energy and momentum results in

$$\Delta E = \hbar \left(\omega_{pt} \pm \omega_{pn} \right), \tag{5.52}$$

$$\Delta k = k_{pt} \pm k_{pn} \simeq \pm k_{pn}, \tag{5.53}$$

where $\hbar\omega_{pt}$ is the energy of an absorbed photon, and $\hbar\omega_{pn}$ is the energy of an absorbed $(+)$ or emitted $(-)$ phonon.

Note that the absorption of the photon alone puts the electron in the band gap, where, based on the previous theory, there are no electron states. A more sophisticated analysis would show that virtual states can exist within the bandgap. These states don't correspond

†It is worthwhile to note that the idea of energy level transitions explains why materials are opaque at some frequencies and transparent at others. If the incident photon energy does not result in any energy transitions (and here we need to consider not just the interaction of photons and electrons, but also the interaction of photons and other quasi particles such as phonons), then the photon is not absorbed. The photon can therefore pass through the material, i.e., the material is transparent to that energy. This explains, for example, why a material that is opaque to visible light will usually be relatively transparent to lower frequency (lower-energy) electromagnetic waves, and thus, although we can't see through walls, we can receive cell phone calls in buildings.

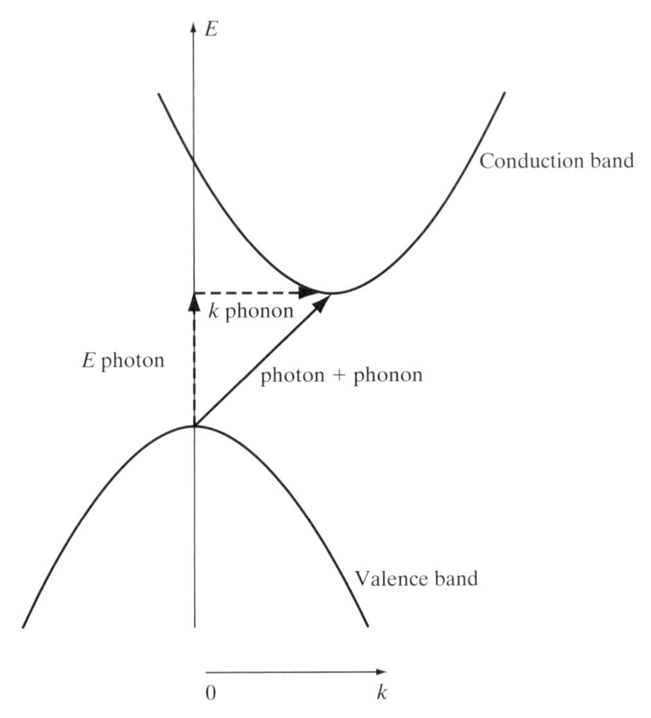

Figure 5.31 Transition in an indirect semiconductor involving a phonon and a photon. Because of the small wavevector associated with the photon, a phonon is required to provide the necessary change in wavevector.

to propagating waves, but decay exponentially with time, so that the electron must interact with a phonon very quickly in order to scatter into the conduction band. This makes the transition rate for indirect transitions much smaller than for direct transitions.

Aside from bandedge-to-bandedge absorption as considered earlier, even in indirect semiconductors electrons can be elevated from the valence band directly to the conduction band at the same k value (i.e., by a vertical transition), although obviously this will occur at relatively high energies (higher then the bandgap energy). Although we won't discuss the transition rate in detail, there is a way of calculating how likely a transition is to occur. This is given by *Fermi's golden rule*, which states that the transition rate is proportional to a factor relating to the wavevectors of the two states in question, and to the joint density of states. (Density of states is discussed in Section 8.1.) In particular, over portions of the E–k plot where conduction and valence bands are essentially parallel, the same photon energy E can result in transitions for many different k values (thus leading to a large joint density of states), and the transition rate will be relatively large. For example, it can be seen from Fig. 5.12 on page 143 that silicon has a conduction and valence band that are approximately parallel, although the separation is on the order of 3–4 eV. Thus, one would expect absorption to greatly increase in the vicinity of photon energies of 3–4 eV, and this is indeed found to be the case. The direct absorption in this energy range is much larger than the indirect absorption near the bandgap energy.

Excitons. We have considered band transitions in semiconductors, and it was implicitly assumed that the process of absorption (of a photon, thermal energy, etc.) created a free electron and a free hole, each of which can contribute to conduction. There is another effect that is worth mentioning, primarily because of its importance in quantum-confined structures such as carbon nanotubes and quantum dots. The basic idea is that after an electron transition, it is possible for the electron and the created hole to be bound together by their mutual Coulomb attraction, forming a quasi particle known as an *exciton*. (See Chapter 9 for a discussion of excitons in quantum-confined structures.)

The two-particle electron-hole exciton can be modeled like the two-particle hydrogen atom considered in Section 4.6.1. However, unlike the hydrogen atom, which consists of one proton and one electron (having greatly different masses) in empty space, here the bound electron–hole pair moves through a material characterized by relative permittivity[†] ε_r, as depicted in Fig. 5.32.

Considering the formulas for energy and Bohr radius of the hydrogen atom, (4.115) and (4.117), and substituting $\varepsilon_r \varepsilon_0$ in place of ε_0 and the reduced mass m_r^*,

$$m_r^* = \frac{m_e^* m_h^*}{m_e^* + m_h^*}, \tag{5.54}$$

in place of m_e, then the binding energy and radius of the ground state exciton are given by

$$E = -\frac{m_r^* q_e^4}{8\varepsilon_r^2 \varepsilon_0^2 h^2 n^2} = -\frac{m_r^*}{m_e \varepsilon_r^2} R_Y = -\frac{m_r^*}{m_e \varepsilon_r^2} 13.6 \text{ eV}, \tag{5.55}$$

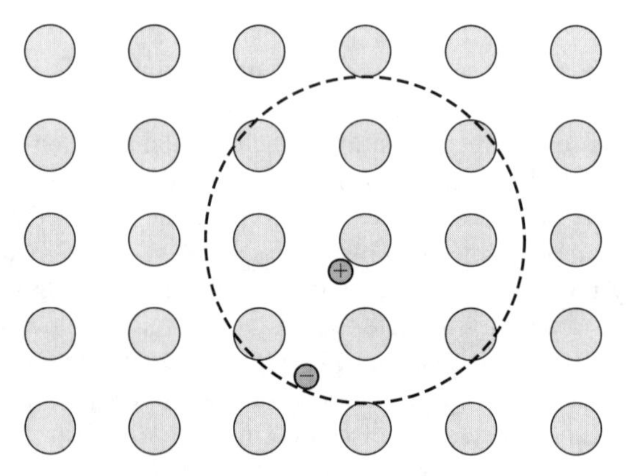

Figure 5.32 Depiction of a bound electron–hole pair known as an exciton. The exciton radius is much larger than the lattice constant.

[†]The macroscopic model of permittivity is appropriate here since the average separation of the electron and the hole is much greater than the lattice constant.

$$a_{ex} = \frac{4\pi\varepsilon_r\varepsilon_0\hbar^2}{m_r^* q_e^2} = \frac{\varepsilon_r m_e}{m_r^*} a_0 = \frac{\varepsilon_r m_e}{m_r^*}\left(0.53\ \text{Å}\right), \tag{5.56}$$

where R_Y is the Rydberg energy and a_0 is the Bohr radius.

As an example, for GaAs ($\varepsilon_r = 13.3$), using an average of the heavy and light hole masses, $m_r^* = 0.0502 m_e$, we find that

$$E = -3.86\ \text{meV}, \tag{5.57}$$

$$a_{ex} = 265 a_0 = 14\ \text{nm}. \tag{5.58}$$

The binding energy of the pair, E, can be easily overcome by thermal effects (e.g., $k_B T \simeq 25$ meV at room temperature), thus breaking the exciton into free electrons and holes. Therefore, in bulk materials, exciton effects are usually only observed at very low temperatures,† and for relatively pure samples (since impurities such as dopants tend to screen the Coulomb interaction, much as occurs in conductors).

To demonstrate the concept of optical absorption by bandgap transitions and excitons, the absorption coefficient‡ versus incident photon energy for GaAs is shown in Fig. 5.33 for various temperatures. At room temperature, one can observe the lack of absorption below

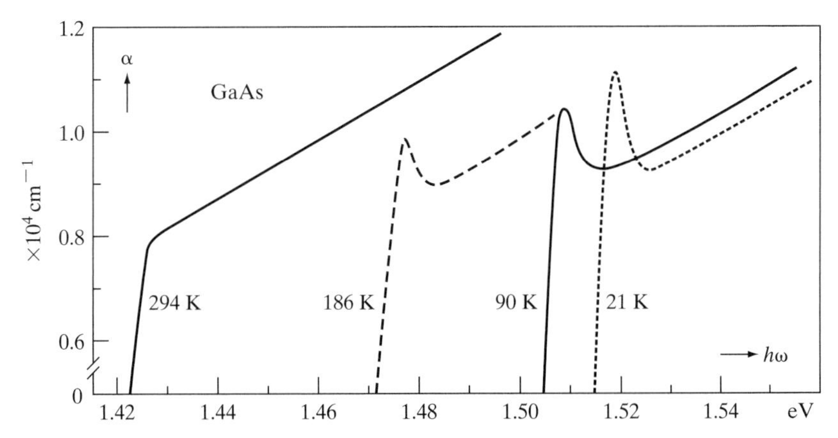

Figure 5.33 Absorption coefficient versus incident photon energy for GaAs at various temperatures. At room temperature, one can observe the lack of absorption below the bandgap. At low temperatures, the peaks at the onset of absorption are due to excitons. (Based on Figure 6.8 (page 55) from *Band Theory and Electronic Properties of Solids* by John Singleton, Oxford University Press, 2001. Reprinted with permission from John Singleton and Oxford University Press. Data from M.D. Sturge, *Phys. Rev.* **127**, 768 (1962).)

†Here we consider "free" excitons, known as *Wannier–Mott excitons*, which occur in semiconductors. Another type, known as *Frenkel excitons*, have a much smaller radius and can be stable at room temperature, although these are not of interest here.

‡The absorption coefficient is defined to be the fraction of power absorbed per unit length of a material.

the bandgap ($E_g = 1.43$ eV). At low temperatures, the peaks at the onset of absorption are due to the creation of excitons.

5.5 GRAPHENE AND CARBON NANOTUBES

Since the relatively recent discovery of *carbon nanotubes* (CNs), there has been an enormous amount of research into their fundamental properties, and great excitement concerning their possible applications. Electronic applications of CNs will be discussed later (e.g., in Sections 6.3.1, 7.3.1 and 10.3), and here we will concentrate on their band structure.

A single-wall carbon nanotube (SWNT) is, roughly speaking, a rolled-up sheet of graphene, which is a mono-atomic layer of graphite. CNs typically have radius values of a few nanometers, and lengths (so far) up to centimeters. Multiwalled carbon nanotubes (MWNTs) are also common, and other related structures exist, such as nanotube ropes (bundles of nanotubes), although here we will focus on SWNTs.

5.5.1 Graphene

At an atomic level, graphene has the periodic honeycomb structure shown in Fig. 5.34, where the small circles denote the location of carbon atoms and the lines represent carbon–carbon bonds. The depicted lattice basis vectors are $\mathbf{a}_1 = \left(\sqrt{3}\mathbf{a}_x + \mathbf{a}_y\right) a/2$ and $\mathbf{a}_2 =$

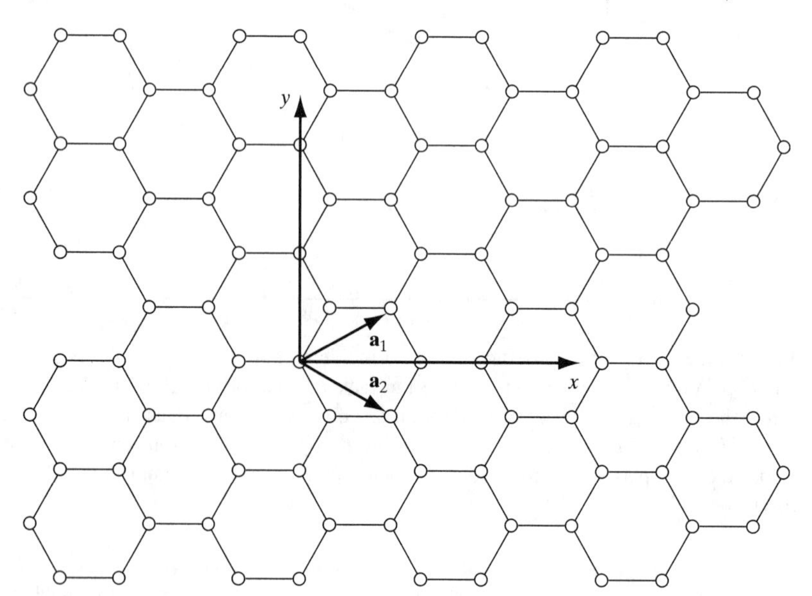

Figure 5.34 Graphene sheet (single layer of graphite). The small circles denote the location of carbon atoms. Lattice basis vectors are $\mathbf{a}_1$ and $\mathbf{a}_2$, as shown.

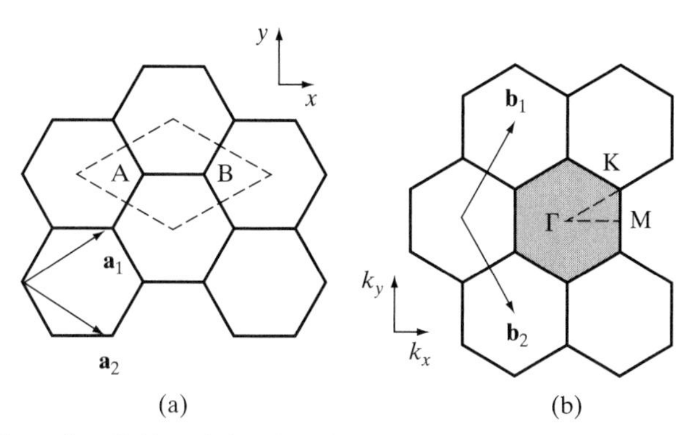

Figure 5.35 (a) The unit cell (dotted rhombus) in graphene. (b) The first Brillouin zone in graphene (shaded region). Both the real-space and reciprocal space structure in graphene consists of hexagons. (Based on a figure in *Physical Properties of Carbon Nanotubes* by R. Saito, G. Dresselhaus, and M.S. Dresselhaus. Singapore: World Scientific Publishing Co. Pte. Ltd., 1998. Used by permission.)

$\left(\sqrt{3}\mathbf{a}_x - \mathbf{a}_y\right) a/2$, where $a = \sqrt{3}b$ and $b = 0.142$ nm is the interatomic distance between carbon atoms in graphene.

Since graphene is a two-dimensional periodic material, it has an energy band structure similar to the three-dimensional crystalline solids discussed previously. The unit cell and first Brillouin zone are shown in Fig. 5.35, where the high-symmetry points in the Brillouin zone are

$$\Gamma = (0, 0), \quad \mathbf{K} = \left(\frac{2\pi}{\sqrt{3}a}, \frac{2\pi}{3a}\right), \quad \mathbf{M} = \left(\frac{2\pi}{\sqrt{3}a}, 0\right). \tag{5.59}$$

The Fermi surface is really a collection of points, the six points of the hexagonal Brillouin zone, where $E = E_F = 0$. (See problem 5.21.)

The most important bands arise from the so-called $\pi-$orbitals, and for these bands, the two-dimensional E–k relationship is well approximated by

$$E\left(k_x, k_y\right) = \frac{\mp\gamma_0 w\left(k_x, k_y\right)}{1 \pm s w\left(k_x, k_y\right)}, \tag{5.60}$$

where

$$w\left(k_x, k_y\right) = \sqrt{1 + 4\cos\left(\frac{\sqrt{3}k_x a}{2}\right)\cos\left(\frac{k_y a}{2}\right) + 4\cos^2\left(\frac{k_y a}{2}\right)}, \tag{5.61}$$

$s \simeq 0.129$, $\gamma_0 \simeq 3$ eV, and where the upper and lower signs correspond to the bonding/antibonding bands. (See Section 5.4.2.) In the important vicinity of the Fermi points, this can be approximated as

$$E\left(k_x, k_y\right) = \pm\gamma_0 w\left(k_x, k_y\right), \tag{5.62}$$

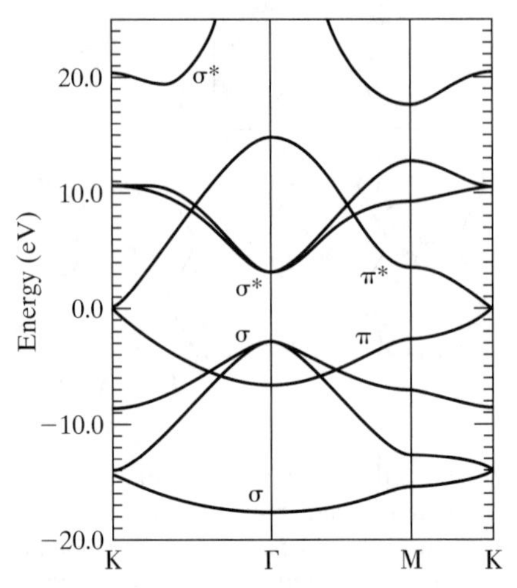

Figure 5.36 Calculated bandstructure of the π- and σ-bands in graphene. The Γ point is $\mathbf{k} = (0, 0)$, and the K and M points are shown in Fig. 5.35 (Based on a figure in *Physical Properties of Carbon Nanotubes* by R. Saito, G. Dresselhaus, and M.S. Dresselhaus. Singapore: World Scientific Publishing Co. Pte. Ltd., 1998.)

and from either (5.60) or (5.62), it is easy to plot the energy band behavior of the π electrons. (See problem 5.21.) The π orbitals arise from electrons that are weakly bound to the carbon atoms. It is these electrons that are of most importance in electronic applications, since a small amount of energy can free them for conduction. Electrons more strongly held to the carbon atoms are known as σ electrons, and these play an important role at higher energies. The band structure of both π- and σ-electrons in graphene is shown in Fig. 5.36.

It is particularly important to note the nearly linear dispersion of the π bands near the K point. Hence, rather then the usual parabolic dispersion for an electron in free space or in a material, $E = \hbar^2 k^2/2m^*$, dispersion for electrons near the K point in graphene is

$$E = v_F \hbar k, \tag{5.63}$$

where $v_F \simeq 9.71 \times 10^5$ m/s is the Fermi velocity of electrons (to be discussed later). For photons, from (2.25),

$$E = c\hbar k, \tag{5.64}$$

where c is the speed of light, and so near the π-band crossing point electrons act more like photons than particles with mass. In this case the electrons are called *Dirac fermions*, and their properties are of interest in recently-developed graphene devices.

5.5.2 Carbon Nanotubes

A carbon nanotube is formed by wrapping the graphene sheet into a cylinder,† as depicted in Fig. 5.37, by connecting points O and A and B and B'. The circumference of the tube is related to the length of the *chiral vector*, $\mathbf{C}_h = n\mathbf{a}_1 + m\mathbf{a}_2$, where n, m are integers.

†Carbon nanotubes form naturally in, for example, the arc discharge of carbon electrodes, and are not made by literally rolling graphene sheets into cylinders.

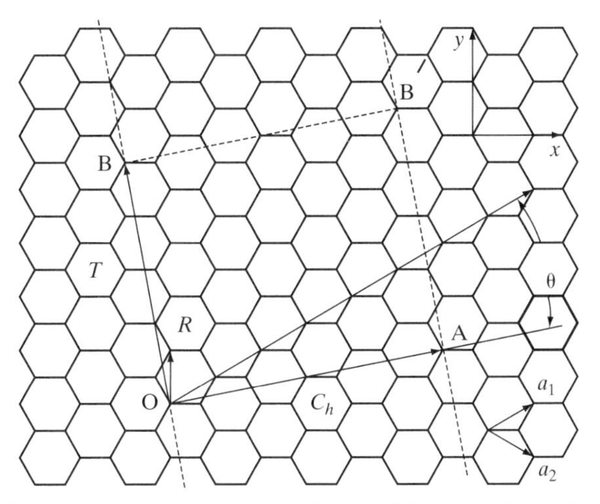

Figure 5.37 Depiction of forming a carbon nanotube by rolling a graphene sheet into a tube. (Based on a figure in *Physical Properties of Carbon Nanotubes* by R. Saito, G. Dresselhaus, and M.S. Dresselhaus. Singapore: World Scientific Publishing Co. Pte. Ltd., 1998. Used by permission.)

It is obvious that the CN cylinder can be formed by wrapping the graphene sheet along any preferred axis. If the cylinder axis is the x axis in Fig. 5.34, the resulting tube is called a *zigzag* CN because at an open end of the tube the carbon–carbon bonds would form a zigzag pattern. If the cylinder axis is the y axis in Fig. 5.34, the resulting tube is called an *armchair* CN, and if the cylinder axis is neither the x nor the y axis as shown, the resulting nanotube is called a *chiral* CN. Thus, carbon nanotubes can be characterized by the dual index (n, m), where $(n, 0)$ for zigzag CNs, (n, n) for armchair CNs, and (n, m), $0 < m \neq n$, for chiral nanotubes, as depicted in Fig. 5.38.

The values of n and m denote, respectively, the number of unit vectors $\mathbf{a}_1$ and $\mathbf{a}_2$ required to make the tube. The resulting cross-sectional radius of a carbon nanotube is given by†

$$r = \frac{|\mathbf{C}_h|}{2\pi} = \frac{\sqrt{3}}{2\pi} b\sqrt{n^2 + nm + m^2}. \tag{5.65}$$

An armchair, zigzag, and chiral CN are shown in Fig. 5.39. Also, shown in the figure is a multi-wall CN.

Since graphene is a two-dimensional periodic structure, upon forming an infinite tube, we have a periodic structure in the axial direction and a finite structure in the transverse direction. It can be shown that the lattice constants are $a_{ac} = a = \sqrt{3}b$ for armchair tubes and $a_{zz} = \sqrt{3}a = 3b$ for zigzag tubes, and, therefore, the edge of the first Brillouin zone occurs at $k_y = \pi/a_{ac}$ for armchair tubes, and $k_x = \pi/a_{zz}$ for zigzag tubes. In the transverse direction, the wavenumber becomes quantized by the finite circumference of the tube, and

†The interatomic distance $b = 0.142$ nm in graphene becomes slightly larger in a carbon nanotube, and the value $b = 0.144$ nm is often used.

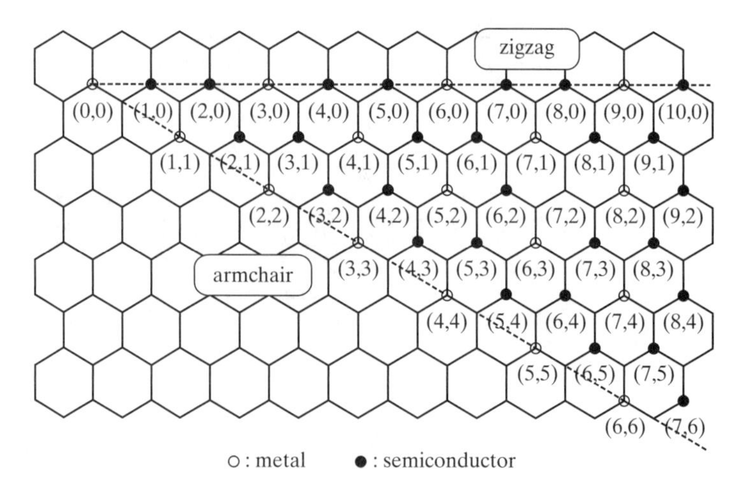

Figure 5.38 Forming armchair, zigzag, and chiral carbon nanotubes, denoted by the index (n, m). Metal and semiconducting tubes are denoted by hollow and solid circles, respectively. (Based on a figure in *Physical Properties of Carbon Nanotubes* by R. Saito, G. Dresselhaus, and M.S. Dresselhaus. Singapore: World Scientific Publishing Co. Pte. Ltd., 1998. Used by permission.)

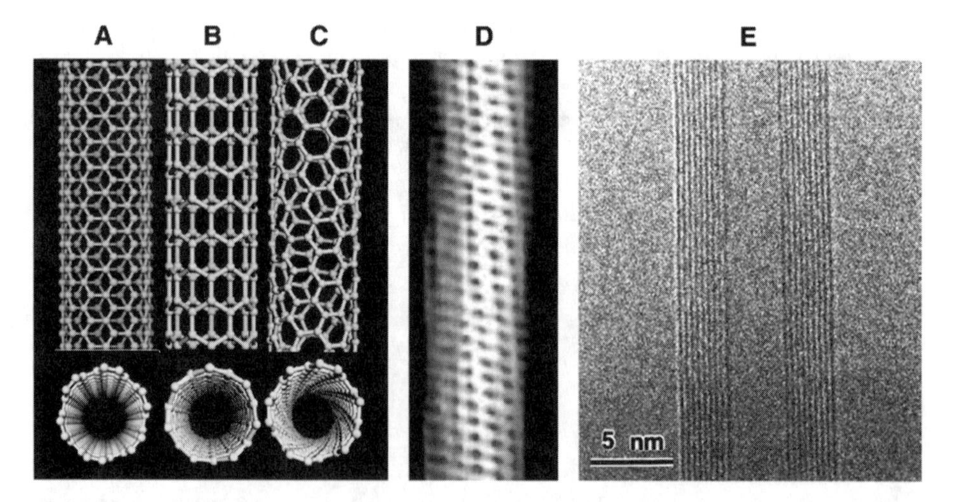

Figure 5.39 (A) armchair CN, (B) zigzag CN, (C) chiral CN, (D) TEM image of a 1.3 nm diameter chiral CN, (E) TEM image of a multi-wall CN, consisting of nine concentric single-wall CNs. (Courtesy Shenzhen Nano-Technologies Port Co., Ltd.)

is given by

$$k_{\perp} = k_{x,q} = \frac{2\pi q}{n3b}, \qquad q = 1, 2, \ldots, 2n \tag{5.66}$$

for the armchair tube $(m = n)$, (see problem 5.23) and

$$k_{\perp} = k_{y,q} = \frac{2\pi q}{n\sqrt{3}b}, \qquad q = 1, 2, \ldots, 2n \tag{5.67}$$

for the zigzag tube. The $E - k$ relations for the π electrons in a nanotube result from plugging (5.66) or (5.67) into (5.62), such that for armchair tubes we have[†]

$$E_{ac}\left(k_y\right) = \pm\gamma_0\sqrt{1 + 4\cos\left(\frac{\pi q}{n}\right)\cos\left(\frac{k_y a}{2}\right) + 4\cos^2\left(\frac{k_y a}{2}\right)}, \tag{5.68}$$

$-\pi < k_y a < \pi$, $q = 1, 2, \ldots, 2n$, and for zigzag tubes,

$$E_{zz}\left(k_x\right) = \pm\gamma_0\sqrt{1 + 4\cos\left(\frac{\sqrt{3}k_x a}{2}\right)\cos\left(\frac{\pi q}{n}\right) + 4\cos^2\left(\frac{\pi q}{n}\right)}, \tag{5.69}$$

$-\pi < k_x\sqrt{3}a < \pi$, $q = 1, 2, \ldots, 2n$. These are approximate relations, since, among other things, the effect of tube curvature on atomic bonds is ignored, although (5.68) and (5.69) turn out to be generally quite accurate. The dispersion behavior is considered in problem 5.24, although in general it is found that carbon nanotubes can be either metallic or semiconducting, depending on their geometry (i.e., on n, m).[‡] Armchair CNs are always metallic (they exhibit no energy bandgap), as are zigzag CNs with $n = 3p$, where p is an integer. Chiral tubes can be either metallic or semiconducting depending on the (n, m) values, and a general rule is that tubes where $(2n + m)/3$ is an integer are metallic.[§] Semiconducting tubes have energy bandgaps given by

$$E_g \simeq \frac{\gamma_0 a}{2\sqrt{3}r} \simeq \frac{0.383}{d_t}\text{ eV}, \tag{5.70}$$

where d_t is the tube radius in nm.

Carbon nanotubes are held together by the carbon–carbon bonds, and thus they exhibit extraordinary strength. In fact, CNs are many times stronger than steel. They also have very high thermal conductivity and stiffness. Their rather extraordinary properties are in part due to the fact that they can be grown to be nearly defect free. (However, see the discussion in Section 10.3.) For example, metals tend to exhibit failure well below their theoretical limits,

[†]For graphene, $\gamma_0 \simeq 3$ eV. For carbon nanotubes, the corresponding value changes due to curvature effects, and is usually taken to be in the range 2.5–3.0 eV.

[‡]Graphene itself is a *semimetal* (also called a zero bandgap semiconductor), exhibiting properties between a metal and a semiconductor. This is why, merely depending on how the cylinder is wrapped, the resulting tube can be either metallic or semiconducting. Semiconducting CNs have bandgaps ranging from a few meV up to, on the order of, an eV, and metallic CNs tend to have very high conductivities.

[§]The described characterization of tubes as being metallic or semiconducting is based on the simple idea of rolling up a graphene sheet without, essentially, changing its atomic structure. This is valid for relatively large-radius tubes (above perhaps a nanometer), although for small-radius tubes the large radius of curvature hybridizes the π– and σ–orbitals, and deviations to the presented theory are encountered. For example, whereas a (5, 0) tube should be semiconducting based on the simple analysis, it is actually found to be metallic. Moreover, some zigzag tubes that should be metallic by the simple mode are actually semiconducting. See, e.g., [16].

Furthermore, the analysis assumes infinitely long tubes. Finite length tubes having lengths less then approximately 10 nm tend to act more like quantum dots (zero-dimensional structures) rather than quantum wires (one-dimensional structures).

due to defects. High quality (nearly molecularly perfect) CNs can be formed that perform near to their theoretical limit, which results in extremely attractive mechanical and electrical properties. Carbon nanotubes are the stiffest known fibers, and exhibit the highest tensile strength of any known material. The Young's modulus of CNs have been measured to be more than an order of magnitude larger than that for steel. Furthermore, carbon nanotubes can carry very high current densities (much higher than typical metals), at 10^9 A/cm^2 or more, without melting. For comparison, copper is generally limited to 10^6 A/cm^2 due to heating and electromigration effects.

For the semiconducting tubes, electronic properties can be controlled by doping, as in a conventional semiconductor. Doping can be introduced chemically by exposing the CN to certain elements, or by inserting molecules inside the tube. One inherent difference between semiconducting CNs and ordinary semiconducting materials is that at room temperature, the undoped CN tends to be a p-type material. It was first thought that this behavior was due to chemical contamination, such as adsorbed oxygen. However, it is now thought that the origin of the p-type behavior is due to what is known as *self-doping*, caused by the curvature of the tube at the nanoscale and the commensurate effect on atomic bonding.

Carbon nanotubes can currently be fabricated using a variety of techniques including carbon arc discharge, laser evaporation, and chemical vapor deposition. Often, however, a mixture of semiconducting and metallic tubes are produced, which must be separated and isolated for use. Making good electrical contact between carbon nanotubes and electrodes can be problematic, as is positioning of the tubes in device fabrication.

5.6 MAIN POINTS

This chapter presents some important concepts from solid state physics, and, in particular, the formation of energy bands in periodic structures. After studying this chapter you should know

- the principles of various crystal structures;
- the effect of a periodic potential on electron properties, and the Kronig–Penney model;
- the band theory of solids;
- the concept of effective mass, and what effective mass accounts for, including the use of effective mass in Schrödinger's equation;
- the effect of an electric field or of an applied electrical potential on energy band structure;
- energy band models for typical semiconductors, including the concept of direct and indirect bandgaps;
- the interaction of electromagnetic energy with an energy band system, including the role of vibrational modes (phonons);
- the basic π band structure of graphene;
- the band structure of π electrons in carbon nanotubes.

5.7 PROBLEMS

1. To gain an appreciation of the important role of surface effects at the nanoscale, consider building up a material out of bcc unit cells. (See Section 5.1.) For one bcc cube, there would be nine atoms, eight on the outside and one interior, as depicted on page 134. If we constrain ourselves to only consider cubes of material, the next largest cube would consist of eight bcc unit cells, and so on. If one side of the bcc unit cell is 0.5 nm, how long should the material's side be in order for there to be more interior atoms than surface atoms?
2. Consider the Kronig–Penney model of a material with $a_1 = a_2 = 5$ Å and $V_0 = 0.5$ eV. Determine numerically the starting and ending energies of the first allowed band.
3. Use the equation of motion (5.34) to show that the period of Bloch oscillation for a one-dimensional crystal having lattice period a is

$$\tau = \frac{h}{e\mathcal{E}a}, \tag{5.71}$$

where $\mathcal{E}$ is the magnitude of the applied electric field.

4. Determine the probability current density (A/m) from (3.187) for the Bloch wavefunction

$$\psi(x) = u(x)\, e^{ikx} e^{-i\omega t}, \tag{5.72}$$

where u is a time-independent periodic function having the period of the lattice,

$$u(x) = u(x + a). \tag{5.73}$$

5. If an energy–wavevector relationship for a particle of mass m has the form

$$E = \frac{\hbar^2}{3m} k^2, \tag{5.74}$$

determine the effective mass. (Use (5.29).)

6. If the energy–wavenumber relationship for an electron in some material is

$$E = \frac{\hbar^2}{2m} \cos(k), \tag{5.75}$$

determine the effective mass and the group velocity. (Use (5.29).) Describe the motion (velocity, direction, etc.) of an electron when a d.c. (constant) electric field is applied to the material, such that the electric field vector points right to left (e.g., an electron in free space would then accelerate towards the right). In particular, describe the motion as k varies from 0 to 2π. Assume that the electron does not scatter from anything.

7. If the energy–wavenumber relationship for an electron in some material is

$$E = E_0 + 2A\cos(ka), \tag{5.76}$$

determine the electron's position as a function of time. Ignore scattering.

8. Consider an electron in a perfectly periodic lattice, wherein the energy–wavenumber relationship in the first Brillouin zone is

$$E = \frac{\hbar^2 k^2}{5m_e}, \tag{5.77}$$

where m_e is the mass of an electron in free space. Write down the time-independent effective mass Schrödinger's equation for one electron in the first Brillouin zone, ignoring all interactions except between the electron and the lattice. Define all terms in Schrödinger's equation.

9. Assume that a constant electric field of strength $\mathcal{E} = -1$ kV/m is applied to a material at $t = 0$, and that no scattering occurs.

(a) Solve the equation of motion (5.34) to determine the wavevector value at $t = 1, 3, 7,$ and 10 ns.

(b) Assuming that the period of the lattice is $a = 0.5$ nm, determine which Brillouin zone the wavevector is in at each time. If the wavevector lies outside the first Brillouin zone, map it into an equivalent place in the first zone.

10. Using the hydrogen model for ionization energy, determine the donor ionization energy for GaAs ($m_e^* = 0.067m_e$, $\varepsilon_r = 13.1$).

11. Determine the maximum kinetic energy that can be observed for emitted electrons when photons having $\lambda = 232$ nm are incident on a metal surface with work function 5 eV.

12. Photons are incident on silver, which has a work function $e\phi = 4.8$ eV. The emitted electrons have a maximum velocity of 9×10^5 m/s. What is the wavelength of the incident light?

13. In the band theory of solids, there are an infinite number of bands. If, at $T = 0$ K, the uppermost band to contain electrons is partially filled, and the gap between that band and the next lowest band is 0.8 eV, is the material a metal, an insulator, or a semiconductor?

14. In the band theory of solids, if, at $T = 0$ K, the uppermost band to have electrons is completely filled, and the gap between that band and the next lowest band is 8 eV, is the material a metal, an insulator, or a semiconductor? What if the gap is 0.8 eV?

15. Describe in what sense an insulator with a finite band gap cannot be a perfect insulator.

16. Draw relatively complete energy band diagrams (in both real space and momentum space) for a p-type indirect bandgap semiconductor.

17. For an intrinsic direct bandgap semiconductor having $E_g = 1.72$ eV, determine the required wavelength of a photon that could elevate an electron from the top of the valence band to the bottom of the conduction band. Draw the resulting transition on both types of energy band diagrams (i.e., energy–position and energy–wavenumber diagrams).

18. Determine the required phonon energy and wavenumber to elevate an electron from the top of the valence band to the bottom of the conduction band in an indirect bandgap semiconductor. Assume that $E_g = 1.12$ eV, the photon's energy is $E_{pt} =$

0.92 eV, and that the top of the valence band occurs at $k = 0$, whereas the bottom of the conduction band occurs at $k = k_a$.

19. Calculate the wavelength and energy of the following transitions of an electron in a hydrogen atom. Assuming that energy is released as a photon, using Table 1, on page 4 classify the emitted light (e.g., x-ray, infrared (IR), etc.).

(a) $n = 2 \rightarrow n = 1$
(b) $n = 5 \rightarrow n = 4$
(c) $n = 10 \rightarrow n = 9$
(d) $n = 8 \rightarrow n = 2$
(e) $n = 12 \rightarrow n = 1$
(f) $n = \infty \rightarrow n = 1$

20. Excitons were introduced in Section 5.4.5 to account for the fact that sometimes when an electron is elevated from the valence band to the conduction band, the resulting electron and hole can be bound together by their mutual Coulomb attraction. Excitonic energy levels are located just below the band gap, since the usual energy to create a free electron and hole, E_g, is lessened by the binding energy of the exciton. Thus, transitions can occur at

$$E = E_g - \frac{m_r^*}{m_e \varepsilon_r^2} 13.6 \text{ eV}, \tag{5.78}$$

where E_g is in electron volts.†

(a) For GaAs, determine the required photon energy to create an exciton. For m_r^*, use the average of the heavy and light hole masses.
(b) The application of a d.c. electric field tends to separate the electron and the hole. Using Coulomb's law, show that the magnitude of the electric field between the electron and the hole is

$$|\mathcal{E}| = \left(\frac{m_r^*}{m_e}\right)^2 \frac{2}{\varepsilon_r^3 \, |q_e|} \frac{R_Y}{a_0}. \tag{5.79}$$

(c) For GaAs, determine $|\mathcal{E}|$ from (5.79). Determine the magnitude of an electric field that would break apart the exciton.

21. The E–k relationship for graphene is given by (5.62). The Fermi energy for graphene is $E_F = 0$, and the first Brillouin zone forms a hexagon (as shown in Fig. 5.35), the six corners of which correspond to $E = E_F = 0$. The six corners of the first Brillouin zone are located at

$$k_x = \pm\frac{2\pi}{\sqrt{3}a}, \quad k_y = \pm\frac{2\pi}{3a}, \tag{5.80}$$

and

$$k_x = 0, \quad k_y = \pm\frac{4\pi}{3a}. \tag{5.81}$$

†The quantity 13.6 should really be replaced by $13.6/n^2$, where n is the energy level of the exciton. Here we consider the lowest level exciton ($n = 1$), which is dominant.

(a) Verify that at these points, $E = E_F = 0$.

(b) At the six corners of the first Brillouin zone, $|\mathbf{k}| = 4\pi/3a$. Make a two-dimensional plot of the E–k relationship for k_x, k_y extending a bit past $|\mathbf{k}|$. Verify that the bonding and antibonding bands touch at the six points of the first Brillouin zone hexagon, showing that graphene is a semimetal (sometimes called a zero bandgap semiconductor). Also make a one-dimensional plot of $E\left(0, k_y\right)$ for $-|\mathbf{k}| \leq k_y \leq |\mathbf{k}|$, showing that the bands touch at $E = 0$ at $k_y = \pm 4\pi/3a$.

22. What is the radius of a $(19, 0)$ carbon nanotube? Repeat for a $(10, 10)$ nanotube. Consider an $(n, 0)$ zigzag carbon nanotube that has radius 0.3523 nm. What is the value of the index n?

23. Since carbon nanotubes are only periodic along their axis, the transverse wavenumber becomes quantized by the finite circumference of the tube. Derive (5.66) and (5.67) by enforcing the condition that an integer number q of transverse wavelengths must fit around the tube ($k_\perp = 2\pi/\lambda_\perp$).

24. Using (5.68) and (5.69), plot the dispersion curves for the first eight bonding and antibonding bands in a $(5, 5)$, $(9, 0)$, and $(10, 0)$ carbon nanotube. Let the axial wavenumber vary from $k = 0$ to $k = \pi/a_{ac}$ for the armchair tube, and from $k = 0$ to $k = \pi/a_{zz}$ for the zigzag tube. Comment on whether each tube is metallic or semiconducting, and identify the band (i.e., the q value) that is most important. If the tube is semiconducting, determine the approximate band gap.

Part II

Single-Electron and Few-Electron Phenomena and Devices

In previous chapters, Schrödinger's equation and the principles of quantum physics were developed, with an emphasis on single particles (primarily electrons) and collections of noninteracting particles in different spatial regions. The remainder of the text is divided into two parts, and presents some basic nanoelectronic applications of these principles. In the next part, we will be concerned with physical phenomena associated with single electrons, or small numbers of electrons (perhaps, say, 10^0–10^5 electrons). The main emphasis is on electrons confined to nanoscopic spaces, such as quantum dots, and devices constructed from quantum dots and "charge islands." Nanoelectronics principles are developed for the so-called "single-electron" devices, including the single-electron transistor, after the important concept of Coulomb blockade has been discussed. Although most single-electron devices are at an early stage of development, especially in the area of manufacturability, they offer the potential benefits of ultralarge scale integration, with device dimensions on the order of nanometers. They also may exhibit very low power dissipation, and high speed. All of these positive attributes arise from the need to move only single electrons, or small groups of electrons, through devices.

The use of the term "single-electron" device merits some discussion. In conventional microelectronics, currents are typically on the order of 1 μA to 1 mA, corresponding to the movement of 6.25×10^6–6.25×10^9 electrons per microsecond. This occurs through a device perhaps 100 nm in length. Even considering devices at the upper limit of optical lithography, perhaps on the order of 10^5 electrons are involved in performing, for example, a digital operation. In the following chapters, conversely so-called "single-electron" devices are studied. In fact, this is a bit of a misnomer, and in the literature the term "single-electron precision" device has been suggested as a more descriptive name. This allows for the fact that usually much more than one electron is involved, although the number may be relatively small, perhaps 10–10, 000 electrons. It is important to note that these devices are typically sensitive to the transfer of a single electronic charge, and therefore they can operate by manipulating an extremely small number of electrons. However, this generally positive attribute has it "dark side" as well. For example, if a device is sensitive to the movement of a single electronic charge, then the presence of a single charge impurity, in, say, an oxide layer, may drastically influence device operation.

Chapter

6

TUNNEL JUNCTIONS AND APPLICATIONS OF TUNNELING

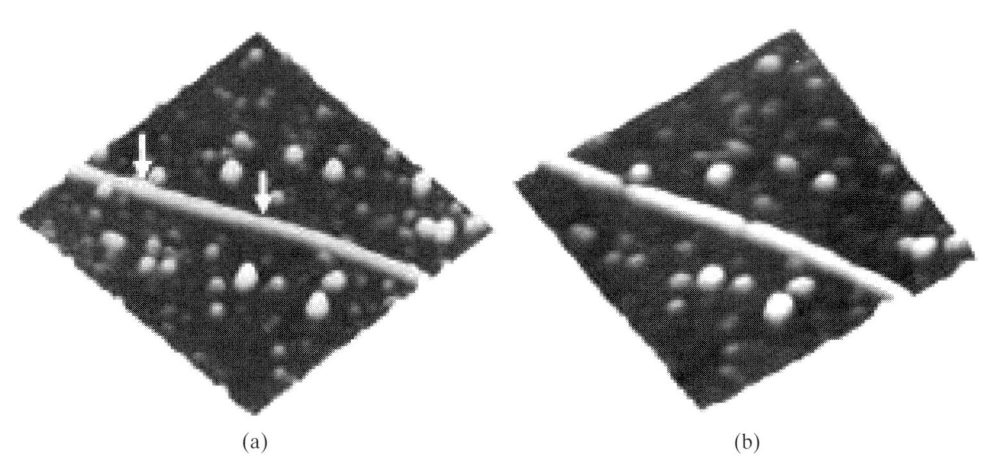

(a) (b)

Quantum dot formed by cutting a carbon nanotube. Transport through the dot is via tunneling. Image size is 500 × 500 nm^2, and the height of the tube is 3 nm. (From Park, J.-Y., et al., "Electrical Cutting and Nicking of Carbon Nanotubes Using an Atomic Force Microscope." *Appl. Phys. Lett.*, 80 (2002): 4446. © 2002, American Institute of Physics.)

In Chapter 4, the general concept of a quantum dot was introduced. A very important aspect to consider is the connection of the quantum dot to the "outside world," or, alternatively, the interaction of a quantum dot with the outside world. In the second case, we often want to "interrogate" the dot remotely. For example, we may want the dot to glow (i.e., emit photons) when illuminated with radiation, in order to be used as a marker, for, say, locating a cancer cell to which the dot is attached. We may even want the dot to cause sufficient heating of the cell to kill it, which has, in fact, already been used in the treatment of skin cancer. This type of application will be further described in Chapter 9.

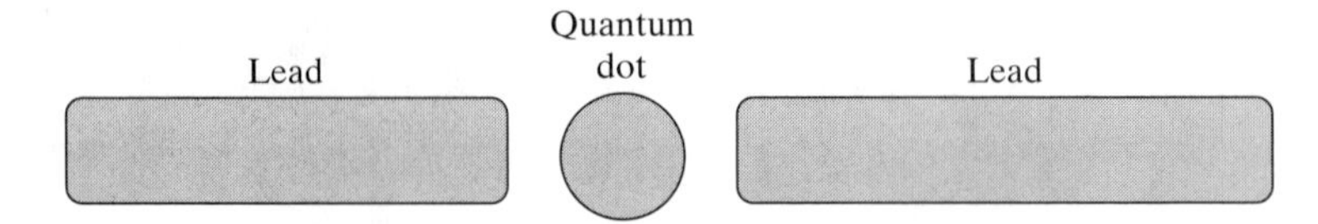

Figure 6.1 Nano-object coupled to external leads.

In this chapter and the next, we want to consider the first scenario, where we "connect" a quantum dot to wires via tunnel junctions, in order to form an electronic device such as a transistor. Not only is this a practical issue, but also, as it turns out, it is one that leads to interesting phenomena and useful applications. In particular, we want to study a method of "communicating" with a nanoscopic object by bringing electrical leads into close proximity to, but not making contact with, the object. This is depicted schematically in Fig. 6.1, which shows a quantum dot separated from two leads by an insulating region.

Although the leads do not contact the object, d.c. electrical current can pass through the system if the gap between the leads and the dot is sufficiently small, despite the fact that the gap is modeled as a perfect insulator. Indeed, the connection between the object and the outside world (i.e., the leads) is by a process known as *quantum tunneling*, or simply tunneling, and in this chapter we consider tunneling in a general sense. In the next chapter, the concept of tunneling is applied to the interaction of the dot and electrical leads, and to related structures such as the single electron transistor.

6.1 TUNNELING THROUGH A POTENTIAL BARRIER

The topic of tunneling is very important for nanoelectronic devices, and is used fruitfully in a large number of applications. We will consider some of these in the next few sections; however, we first consider a general tunneling problem.

To investigate tunneling, consider a particle such as an electron, incident from the left on a potential energy barrier, as depicted in Fig. 6.2.

The potential energy profile is given by

$$V = \begin{cases} V_0, & 0 \leq x \leq a, \\ 0, & x < 0, \; x > a, \end{cases} \tag{6.1}$$

which models, for example, the energy profile in a metal–insulator–metal junction, as discussed subsequently. Other, qualitatively similar potential energy profiles model an electron bound to an atom or molecule, an electron bound to a quantum dot, and similar confinement structures. We make the assumption that the barrier does not contain any scattering objects, so that particles transverse the barrier coherently.† This assumption allows one to

†Certain collisions between particles would lead to incoherent transport, as further discussed in Chapter 10.

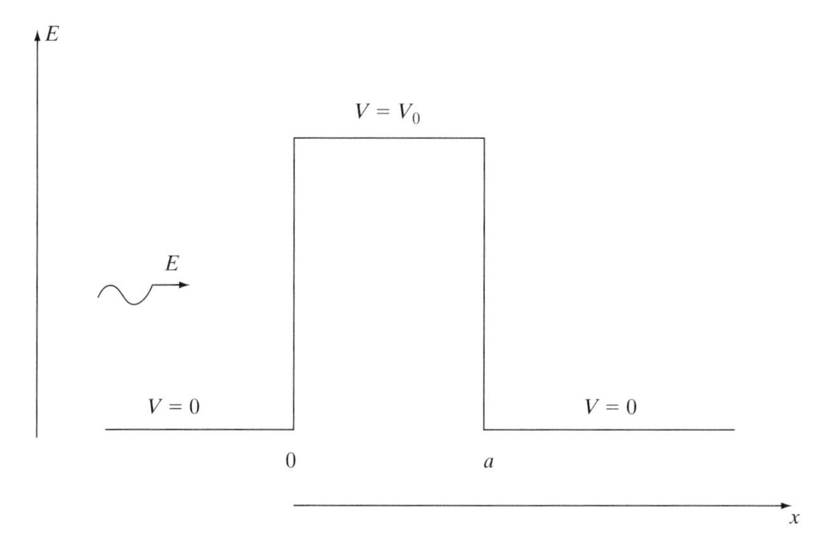

Figure 6.2 Particle with energy E incident on a potential barrier.

solve Schrödinger's equation with boundary/continuity conditions applied at the barrier's interfaces, rather than at various points inside the barrier.

We are interested in two cases; when the particle has total energy $E > V_0$, and when $E < V_0$. According to classical physics, for $E > V_0$, the particle will simply move past the potential barrier. This would happen with 100 percent certainty. (Its transmission coefficient would be unity, and its reflection coefficient would be zero.) For the case $E < V_0$, the particle would be reflected from the barrier with 100 percent certainty. (Its transmission coefficient would be zero, and its reflection coefficient would be unity.) However, quantum mechanics shows a more complicated behavior.

Starting with Schrödinger's equation,

$$\left(-\frac{\hbar^2}{2m^*}\frac{d^2}{dx^2} + V(x)\right)\psi(x) = E\psi(x), \tag{6.2}$$

we note that it is difficult to solve (6.2) when $V(x)$ varies as a function of position (i.e., in this case, Schrödinger's equation is a nonconstant coefficient differential equation). If $V(x)$ is a piecewise constant function, the usual method to avert this problem is to solve Schrödinger's equation separately in each region where V is constant, and to connect the solutions using the boundary conditions for the wavefunction, (3.143).

Proceeding in this manner, in region I ($x < 0$) where $V = 0$, we find that Schrödinger's equation is

$$-\frac{\hbar^2}{2m^*}\frac{d^2}{dx^2}\psi_1(x) = E\psi_1(x), \tag{6.3}$$

which has solutions

$$\psi_1(x) = Ae^{ik_1x} + Be^{-ik_1x}, \tag{6.4}$$

where

$$k_1^2 = \frac{2m^* E}{\hbar^2}. \tag{6.5}$$

In region II ($0 \le x \le a$) where $V = V_0$, Schrödinger's equation is

$$\left(-\frac{\hbar^2}{2m^*}\frac{d^2}{dx^2} + V_0\right)\psi_2(x) = E\psi_2(x), \tag{6.6}$$

which has solutions

$$\psi_2(x) = Ce^{ik_2 x} + De^{-ik_2 x}, \tag{6.7}$$

where

$$k_2^2 = \frac{2m^*(E - V_0)}{\hbar^2}. \tag{6.8}$$

Note that for the case $E \lesseqgtr V_0$,

$$k_2^2 = \frac{2m^*(E - V_0)}{\hbar^2} \lesseqgtr 0, \tag{6.9}$$

and so k_2 is either pure imaginary or real valued.

Lastly, in region III ($x > a$), Schrödinger's equation is

$$-\frac{\hbar^2}{2m^*}\frac{d^2}{dx^2}\psi_3(x) = E\psi_3(x), \tag{6.10}$$

which has solutions

$$\psi_3(x) = Fe^{ik_3 x} + Ge^{-ik_3 x}, \tag{6.11}$$

where

$$k_3^2 = \frac{2m^* E}{\hbar^2} = k_1^2. \tag{6.12}$$

Since there is no potential disturbance to reflect the wave after it reaches region III, $G = 0$.

Therefore, we have

$$\begin{aligned} \psi_1(x) &= Ae^{ik_1 x} + Be^{-ik_1 x}, \\ \psi_2(x) &= Ce^{ik_2 x} + De^{-ik_2 x}, \\ \psi_3(x) &= Fe^{ik_1 x}. \end{aligned} \tag{6.13}$$

Note that since in region III $|\psi_3|^2$ is constant, the particle is equally likely to be found at any point in this region.

The boundary conditions (3.143), and continuity of ψ and ψ' at $x = 0$ and at $x = a$, lead to

$$\frac{B}{A} = \frac{\left(k_1^2 - k_2^2\right)\left(1 - e^{i2ak_2}\right)}{(k_1 + k_2)^2 - (k_1 - k_2)^2 e^{i2ak_2}},$$
$$\frac{F}{A} = \frac{4k_1 k_2 e^{i(k_2 - k_1)a}}{(k_1 + k_2)^2 - (k_1 - k_2)^2 e^{i2ak_2}}. \tag{6.14}$$

We can define a *tunneling probability* as

$$T = \left|\frac{F}{A}\right|^2 = \frac{4E(E - V_0)}{V_0^2 \sin^2(k_2 a) + 4E(E - V_0)}, \tag{6.15}$$

which is obviously the modulus squared of the ratio of the transmitted to incident wavefunctions. We can define a *reflection probability* as

$$R = \left|\frac{B}{A}\right|^2 = \frac{V_0^2 \sin^2(k_2 a)}{V_0^2 \sin^2(k_2 a) + 4E(E - V_0)}, \tag{6.16}$$

which is the modulus squared of the ratio of the reflected to incident wavefunctions.

If $E < V_0$, then, classically, the particle would be turned back by the barrier ($T = 0$, $R = 1$), whereas, classically, for $E > V_0$ the particle would move unimpeded past the barrier ($T = 1$, $R = 0$). However, consider the plot of $T(E)$ versus E, as shown in Fig. 6.3. It can be seen that the classical values are limiting cases for $E \ll V_0$ and $E \gg V_0$. In general, however, for most values of energy $0 < T < 1$, meaning that there is some nonzero probability that the electron will be transmitted through the barrier. For larger values of

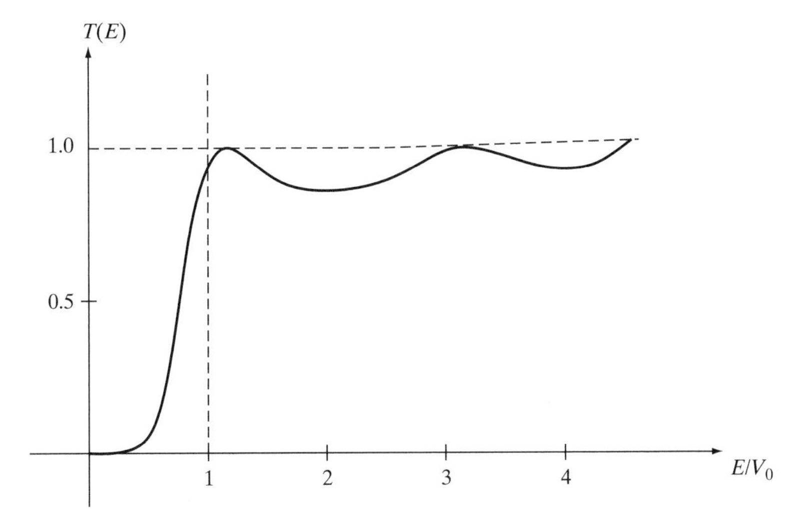

Figure 6.3 Tunneling probability versus energy for a potential energy barrier.

energy, it is more likely that the electron will be transmitted through the barrier, but even for relatively large energies the electron may be reflected by the barrier.

If the electron is transmitted through the barrier in the case $E < V_0$, the electron is said to *tunnel through the barrier* for the following reason. The total energy of the electron is the sum of kinetic and potential energies,

$$E = E_{KE} + E_{PE} \tag{6.17}$$

$$= E_{KE} + V_0, \quad \text{inside the barrier,} \tag{6.18}$$

$$= E_{KE} + 0, \quad \text{outside the barrier.} \tag{6.19}$$

If $E < V_0$, then inside the barrier,

$$E = E_{KE} + V_0 < V_0, \tag{6.20}$$

which indicates that the electron's kinetic energy would be negative (i.e., $(1/2)\,mv^2 < 0$), as discussed for a similar situation in Section 4.5.1. According to classical physics, kinetic energy cannot be negative, and therefore, classically, for $E < V_0$ the electron cannot be found inside the barrier. Thus, classically, the electron cannot get to the other side of the barrier. However, it is well known from experiments that the electron can indeed cross the barrier, with probability (6.15), and so the electron is said to *tunnel through the barrier* in order to get to the other side. For students familiar with electromagnetics and optics, it can be appreciated that particle tunneling is analogous to evanescent wave propagation (as occurs in total internal reflection, sections of below cutoff waveguides, etc.), although that topic will not be addressed here.

It can be seen that for $E/V_0 > 1$, there is a series of transmission resonances where $T = 1$. Rewriting (6.15) as

$$T = \left(1 + \frac{V_0^2}{4E\,(E - V_0)} \sin^2(k_2 a)\right)^{-1}, \tag{6.21}$$

we can see that full transmission (i.e., $T = 1$) will occur when

$$\sin(k_2 a) = 0, \tag{6.22}$$

that is, when

$$k_2 a = n\pi, \qquad n = 0, 1, 2, \ldots . \tag{6.23}$$

At these points, the internal reflections "bouncing around" in the barrier that lead to left-moving waves exactly cancel, and only the right-moving waves remain. With $k_2 = 2\pi/\lambda_2$, where λ_2 is the wavelength in region II, (6.23) becomes

$$a = n\frac{\lambda_2}{2}, \tag{6.24}$$

so that complete transmission occurs when the barrier thickness is an integral number of half wavelengths. (This situation is called a *transmission resonance*.) The same phenomenon is encountered in classical electromagnetics and optics, where electromagnetic energy can pass through a half-wavelength dielectric sheet without reflection.

An important, interesting special case is when $E < V_0$, and a is sufficiently large. In this case, we have

$$\begin{aligned} \sin\left(\sqrt{\frac{2m_e^*(E-V_0)}{\hbar^2}}a\right) &= \sin\left(i\sqrt{\frac{2m_e^*(V_0-E)}{\hbar^2}}a\right) \\ &= i\sinh\left(\sqrt{\frac{2m_e^*(V_0-E)}{\hbar^2}}a\right) \\ &\underset{a\to\infty}{\to} \frac{i}{2}e^{\sqrt{\frac{2m_e^*(V_0-E)}{\hbar^2}}a} = \frac{i}{2}e^{\alpha a}, \end{aligned} \tag{6.25}$$

where

$$\alpha = \sqrt{\frac{2m_e^*(V_0-E)}{\hbar^2}} > 0, \tag{6.26}$$

so that

$$T \to \frac{16E(V_0-E)}{V_0^2}e^{-2\alpha a}. \tag{6.27}$$

Therefore, the tunneling probability is exponentially decaying as a function of the barrier width a. So, as might be expected, the tunneling probability is low for thick barriers, and increases as the barrier thickness decreases. Table 6.1 shows the tunneling probability for a $V_0 = 0.2$ eV barrier for two different barrier widths. It can be seen that a doubling of the barrier width significantly changes the tunneling probability (in a nonlinear manner).

For the example of tunneling across the gate oxide in a MOSFET, this explains why, as oxide thickness is decreased, tunneling can become a significant problem, leading to non-negligible gate currents (as discussed in more detail later). Of course, tunneling is often a beneficial phenomenon. Flat panel displays make use of field emission (Section 6.3.1), which is a tunneling phenomenon, and the basis of the scanning tunneling microscope (described on page 202) is tunneling. Last, many nanoelectric devices rely on tunneling for their operation, as described in the next chapter.

TABLE 6.1 TUNNELING PROBABILITY FOR A 0.2 eV BARRIER FOR TWO DIFFERENT BARRIER WIDTHS.

E (eV)	T ($a = 1$ nm)	T ($a = 2$ nm)
0.01	8.86×10^{-3}	1×10^{-4}
0.10	0.145	6.11×10^{-3}
0.20	0.432	0.160

6.2 POTENTIAL ENERGY PROFILES FOR MATERIAL INTERFACES

Although we have considered tunneling through a simple potential energy barrier in a general sense, it is informative to consider the nature of the potential energy profile itself. That is, we should consider what the potential energy profile represents, and whether or not it is a realistic model. It turns out that, for our purposes, the tunneling barrier will often be associated with the junction between two different materials. We consider this topic in the next section.

6.2.1 Metal–Insulator, Metal–Semiconductor, and Metal–Insulator–Metal Junctions

First of all, the rectangular barrier shown in Fig. 6.2 on page 185 is a fairly gross approximation to the actual potential energy profile usually seen in real tunnel junctions. For example, consider the interface between metal and vacuum. If we supply enough energy to the material, electrons can escape from the metal surface.† The amount of energy needed to liberate electrons from the metal's surface is the work function, $e\phi$. (See also Section 5.4.) In considering potential energy problems, $e\phi$ can be thought of as merely a material constant. Furthermore, the energy of electrons in the metal, at least the most important electrons, is the Fermi energy, E_F (Section 4.4), which can also be thought of as simply a material constant.‡ Therefore, the metal–vacuum junction can be modeled as shown in Fig. 6.4, where E_{vac} is the vacuum energy, and the energy difference $E_{\text{vac}} - E_F$ is, by definition, the work

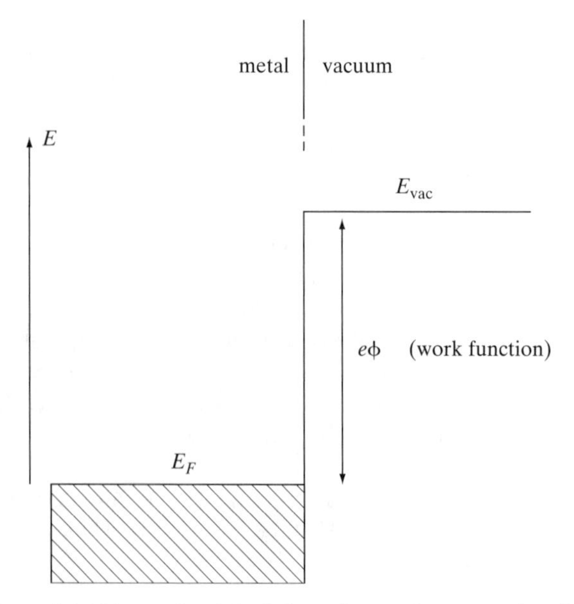

Figure 6.4 Energy band depiction of a metal–vacuum interface.

†This is called *thermionic emission* if we supply thermal energy, *photo emission* if we supply electromagnetic energy (photons), and simply *field emission* if energy is supplied by an electric field.

‡For copper and gold the work function is on the order of 4.5–5.0 eV, and for copper, $E_F \simeq 7$ eV at room temperature.

function $e\phi$. The shading below E_F in the metal indicates that energy levels below E_F are filled with electrons, since E_F is the energy of the most energetic electrons.

Therefore, for example, thermal emission involves supplying sufficient thermal energy such that the electron's energy is raised at least $e\phi$ above the level E_F, and then the electron can escape from the metal. Thus, in this case, the electron effectively goes *over the barrier*, and the electron emerges from the metal with energy greater than or equal to $E_F + e\phi$ (depending on how much energy is supplied).

If an insulator replaces the vacuum, then the work function is replaced with a *reduced (or modified) work function* $e\phi'$, which is the energy required to liberate an electron from the metal's surface into the insulating region (i.e., into the conduction bandedge of the insulator (Section 5.4)). For example, since the electron affinity, $e\chi$, of SiO_2 is 0.9 eV, the modified work function of a metal–SiO_2 junction is $e\phi' = e\phi - e\chi = e\phi - 0.9$. The modified work function for several metal–SiO_2 junctions is shown in Table 6.2.

Furthermore, if the vacuum region is replaced by a semiconductor, the resulting metal–semiconductor junction behaves in a similar manner to the metal–insulator junction, although the energy bands on the semiconductor side become curved, rather than forming straight lines, and the barrier height is approximately one-half of the bandgap energy. For example, Fig. 6.5 shows the energy band diagram for a metal–semiconductor junction before the materials are joined (the semiconductor is assumed to be n-type, and $e\phi_m > e\phi_s$), and Fig. 6.6 shows the junction after the two materials are brought together and thermal equilibrium is established.

The band bending on the semiconductor side is due to the fact that upon contact, charges will flow across the junction (in this case, electrons from the semiconductor will cross into the metal, since $E_c > E_{F,\,\mathrm{metal}}$) until the Fermi levels of the two materials are aligned. For this example, electrons are depleted from the semiconductor near to the interface, resulting in a net positive charge and an upward bending of the energy bands near to, and on the semiconductor side of, the interface. There is no band bending on the metal side, since, for instance, there is no voltage drop in the metal.[†] This results in a barrier (called a

TABLE 6.2 WORK FUNCTION FOR A METAL-VACUUM INTERFACE, AND MODIFIED (REDUCED) WORK FUNCTION FOR A METAL-SiO_2 INTERFACE (FROM [9]).

Metal	$e\phi$ (eV)	$e\phi'$ (eV)
Al	4.1	3.2
Ag	5.1	4.2
Au	5.0	4.1
Cu	4.7	3.8

Goser, K., P. Glösekötter, and J. Dienstuhi (2004). *Nanoelectronics and Nanosystems—From Transistors to Molecular and Quantum Devices*, Berlin: Springer-Verlag.

[†]If the metal is approximated as a perfect conductor, there is no band bending on the metal side. However, if the metal is more accurately modeled as an imperfect conductor, band banding does occur, but is limited to a very small region near the surface of the metal.

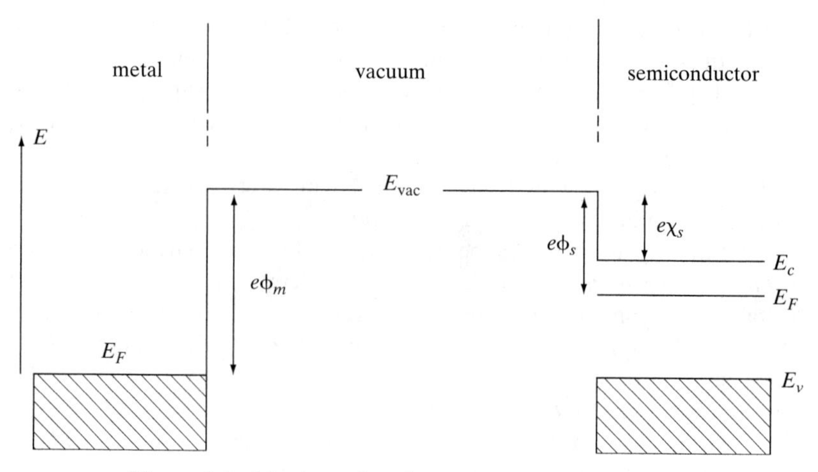

Figure 6.5 Metal–semiconductor junction before contact.

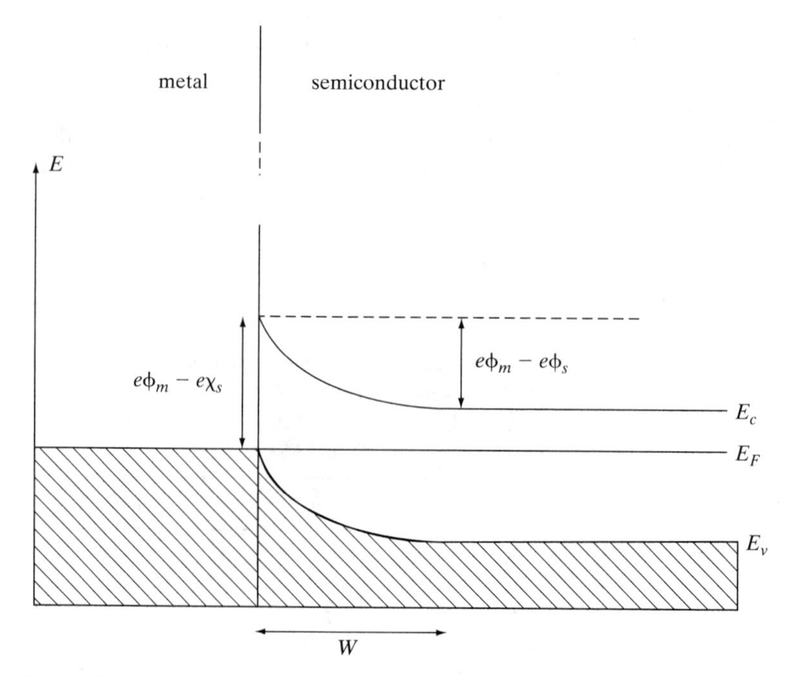

Figure 6.6 Metal–semiconductor junction after contact at thermal equilibrium.

Schottky barrier) to electron flow, with the barrier height given by

$$e\phi_b = e\phi_m - e\chi_s. \tag{6.28}$$

The depleted region is called a *space charge layer*, and has approximate width W. The exact form of the energy band profile in the depletion region must be obtained by solving Poisson's equation for the charge profile, and will not be discussed here.† The resulting junction is called a *Schottky diode*, since upon applying a voltage bias positive with respect to the metal, the barrier will be lowered,‡ allowing large currents to flow metal to semiconductor (i.e., for electrons to cross from the semiconductor to the metal), and applying a voltage bias negative with respect to the metal, the barrier will be raised, impeding current flow.

Many applications require an *ohmic contact*, in which current can flow in either direction with very little resistance. Being able to align the energy levels of the metal and semiconductor would help accomplish this, but interface effects also play a role. Often ohmic contacts are made by heavily doping the semiconductor near the metallic contact.

In summary, the junction between two materials presents a change in potential energy, and results in either a step change in energy, as in the case of a metal–vacuum or metal–insulator junction, or a more complicated energy profile, such as for a metal–semiconductor junction. It can then be seen that the energy profile depicted in Fig. 6.2 on page 185 is a model for a metal–vacuum–metal junction, as shown in Fig. 6.7 (assuming identical metals in thermal equilibrium on either side of the vacuum region).

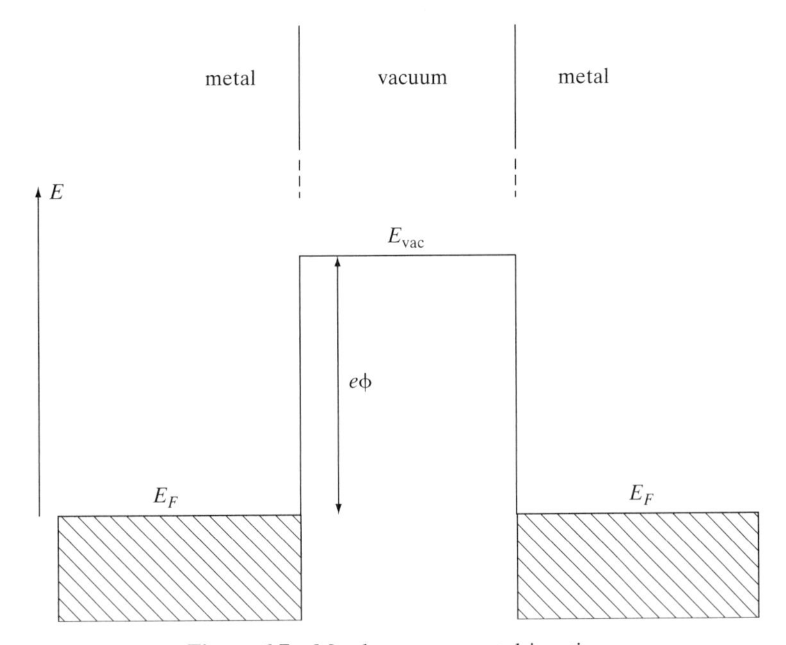

Figure 6.7 Metal–vacuum–metal junction.

†This is standard material in semiconductor device physics texts.

‡See Section 5.4.3 for the effect of an applied voltage on energy bands.

A metal–insulator–metal structure would have a very similar energy profile, although the modified work function of the metals, $e\phi'$, would replace the work function $e\phi$. A metal–semiconductor–metal junction would also lead to a similar profile, although with band bending in the semiconductor.

To consider the effect of an applied field on a material junction, we can examine the metal–vacuum–metal junction in Fig. 6.7. In this case, it is convenient to consider applying a voltage V_0 across the vacuum region, resulting in the electric field magnitude $\mathcal{E}_0 = V_0/d$, where d is the thickness of the vacuum region. Combining this result with the work function then leads to the total potential energy profile

$$e\phi - q_e V_0 x/d \tag{6.29}$$

in the vacuum region. Of course, a similar result holds if the vacuum is replaced by an insulating material, with the resulting band diagram shown in Fig. 6.8. If the insulator were a vacuum, then the barrier height would extend up to the vacuum level, i.e., $e\phi' = e\phi$.

At this point, it is worthwhile to consider the possible tunneling currents that arise in a metal–insulator–metal junction under an applied bias. Three possible currents (I_1, I_2, and I_3) are shown in Fig. 6.8. We assume that in each metal region, all states below E_F are filled and all states above E_F are empty. As such, we must have $I_1 = 0$, since this current would result from the flow of electrons having energy above E_F; however, these energy states are empty. Current I_3 would result from filled states on the left tunneling into already filled states on the right, and so, since this is impossible, $I_3 = 0$ as well. The actual

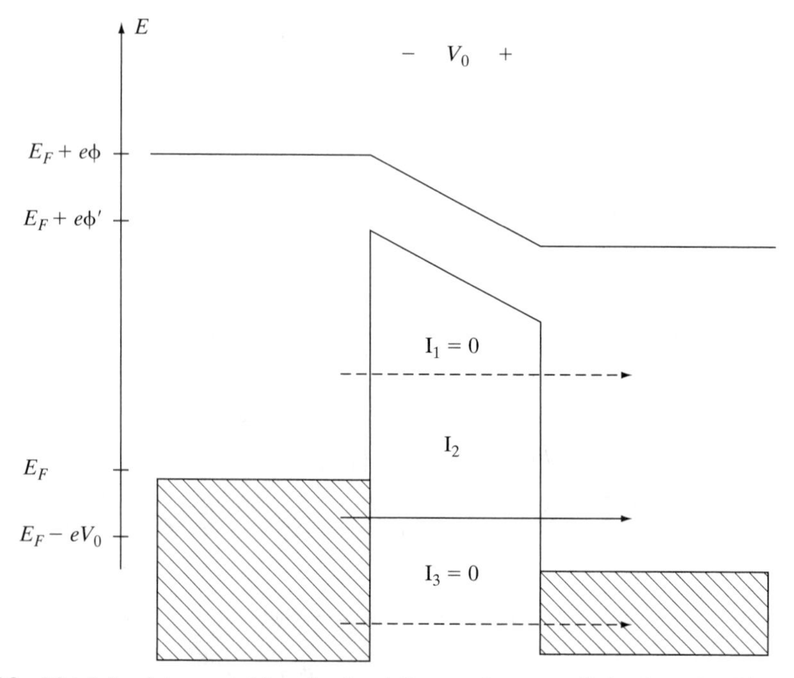

Figure 6.8 Metal–insulator–metal junction band diagram for an applied voltage V_0. Bias is positive on right side, and negative on left side.

tunneling current arises from *filled* states on the left tunneling into *unfilled* states on the right, resulting in tunneling current I_3.

6.3 APPLICATIONS OF TUNNELING

As described briefly earlier, there are many applications of tunneling, especially in the emerging area of nanoelectronics. Tunneling can be used for beneficial purposes, or it can be detrimental to device performance. Next we briefly examine several different aspects of tunneling.

6.3.1 Field Emission

Excepting Fig. 6.8, the energy profiles described in the previous section correspond to the case of no applied field or bias. That is, the energy profile reflects the junction after the materials are joined together, and after thermal equilibrium is established, which equalizes the Fermi levels. However, as might be guessed from Section 5.4.3, the presence of an electric field or voltage modifies the band structure. In particular, we know that an electric field or voltage tilts energy bands, depressing the energy band on the positive side of the field. Therefore, if an electric field or voltage is applied across the junction between a metal and a vacuum, due to (6.29), the potential energy profile becomes triangular, as shown in Fig. 6.9.

As described previously, if energy $E_F + e\phi$ or more is supplied to the structure, an electron can *go over* the energy barrier, as depicted in Fig. 6.10. However, note that as the applied electric field magnitude is increased, the slope of the energy profile in Fig. 6.9 becomes greater, and the triangular barrier becomes thinner. Thus, for a sufficiently large electric field, electrons can easily tunnel *through* the thin barrier, as depicted in Fig. 6.11. This is called field emission, or *cold emission*, since the electrons emerge from the metal with energies lower than $E_F + e\phi$. This is also called *Fowler–Nordheim* (FN) *tunneling*, named after the researchers who investigated field emission in the 1920s. It can be shown that the tunneling probability through the triangular barrier depicted in Fig. 6.9 is[†]

$$T = \exp\left(\frac{-4\sqrt{2m_e^*}}{3\left|q_e\right|\mathcal{E}\hbar}\left(e\phi - (E - E_F)\right)^{3/2}\right), \tag{6.32}$$

[†]This result comes from the Wentzel–Kramers–Brillouin (WKB) approximation of the wavefunction $\psi(x)$ in a region of slowly varying potential energy $V(x)$. This is described in standard textbooks on quantum mechanics, and the main result is that for a tunnel barrier extending from x_1 to x_2,

$$T \simeq e^{-2\int_{x_1}^{x_2} \beta(x)dx}, \tag{6.30}$$

where

$$\beta(x) = \sqrt{\frac{2m^*}{\hbar^2}\left(V(x) - E\right)}. \tag{6.31}$$

When $V(x)$ has a triangular shape, (6.32) results. Although the discontinuity in V violates the condition of $V(x)$ being slowly varying, the result is approximately correct.

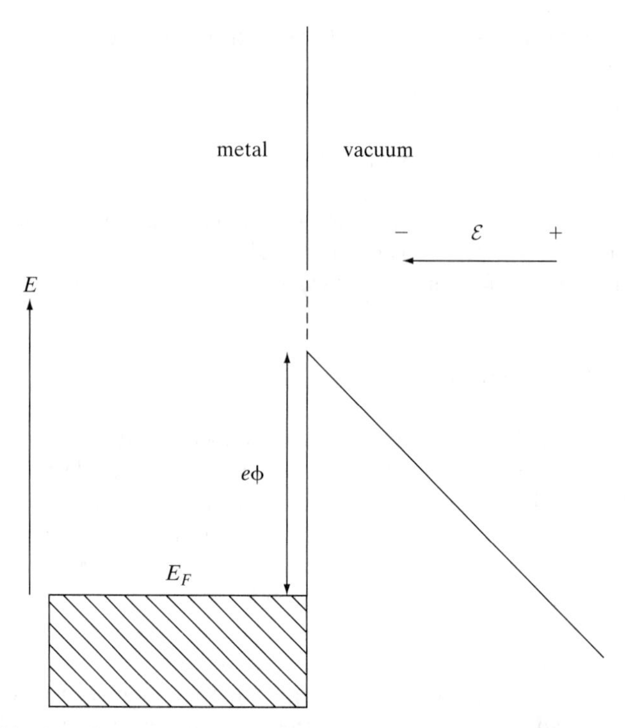

Figure 6.9 Metal–vacuum interface in the presence of an applied electric field $\mathcal{E}$.

where E is the energy of the electron (often the case $E = E_F$ is of interest), $e\phi$ is the barrier height, and $\mathcal{E}$ is the magnitude of the electric field. If the vacuum region is replaced by an insulator, then we use the metal–insulator work function $e\phi'$, rather than $e\phi$.

A good application of Fowler–Nordheim tunneling is to the description of field emission by carbon nanotubes. As described in Section 5.5, carbon nanotubes have nanoscopic radius values, and thus possess an extremely sharp tip that concentrates the electric field to a very small region of space. This strong field enhances tunneling through the vacuum barrier. As an example, an SEM image of the apparatus to measure the field emission I–V characteristics of an individual carbon nanotubes is shown in Fig. 6.12, along with the best fit Fowler–Nordheim result. The field-emitted current begins at approximately 91 V, and saturates around 150 V.

To apply the Fowler–Nordheim result to model this situation, we must obtain the tunneling current. In general, tunneling current is related to the product of the incident electron density multiplied by the tunneling probability, which is then integrated over various states.[†] For the triangular-like barrier presented by field emission, the result is[‡]

$$I = A\frac{1.5 \times 10^{-6}}{e\phi}\mathcal{E}^2 \exp\left(\frac{10.4}{\sqrt{e\phi}}\right)\exp\left(-6.44 \times 10^9 \frac{(e\phi)^{3/2}}{\mathcal{E}}\right), \tag{6.33}$$

[†]The method for determining tunneling current is the same as that described in Section 10.2.3 for a slightly different application; the basic equation for tunneling current is (10.49), where the limits of integration may change to account for the bandstructure of the material.

[‡]See [11]–[14].

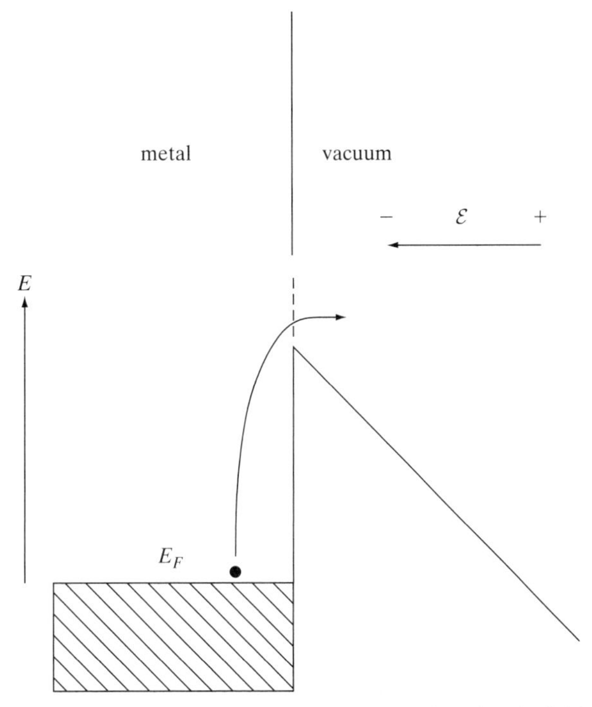

Figure 6.10 Metal–vacuum interface in the presence of an applied electric field $\mathcal{E}$, where an electron at the Fermi level goes over the potential barrier.

where A has the dimension of area (m^2) and the work function $e\phi$ has units of electron volts. The local field $\mathcal{E}$ is related to the applied field V/d by $\mathcal{E} = \gamma V/d$, where the *field enhancement factor* γ quantifies the ability of the emitter to intensify the applied field. For the results shown in Fig. 6.12, $e\phi = 4.9$ eV, $\gamma = 90$, and $A = 5 \times 10^{-16}$ m^2. These values are obtained from the measurement by noting that

$$\ln \frac{I}{V^2} = \ln \left(A \frac{1.5 \times 10^{-6}}{e\phi d^2} \gamma^2 \right) + \left(-6.44 \times 10^9 \frac{(e\phi)^{3/2} d}{\gamma} \frac{1}{V} + \frac{10.4}{\sqrt{e\phi}} \right), \tag{6.34}$$

or

$$\ln \frac{I}{V^2} = c_1 + \left(-c_2 \frac{1}{V} + c_3 \right), \tag{6.35}$$

where c_1–c_3 are constants. Thus, a plot of $\ln\left(I/V^2\right)$ versus $1/V$ should be linear with a negative slope. In the insert of Fig. 6.12, the experimental $\ln\left(I/V^2\right) - (1/V)$ curve is shown, which indeed has the desired behavior. Since d is known for the measurement system, and the approximate value of the work function for the CN is known, γ can be determined from

$$c_2 = 6.44 \times 10^9 \frac{(e\phi)^{3/2} d}{\gamma}, \tag{6.36}$$

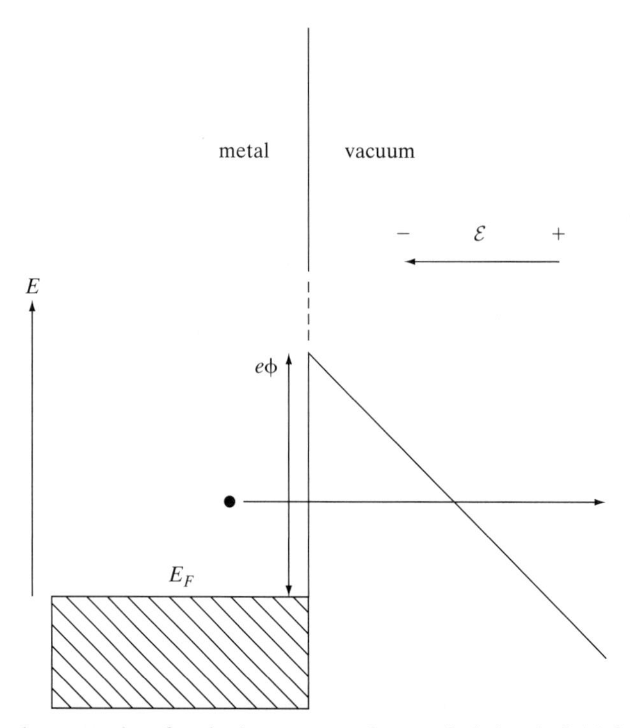

Figure 6.11 Metal–vacuum interface in the presence of an applied electric field $\mathcal{E}$, where an electron at the Fermi level tunnels through the potential barrier.

where c_2 is obtained from the measured slope. Then, A is determined from the constant

$$c_1 = \ln\left(A\frac{1.5 \times 10^{-6}}{e\phi d^2}\gamma^2\right), \tag{6.37}$$

since the constant $c_3 = 10.4/\sqrt{e\phi}$ is known.

The issue of emission stability and device lifetime is a possible concern for CN field emitters. While continuous operation without degradation of CN field emission sources has been demonstrated for time frames exceeding one year, single emitters and CN arrays have been observed to fail for reasons that are not currently well understood. Failure may be due to CN tip damage by large emission currents, although the tube environment (gas concentration, temperature, etc.) seems also to play a role. Nevertheless, it is expected that tunneling-based carbon nanotube flat-panel displays may be commercially available in the near future. Other devices, such as flash memories that use tunneling, have been commercially available for some time.

6.3.2 Gate–Oxide Tunneling and Hot Electron Effects in MOSFETs

Tunneling is a very important aspect in MOSFETs and similar structures, especially as the feature size is reduced. Consider the usual n-type MOSFET structure shown in Fig. 6.13, where s, d, and g indicate the source, drain, and gate, respectively.

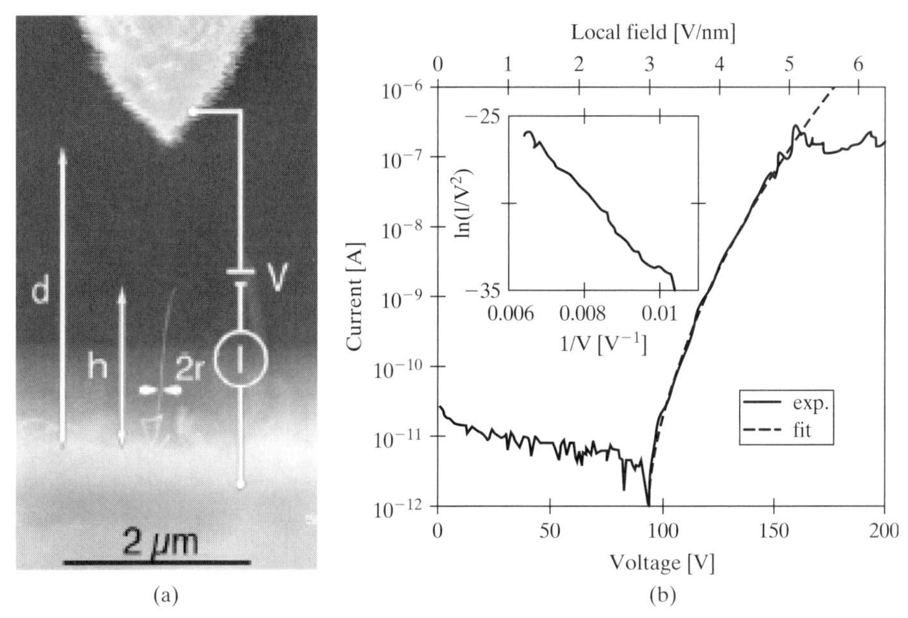

(a) (b)

Figure 6.12 (a) SEM image of the experimental setup for measuring field-emission I–V characteristics of an individual carbon nanotube. The nanotube length is $h = 1.4$ μm, and the tube radius is 7.5 nm, with the anode positioned at $d = 2.65$ nm. (b) Experimental I–V characteristics (solid line) and the best fit Fowler–Nordheim prediction (dashed line; $e\phi = 4.9$ eV, $\gamma = 90$, and $A = 5 \times 10^{-16}$ m^2) for an individual MWNT. The insert of (b) is described in the text. (From Bonard, J. M. et al., "Field Emission of Individual Carbon Nanotubes in the Scanning Electron Microscope," *Phys. Rev. Lett.* 89 (2002): 197602. Copyright 2002, American Physical Society.)

The oxide layer is conventionally formed by oxidizing the silicon substrate, forming an SiO_2 insulating barrier between the gate electrode and the rest of the device.† As shown in Fig. 6.13, there is no conduction channel between the source and the drain (both n-type Si) when the gate voltage V_g is zero; thus, $I_{ds} = 0$ irrespective of V_{ds}. When a positive gate voltage $V_g > 0$ is applied, and has sufficient magnitude,‡ an inversion layer is formed under the gate, connecting the source and drain. The inversion layer is formed since positive voltage V_g pushes away holes, and attracts electrons under the gate, forming, in effect, an n-type channel, as shown in Fig. 6.14. This allows current flow in the induced channel from drain to source upon applying a potential $V_{ds} > 0$. From an energy barrier viewpoint, when $V_g = 0$, there is a large energy barrier between source and drain. As V_g is increased, this barrier is pushed down, eventually below the filled states of the source and drain, and conduction can take place. MOSFET characteristics are reviewed in Appendix C.

Considering the gate–oxide–channel junction, we have something like a metal–insulator–metal junction, and, therefore, the potential energy profile of Figures 6.7 and

†High dielectric constant insulators are being developed for MOS devices that will likely replace SiO_2, allowing for thicker oxide layers resulting in less tunneling.

‡In order to induce a channel, we need $V_g > V_t$, where V_t is called the *threshold voltage*, where typically V_t is on the order of a volt.

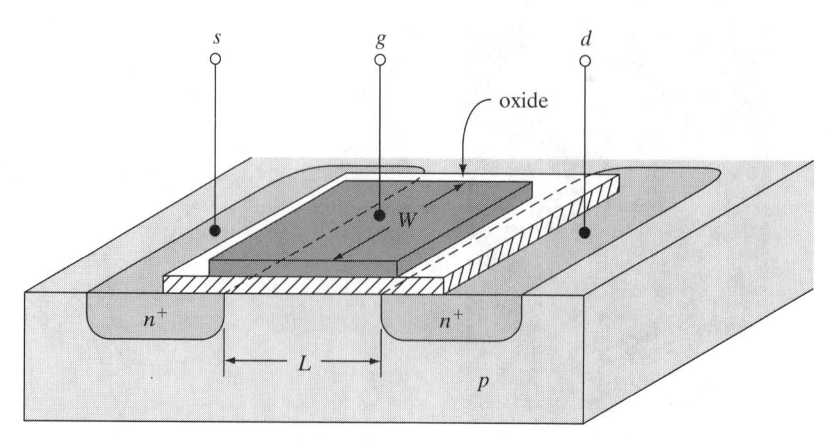

Figure 6.13 Physical structure of an n-type MOSFET. The regions denoted as n^+ are heavily doped, n-type Si, and p denotes the p-type Si substrate. For the gate electrode, typically polycrystalline silicon (polysilicon) is used for a variety of technical reasons.

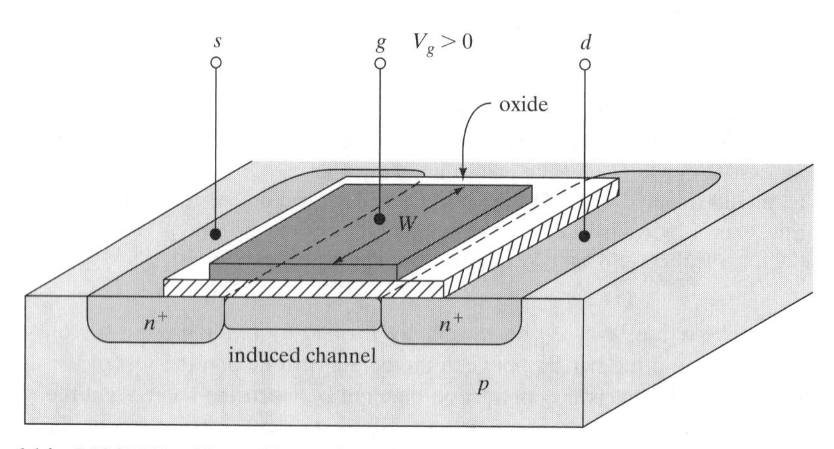

Figure 6.14 MOSFET with positive gate voltage, creating a n-type conducting channel between source and drain.

6.8 approximately applies. That is, for simplicity, we are modeling the inverted p-type semiconductor channel as a metal. A more careful analysis would take into consideration the fact that the channel is actually a semiconductor, and the band diagram would differ a bit on the channel side due to band bending.

Of course, in an ideal classical MOSFET, electrons do not travel between the channel and the gate (i.e., $I_g = 0$) because of the presence of the insulating oxide region. However, in light of our previous discussions, it is obvious that for sufficiently thin oxides, electrons will be able to cross (tunnel through) this energy barrier, leading to† $I_g \neq 0$. There are several

†This tunneling current is in addition to any current that may be present due to the oxide being an imperfect insulator, perhaps due to the presence of trapped charge impurities and other defects.

possible effects relating to quantum tunneling through the oxide, and next we mention two of these.

Hot Electrons. Drain-source current obviously results from accelerating electrons via the source-drain voltage, V_{ds}, which results in a horizontal electric field in the channel. As these electrons are accelerated, they gain kinetic energy. If they gain sufficient kinetic energy, they may tunnel through the oxide.† They may gain a large amount of kinetic energy from either a large V_{ds} and correspondingly large horizontal electric field ($\mathcal{E}_{ds} \propto V_{ds}/L$, where L is the channel length), or from a short channel. Note that this effect is also influenced by the presence of the vertical electric field in the oxide due to V_g. The end result is that electrons that are supposed to transverse the channel and reach the drain may, because of their high kinetic energy, instead tunnel through the oxide and contribute to an undesired gate current, as shown in Fig. 6.15. This is called the *hot-electron effect*, since the highly energetic electrons are considered "hot."

Fowler–Nordheim Tunneling. The second possibility is that, if a strong gate voltage is applied, electrons will be energetic enough from this field alone to become likely to tunnel through the oxide, as depicted in Fig. 6.16.

It is obvious that nonzero gate current can be attributed to a combination of tunneling events (as well as defects, trapped charge states, etc.), all of which lead to significant gate currents if the oxide layer is sufficiently thin. An accurate analysis results from a self-consistent numerical solution of coupled Poisson and Schrödinger's equations. Poisson's equation is used to obtain the potential profile, and an effective mass Schrödinger's equation provides the wavefunction, from which the probability current can be obtained using (3.187). In some cases a Fowler–Nordheim model provides reasonable accuracy. (One major cause

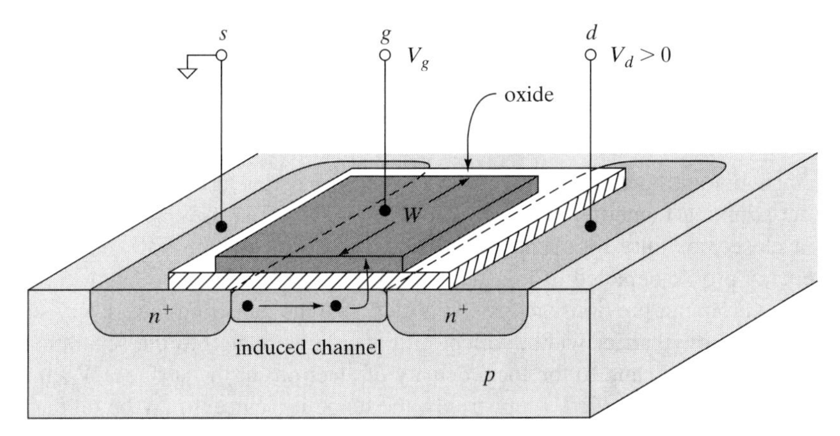

Figure 6.15 Depiction of the hot-electron effect, where electrons transversing the channel gain sufficient energy from the source-drain electric field to tunnel through the oxide energy barrier.

†Recall that tunneling is a random process, and that, for any energy, the electrons may tunnel. However, the tunneling probability goes up if the electrons' energy increases, or if the barrier thickness decreases.

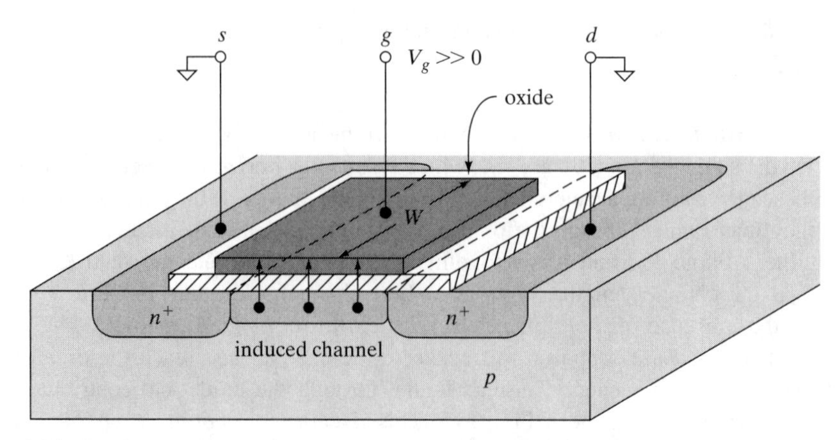

Figure 6.16 Depiction of Fowler–Nordheim tunneling. Strong gate voltage provides electrons in channel with sufficient energy to tunnel across the oxide energy barrier.

of inaccuracy in the Fowler–Nordheim model is that the potential energy profile is assumed, rather then obtained rigorously.)

For typical oxides, thickness values less than approximately 1.5 nm lead to relatively high tunneling rates. However, tunneling can occur even for moderately thick oxides if the gate voltage is high. For example, Fig. 6.17 shows gate current versus gate voltage for an n^+-poly silicon gate (doping concentration is 1×10^{21} cm^{-3}) n-type MOSFET structure, with oxide thickness values of 5, 8, and 10 nm. It can be seen that the Fowler–Nordheim tunneling model† (symbols) agrees well with the measured values (solid curves).

Gate current severely impacts standby power consumption and device functionality. The semiconductor industry is considering a variety of approaches to combat this problem, including using new oxide materials, and different MOSFET structures.

6.3.3 Scanning Tunneling Microscope

An important application of tunneling is to the characterization of material surfaces using the scanning tunneling microscope (STM).‡ The STM uses an extremely fine metallic tip in close proximity to a material surface, as shown in Fig. 6.18, resulting in the potential energy profile depicted in Fig. 6.19 (for the case of no applied bias).

From our previous analysis, it is clear that the tunneling rate is very strongly dependent on the energy barrier width, which, in this case, is related to the separation between the tip and the surface, and to the local density of electrons at the surface. When the tip-to-surface distance is on the order of angstroms, the tunneling current may be on the order of nA, and

†This particular Fowler–Nordheim model is actually somewhat of a hybrid model, and is based on parameters obtained from the self-consistent numerical solution of the coupled Poisson and Schrödinger's equations (assuming $m_e^* \simeq 0.5m_e$ for the electron effective mass in the SiO_2 oxide).

‡In 1986, the Nobel prize in physics was awarded to Gert Binnig and Heinrich Röher at IBM Zürich for their work on developing the STM.

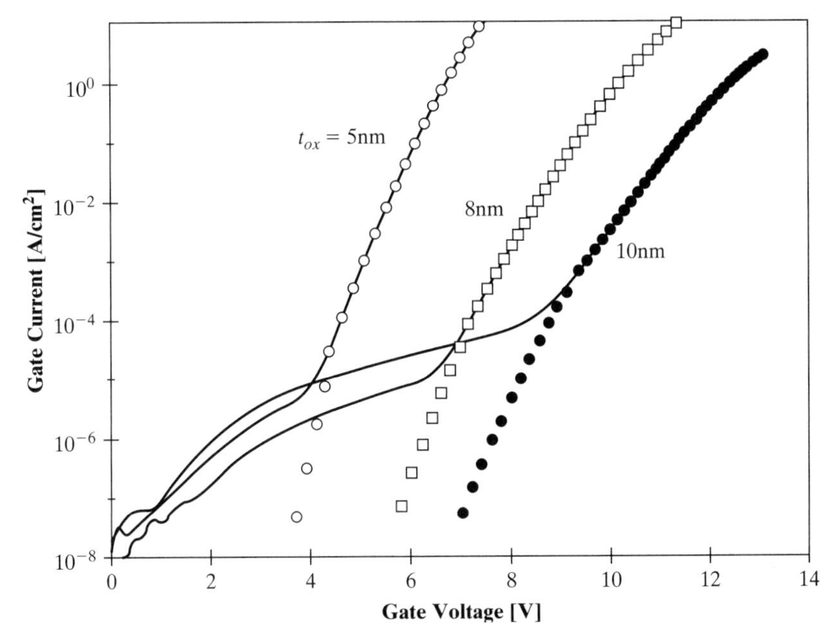

Figure 6.17 Gate tunneling current versus gate voltage for an n^+-poly silicon gate (doping concentration is 1×10^{21} cm^{-3}) n-type MOSFET with oxide thickness values of 5, 8, and 10 nm. The silicon substrate is doped at concentrations of 5×10^{17} cm^{-3}, 3.5×10^{17} cm^{-3}, and 1×10^{-17} cm^{-3}, respectively. The solid curves are measured values, and the symbols are from a Fowler–Nordheim tunneling model. (From Quan, W.-Y., D. M. Kim, and M. K. Cho, "Unified Compact Theory of Tunneling Gate Current in Metal–Oxide-Semiconductor Structures: Quantum and Image Force Barrier Lowering," *J. Appl. Phys.* 92 (2002): 3724. Copyright 2002, American Institute of Physics.)

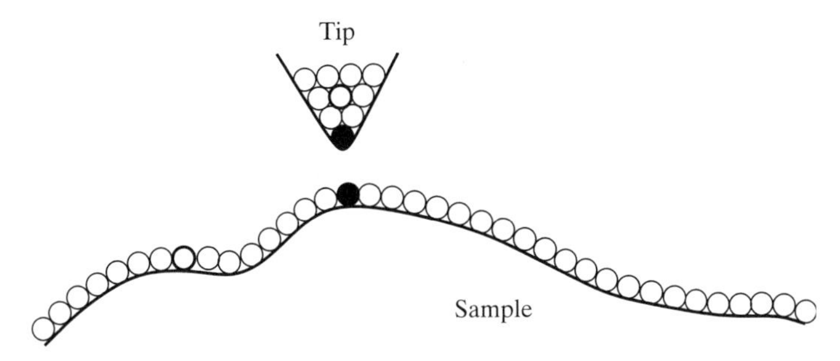

Figure 6.18 Material sample interrogated by probe tip. Most of the tunneling arises between the bottommost atom in the tip, and the nearest atom on the surface, shown in black.

will be very sensitive to the tip-to-surface separation. Thus, as long as the tip position can be controlled to angstrom precision, the surface can be mapped in atomic detail.

It can be appreciated that the STM can function either by moving the tip at a constant height and measuring the change in tunneling current as the tip-to-surface separation varies due to surface features, or by attempting (via feedback) to keep the current constant by

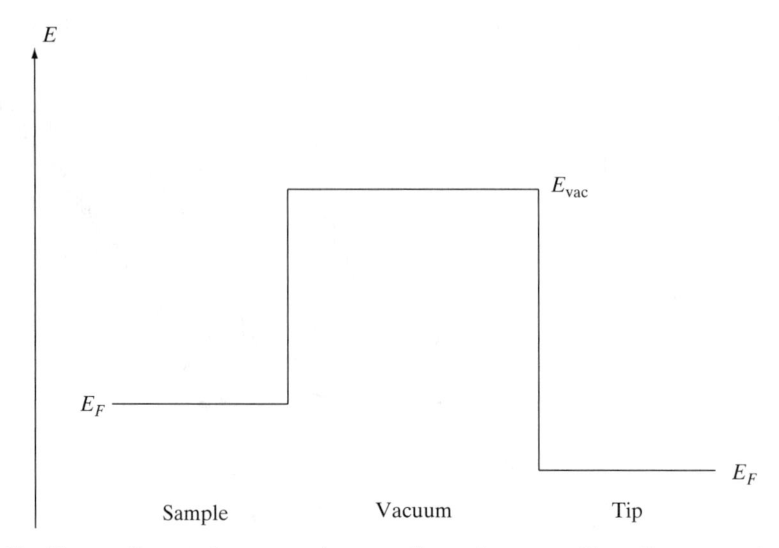

Figure 6.19 Energy diagram for a scanning tunneling microscope. Tunneling occurs between the sample and the tip, surmounting the work function of the material–vacuum interface. Diagram shown for the case of no applied bias.

varying the tip height. In the first method, as the tip moves from location to location on the surface, the amount of tunneling current will become bigger or smaller depending on the local electron density, which is itself related to the positions of the atoms. For example, since the tunneling current falls off exponentially with distance, when the tip is over an atom, the current will be much larger than when the tip is between atoms. However, because of the exponential dependence of the tunneling current, this results in an image in which the atomic peaks look much higher than their actual height. (A true image would result if the tunneling current depended linearly, rather than exponentially, on the sample–tip separation.)

In the constant current method, a piezoelectric crystal is used to vary the tip height, in order to maintain a constant tunneling current. The piezo electric material expands linearly as a function of applied voltage, and, therefore, the voltage needed to expand the crystal (thus moving the tip) to keep the current constant varies linearly with the position of the atoms on the sample. This voltage can be used to record the height of the tip, which is related to the surface features.

Note that the STM image is really an image of the local electron density, and not explicitly a tomographic map of the surface. Therefore, it can be used to image the local density of states, as described further in Section 8.1. For example, if an oxygen atom is located on a metal surface, it will appear as a depression in an STM image, even though the atom is on top of the metal surface. This is because the tip needs to move closer to the surface to maintain the same tunneling current, due to the (roughly) insulting properties of the oxygen atom. Therefore, although the STM itself does not need a vacuum to operate, the STM is often operated in an ultrahigh vacuum to avoid contamination of the sample from the surrounding environment.

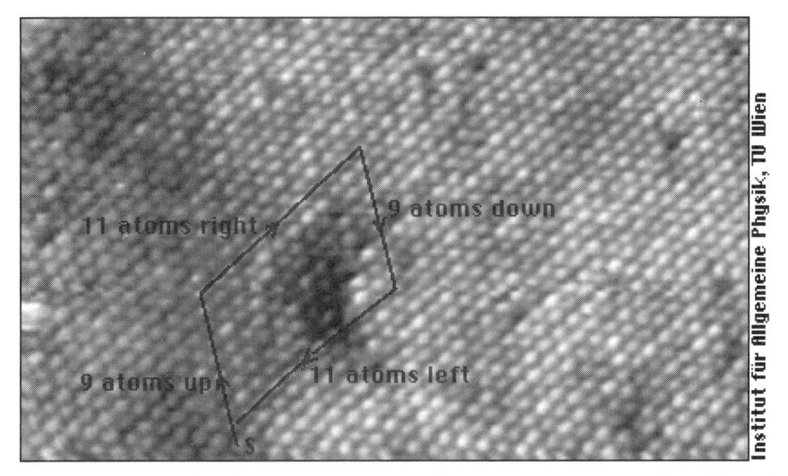

Figure 6.20 STM image of a dislocation in a PtNi alloy. (Courtesy Institut für Allgemeine Physik, Vienna University of Technology.)

Figure 6.21 STM image of a nickel surface. (Image reproduced by permission of IBM Research, Almaden Research Center. Unauthorized use not permitted.)

Figure 6.20 shows an STM image of a dislocation in a PtNi alloy, and Fig. 6.21 shows an STM image of the surface of nickel.

Similar in application, although not in operating principles, is the atomic force microscope (AFM). In AFM, a cantilever with a sharp tip is brought into close proximity to a sample surface. The force between the tip and the sample leads to a deflection of the cantilever according to Hooke's law. The deflection may be measured, for example, using a laser.

6.3.4 Double Barrier Tunneling and the Resonant Tunneling Diode

An interesting case occurs when one considers two barriers separated by a small distance, forming a potential well as shown in Fig. 6.22. This leads to the topic of *resonant tunneling*.

We assume that the barriers are sufficiently thin to allow tunneling, and that the well region between the two barriers is also sufficiently narrow to form discrete (quasi-bound) energy levels, as shown in Fig. 6.23.

The analysis of the double barrier structure shown in Fig. 6.22 is essentially the same as considered at the beginning of Section 6.1, although we now have five regions

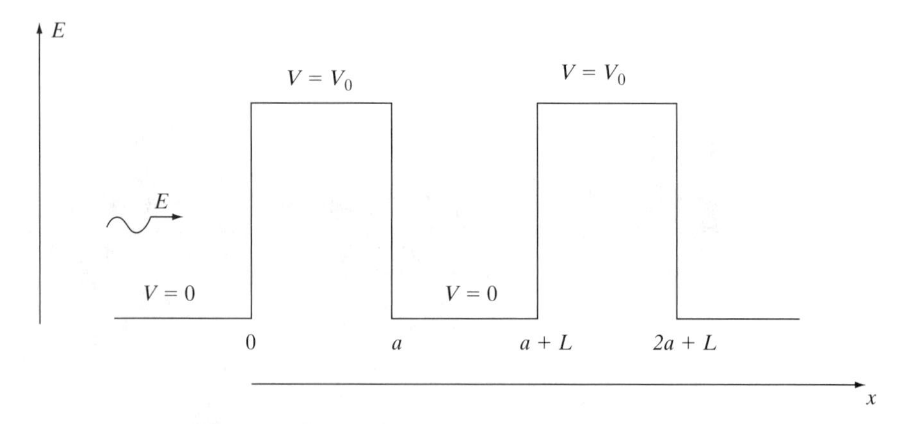

Figure 6.22 Double barrier system forming a potential well.

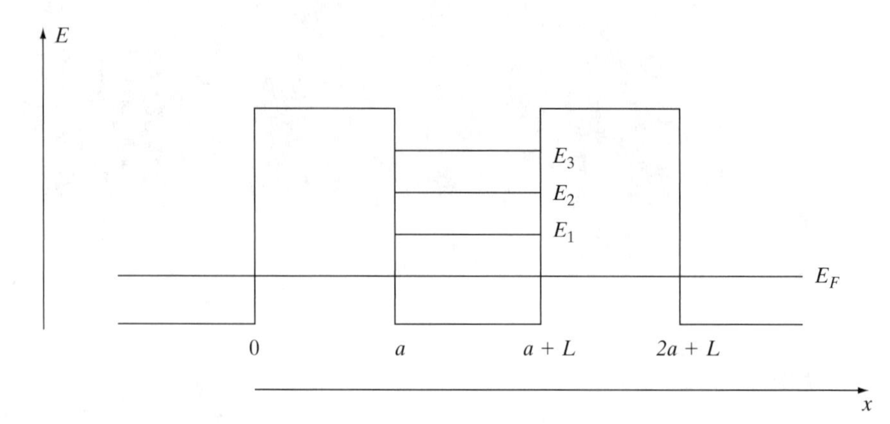

Figure 6.23 Double barrier structure forming a sufficiently narrow well so that energy levels in the well are well quantized. Energy levels E_1, E_2, and E_3 depict quasi-bound states.

to consider, with four interfaces at which to match boundary conditions. Therefore, the analysis is straightforward but tedious. The result for the transmission coefficient of the double symmetric barrier shown in Fig. 6.22 is

$$T = \left(1 + \frac{4R_1}{T_1^2} \sin^2 (k_1 L - \theta)\right)^{-1}, \tag{6.38}$$

where T_1 and R_1 are the transmission and reflection coefficients for a single barrier of width a, given by (6.15) and (6.16), respectively; L is the length of the well between the barriers, and

$$\tan\theta = \frac{2k_1 k_2 \cos (k_2 a)}{\left(k_1^2 + k_2^2\right) \sin (k_2 a)}, \tag{6.39}$$

where

$$k_1^2 = \frac{2m_e^* E}{\hbar^2}, \quad k_2^2 = \frac{2m_e^* (E - V_0)}{\hbar^2}. \tag{6.40}$$

From (6.38), it is easy to see that the transmission probability becomes unity when

$$\sin (k_1 L - \theta) = 0, \tag{6.41}$$

that is, when

$$k_1 L - \theta = n\pi, \quad n = 0, 1, 2, \ldots . \tag{6.42}$$

It turns out these transmission peaks ($T = 1$) will occur when the energy of the incoming electron wave (E) coincides with the energy of one of the quasi-bound states formed by the well. To see this for a simple special case, assume that $V_0 \gg E$, such that $|k_2| \gg |k_1|$. Then, $\tan\theta \to 0$ such that $\theta \to 0$, leading to

$$k_1 L = n\pi.$$

Using the expression for k_1, we see that

$$E = E_n = \frac{\hbar^2}{2m_e^*}\left(\frac{n\pi}{L}\right)^2, \tag{6.43}$$

which is exactly the same as the result for the quantized energy levels in a one-dimensional quantum well (4.35).

The double barrier tunnel junction has important applications to a device known as a *resonant tunneling diode*. The operation of these diodes can be appreciated from considering the influence of bias on the energy band diagrams for the double barrier system. We make use of the fact that when the incident energy E is very different from the energy of a quasi-bound state E_n, transmission will be low, and as $E \to E_n$, transmission will increase, becoming a maximum when $E = E_n$. For example, assume that incident electrons have energy E, and that, at first, all of the quasi-bound states E_n lie above E, as shown in Fig. 6.24.

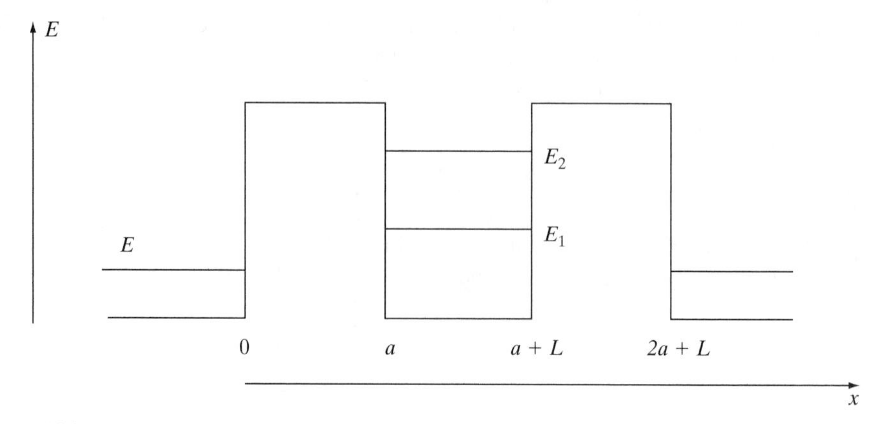

Figure 6.24 Double barrier junction with no applied bias. E is the energy of the incident electron, and $E_{1,2}$ are the energy levels of the quasi-bound states in the well.

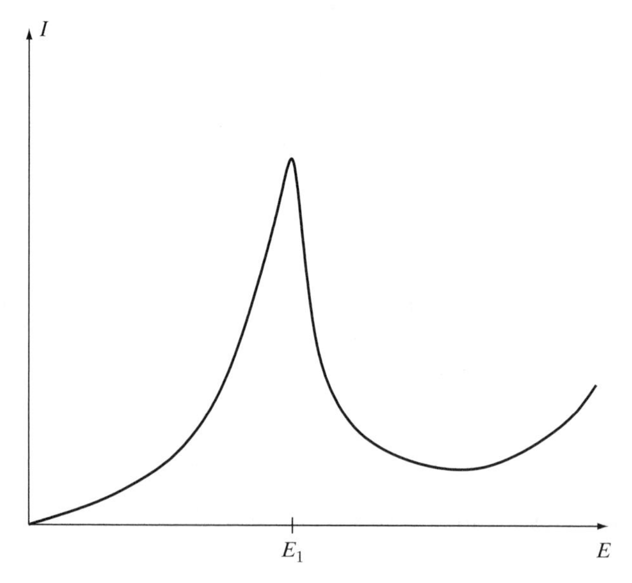

Figure 6.25 Current–energy characteristic for a resonant tunneling junction, where E is the energy of the incident electron and E_1 is the energy level of the first quasi-bound state in the well.

As E increases, tunneling will increase, reaching a peak when $E = E_1$. After that point, a further increase in E will result in a decreasing current, as shown in Fig. 6.25. This decrease of current with an increase of bias is called *negative resistance*. Further peaks and valleys will occur as $[E]$ approaches, and then moves past, other quasi-bound states.

A typical structure is made by using n-type GaAs for the regions to the left and right of both barriers, intrinsic GaAs for the well region, and AlGaAs or AlAs for the barrier material. Tunneling is controlled by applying a bias voltage across the device. For the case of no applied bias, the energy band diagram is similar to that shown in Fig. 6.24.

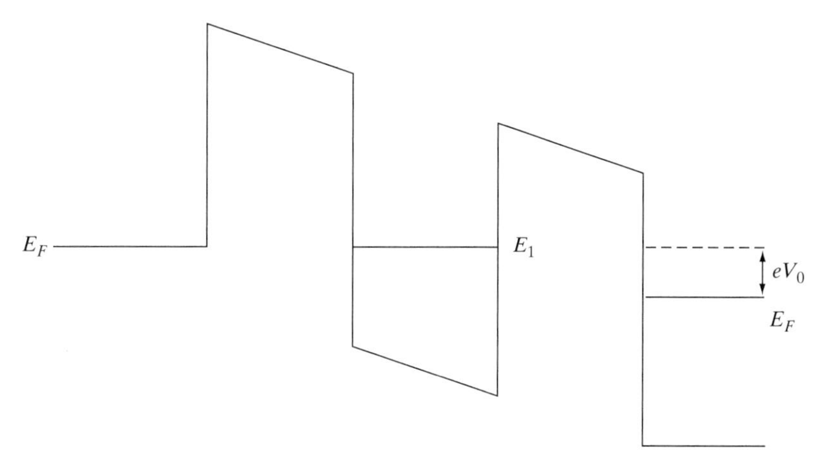

Figure 6.26 Double barrier junction under the action of an applied bias.

For an applied bias V_0, positive on the right side of the double junction, an appreciable current begins to flow when the quasi-bound state is pulled down to the Fermi level of the left region, as shown in Fig. 6.26. Current reaches a maximum when the level of the quasi-bound state is equal to the conduction bandedge of the left region.

Superlattice. Going beyond a double barrier structure is a periodic array of barriers, forming what is known as a *superlattice*. If enough layers are used, this structure results in periodic behavior (and associated band formation), as was found for crystalline materials in Chapter 5. The result is a one-dimensional artificial crystal, although the period is much longer than that found for a crystal (whose periodicity is governed by interatomic spacing).

As a simple approximate analysis, the Kronig–Penney model (Fig. 5.7 on page 138) can be used, where the barrier thickness a_2 and the well thickness a_1 correspond to the thickness of the material layers. Typically, a large bandgap and a small band gap material is used (i.e., alternating layers of a large bandgap material having thickness a_2, and a small bandgap material having thickness a_1). The analysis presented in Section 5.3 is appropriate for an infinite superlattice, or for a sufficiently long structure. The analysis of a finite superlattice can be done using the method described in Section 6.1, although efficient matrix methods can be used to avoid generating large numbers of simultaneous equations.

As a rough approximation, we can consider the energy levels formed by one isolated well. Then, using the ideas of the interacting systems model presented in Section 5.4.2, we see that for a collection of N interacting wells, the energy levels split into N discrete levels, and, as N becomes large, energy bands are developed. In this way, an artificial material can be made with specific band characteristics. Bloch oscillations can be sustained in superlattices, since the band structure can be precisely controlled. (An important aspect is the size of the first Brillouin zone.) Superlattice structures have been used as infrared oscillators and detectors, in various optical and quasi-optical devices, in transistor applications, and in a host of other areas, including heterostructure fabrication and material coatings. Heterostructures are introduced in Chapter 9.

6.4 MAIN POINTS

In this chapter, we have examined particle tunneling through barriers, and implications and applications of tunneling. In particular, after studying this chapter, you should understand

- quantum particle tunneling through simple energy barriers;
- energy profiles as models of material interfaces, and what material junctions are represented by rectangular and triangular energy profiles;
- field emission, and applications to carbon nanotube emitters;
- gate-oxide tunneling in MOSFETs;
- principles of the scanning tunneling microscope;
- tunneling through double barriers, and applications to the resonant tunneling diode;
- the idea of a superlattice, and how it can behave as an artificial crystal.

6.5 PROBLEMS

1. Plot the tunneling probability versus electron energy for an electron impinging on a rectangular potential barrier (Fig. 6.2, page 185) of height 3 eV and width 2 nm. Assume that the energy of the incident electron ranges from 1 eV to 10 eV.

2. Plot the tunneling probability verses barrier width for a 1 eV electron impinging on a rectangular potential barrier (Fig. 6.2, page 185) of height 3 eV. Assume that the barrier width varies from 0 nm to 3 nm.

3. A 6 eV electron tunnels through a 2-nm-wide rectangular potential barrier with a transmission coefficient of 10^{-8}. The potential energy is zero outside of the barrier, and has height V_0 in the barrier. What is the height V_0 of the barrier?

4. Can humans tunnel? Consider running 1 m/s (assume that you have no potential energy) at an energy barrier of 50 J that is 1 m thick.
(a) If you weigh (i.e., your mass is) 50 kg, determine the probability that you will tunnel through the barrier.
(b) Consider that in order for tunneling to have a reasonably large probability of occurring, $k_2 a$ can't be too large in magnitude. Discuss the conditions that would result in this happening.

5. Referring to the development of the tunneling probability through a potential barrier, as shown in Section 6.1, apply the boundary conditions (3.143) to (6.13) to obtain (6.14).

6. Consider the metal–insulator junction shown in Fig. 6.4 on page 190 . Solve Schrödinger's equation in each region (metal and insulator), and derive tunneling and reflection probabilities analogous to (6.15)–(6.16) for this structure.

7. Consider a metal–insulator–metal junction, as shown in Fig. 6.7 on page 193, except assume two different Fermi levels.

(a) Draw the expected band diagram upon first bringing the metals into close proximity.

(b) Because of the difference in Fermi levels, tunneling will occur (assuming that the barrier between the metals is thin), and will continue until a sufficient voltage is built up across the junction, equalizing the Fermi levels. This internal voltage is called the *built-in voltage*. Draw the energy band diagram showing the built-in voltage in this case.

8. Draw the potential energy profile for a metal–vacuum–metal structure when a voltage V_0 is applied across the vacuum region.

9. Determine the tunneling probability for an Al-SiO_2-Al system, if the SiO_2 width is 1 nm and the electron energy is 3.5 eV. Repeat for an SiO_2 width of 2 nm, 5 nm, and 10 nm.

10. Use the WKB tunneling approximation (6.30) to determine the tunneling probability for the rectangular barrier depicted in Fig. 6.2 on page 185.

11. Plot the tunneling probability versus electron energy for an electron impinging on a triangular potential barrier (Fig. 6.9, page 196), where $e\phi = 3$ eV and the electric field is 10^9 V/m. Assume that the energy difference $(E - E_F)$ ranges from 0 to 3 eV.

12. Consider the double barrier structure depicted in Fig. 6.22 on page 206.

(a) Plot the tunneling probability versus electron energy for an electron impinging on the double barrier structure. The height of each barrier is 0.5 eV, each barrier has width $a = 2$ nm, and the well has width 4 nm. Assume that the energy of the incident electron ranges from 0.1 eV to 3 eV, and that the effective mass of the electron is $0.067m_e$ in all regions.

(b) Verify that the first peak of the plot occurs at an energy approximately given by the first discrete bound state energy of the finite-height, infinitely-thick-walled well formed by the two barriers. Use (4.83) adopted to this geometry, i.e.,

$$k_2 \tan(k_2 L/2) = k_1, \tag{6.44}$$

where $L = 4$ nm and

$$k_2 = \sqrt{\frac{2m_e^* E}{\hbar^2}}, \quad k_1 = \sqrt{\frac{2m_e^* (V_0 - E)}{\hbar^2}}. \tag{6.45}$$

13. Derive the tunneling probability (6.38) for the double barrier junction depicted in Fig. 6.22 on page 206.

14. Research how tunneling is used in flash memories, and describe one such commercial flash memory product.

15. Research how field emission is used in displays, and summarize the state of display technology based on field emission.

16. Explain how a negative resistance device can be used to make an oscillator.

Chapter

7

COULOMB BLOCKADE AND THE SINGLE-ELECTRON TRANSISTOR

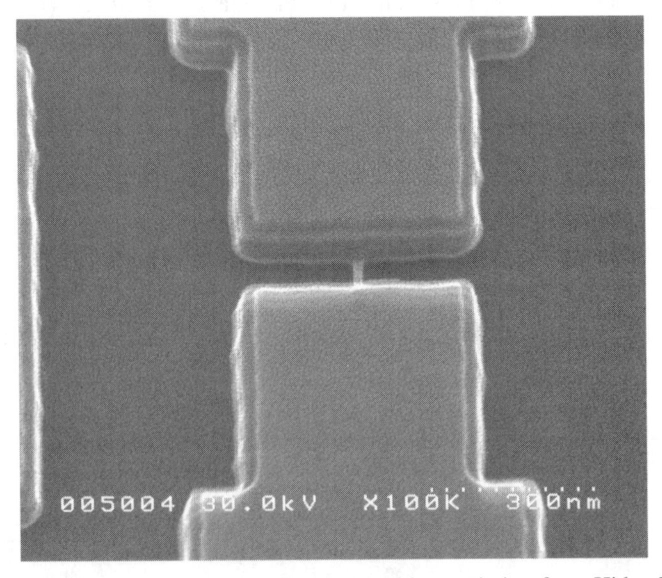

AFM image of a single electron transistor. (Reprinted with permission from Hideo Namatsu, Ph.D., *Journal of Vacuum Science & Technology B: Microelectronics and Nanometer Structures*, January 2003, Volume 21, Issue 1, pages 1–5. © 2007, American Institute of Physics.)

7.1 COULOMB BLOCKADE

With the basics of tunneling described in the preceding chapter, we can finally examine the "connection" between a quantum dot and electrical leads, depicted in Fig. 7.1. Perhaps not surprisingly at this point, it turns out that tunneling is the process by which current can flow from lead to lead through the quantum dot. In this type of situation, the quantum dot, which is merely a very small material region, is also called a *quantum island* or *Coulomb island*. We will model the quantum dot and the exterior leads using the classical concept of capacitance, and consider electron conduction via tunneling, and so we use a

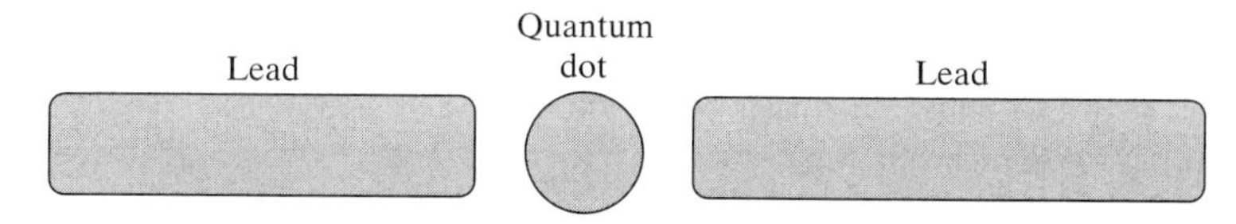

Figure 7.1 Quantum dot coupled to leads. Charge transport through the dot is via tunneling.

mixed classical–quantum model. As we will see, one of the most fundamental effects in nanoelectronics is related to the significant change in energy when a single electron is transferred into a nanoscopic material region, such as a quantum dot, leading to what is known as *Coulomb blockade*.

Early studies of Coulomb blockade arose out of the consideration of electron transport through materials consisting of granular metallic particles. It was realized that current flow through the material could be inhibited if the capacitance of the particle was sufficiently small. However, only more recently has the phenomenon been extensively studied using modern experimental methods. In the following, we first consider a generic tunnel junction consisting of two conductors separated by an insulating gap, after which we examine more complicated structures, leading up to the study of a single-electron transistor.

Recall that for two conductors separated by an insulator, charge and voltage are proportional,

$$Q = CV, \tag{7.1}$$

where Q is the charge on the conductors ($+Q$ on one conductor and $-Q$ on the other), C is the capacitance,† and V is the d.c. voltage between the conductors.

The electrostatic energy stored in a capacitor is

$$E = \frac{1}{2}CV^2 = \frac{Q^2}{2C}, \tag{7.2}$$

which is the energy required to separate the charges initially upon applying the potential across the conductors (i.e., the work that must be done by a source to establish the charge configuration on the capacitor surfaces).

The simplest capacitor is formed by two parallel plates of area A and plate separation d. For this configuration,‡

$$C = \frac{\varepsilon A}{d}, \tag{7.3}$$

where ε is the permittivity of the material between the plates. The capacitance values for several parallel-plate capacitors are given in Table 7.1, where we assume a vacuum between

†As described in basic electrical engineering textbooks, capacitance is a geometric proportionality constant, depending on the size and shape of the conductors, their relative orientation, and the material surrounding the conductors.

‡This equation is approximate, and ignores field fringing.

TABLE 7.1 PROPERTIES ASSOCIATED WITH SEVERAL DIFFERENT CAPACITORS. THE LAST COLUMN IS THE CHANGE IN ENERGY DUE TO THE TRANSFER OF A SINGLE ELECTRON THROUGH THE CAPACITOR.

	A (nm^2)	d (nm)	C (F)	ΔE (eV)
nanocapacitor	5^2	2	1.1×10^{-19}	0.73
nanocapacitor	80^2	1	5.7×10^{-17}	0.0014
mm capacitor	$(5 \times 10^6)^2$	2×10^6	1.1×10^{-13}	7.3×10^{-7}
μ capacitor	$(5 \times 10^3)^2$	2	1.1×10^{-13}	7.3×10^{-7}

the capacitor plates ($\varepsilon = \varepsilon_0$). The first two capacitors have very small nanoscale dimensions, the third capacitor listed in the table is completely macroscopic, and last one has micron plate size and nanoscopic thickness. As can be seen, the nanoscale capacitors have extremely small values of capacitance.

For nanoscale dimensions, the small values of capacitance lead to the interesting observation that the transfer of a single electron results in an appreciable energy change. For example, for the nanoscale capacitor having $C = 1.1 \times 10^{-19}$ F, by (7.2), the transfer of a single electron ($\Delta Q = q_e$) yields a change in energy ΔE that is an appreciable fraction of an electron volt. For the macroscopic capacitor ($C = 1.1 \times 10^{-13}$ F), the change in energy due to the transfer of a single electron is negligible. For nanoscale capacitors, it is this sensitivity to the transfer of an extremely small amount of charge that allows for the possibility of so-called single-electron precision devices. For these small capacitance values, the prefixes femto (f), 10^{-15}, and atto (a), 10^{-18}, are useful.

7.1.1 Coulomb Blockade in a Nanocapacitor

Consider a small capacitor made by creating an insulating layer between two metal surfaces, and assume that the capacitor plates carry charges $\pm Q$, as shown in Fig. 7.2. The initial energy stored in the electrostatic field between the capacitor plates is given by (7.2),

$$E^i = \frac{Q^2}{2C}. \tag{7.4}$$

Note that even though charge itself is quantized (in units of $|q_e|$), the charge Q on the capacitor plates is not quantized. The charge on the capacitor plates (sometimes called a *polarization charge*) merely represents a displacement of electrons relative to a background of positive ions. That is, certain electrons may be found nearer to, or farther from, the material's surface, depending on the situation. These discrete charges can, in this way, contribute a continuous amount of charge to the interface, as depicted in Fig. 7.3.

Therefore, the discreteness of electronic charge does not show up in macroscopic circuits for two reasons. First, because there are so many electrons involved (for example, a 1 mA d.c. current corresponds to the movement of 6.25×10^{15} electrons/second, since A = C/s), and second, because discrete charges can contribute continuous amounts of charge to an interface.

Now let a single electron tunnel through the insulating layer from the negative terminal to the positive terminal, such that charge $Q + q_e$ resides on the top plate, and $-Q - q_e$

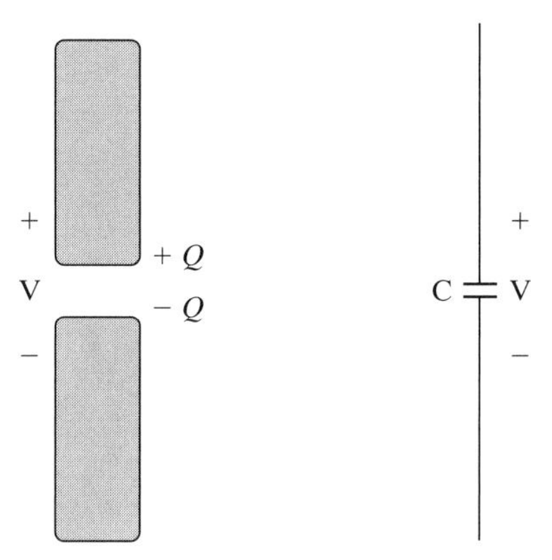

Figure 7.2 Capacitor circuit showing induced charges $\pm Q$ and voltage. The physical structure is depicted on the left, and the circuit model is shown on the right.

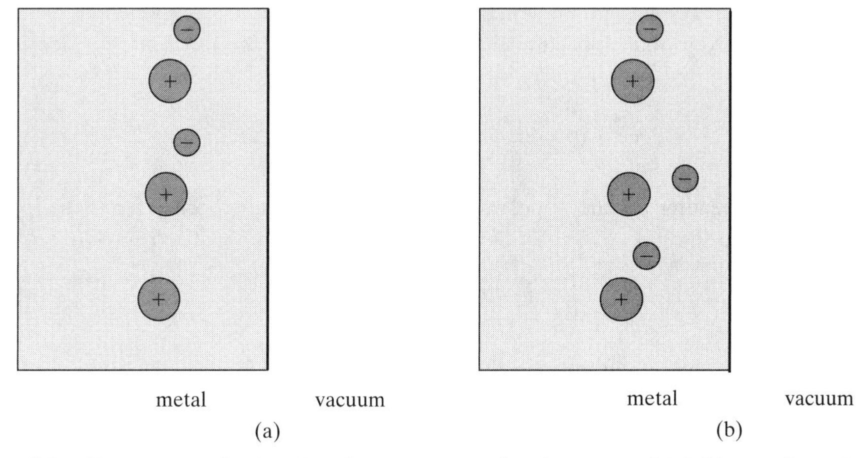

Figure 7.3 Charges near the interface between a metal and vacuum. (a) Initial configuration. (b) Configuration where an electron is nearer to the interface than in (a), contributing some (nonintegral) amount of charge to the surface.

resides on the bottom plate.[†] The energy stored in the field of the capacitor is now

$$E^f = \frac{(Q + q_e)^2}{2C}, \tag{7.5}$$

[†]We assume that tunneling is virtually instantaneous; in reality, tunneling time τ_t is finite, but typically very short. For nanoscopic regions separating metals, τ_t is roughly on the order of 10^{-14} s. For an approximate estimate, assume a tunnel barrier having $L = 10$ nm, and that electrons at the Fermi energy tunnel. It is shown in Section 10.1.2 that electrons at the Fermi energy have an associated Fermi velocity v_F, and for a typical metal $v_F \sim 10^6$ m/s. Therefore, $\tau_t = L/v_F \sim 10 \times 10^{-14}$ s.

Furthermore, charge does not redistribute on the capacitor plates instantaneously, but in some characteristic time τ_p. For a good conductor, typically $\tau_p \simeq 10^{-18}$ s; this time is derived in standard electromagnetics textbooks. Both characteristic times τ_t and τ_p are small enough to ignore in our analysis.

such that the change in energy stored is

$$\Delta E = E^i - E^f = \frac{-q_e (Q + q_e/2)}{C}. \tag{7.6}$$

It must be energetically favorable for the tunneling event to occur, and if we require $\Delta E > 0$, we find that

$$Q > \frac{-q_e}{2} \tag{7.7}$$

for tunneling to occur. In terms of a condition on voltage, from (7.1), we have

$$V > \frac{-q_e}{2C}. \tag{7.8}$$

Recall that $-q_e > 0$.

Energy is conserved, and since the stored energy decreases upon tunneling, the electron ends up above the Fermi level (an increase in kinetic energy) on the other side of the junction. Note that upon tunneling, the voltage over the junction will decrease by $|q_e|/C$ such that

$$V^i = \frac{Q}{C} = \frac{-q_e}{2C} \rightarrow V^f = \frac{Q - |q_e|}{C} = -\frac{|q_e|}{2C}. \tag{7.9}$$

Repeating for the opposite polarity, we find that in order for tunneling to occur, we must have

$$\frac{-q_e}{2C} < V < \frac{q_e}{2C}. \tag{7.10}$$

That is, tunneling current will only flow when a sufficiently large voltage ($|V| > |q_e|/2C$) exists across the capacitor. This effect is called *Coulomb blockade*. Since $q_e V$ is energy,

$$E_c = \frac{q_e^2}{2C} \tag{7.11}$$

is called the *charging energy* of the capacitor. This is the energy required to add a charge to the capacitor, which is, itself, a classical idea based on charge repulsion. The resulting I–V characteristics depicting Coulomb blockade are shown in Fig. 7.4.[†]

For example, in the small yet macroscopic capacitor considered in Table 7.1 on page 214, where $C = 1.1 \times 10^{-13}$ F, we only need $|V| > 0.73$ μV for tunneling to occur. Note, however, that regardless of the voltage, capacitor plate separation d must be adequately small such that tunneling can take place, which is the difference between the last two capacitors in Table 7.1. For the nanoscale capacitor with $C = 5.7 \times 10^{-17}$ F, we need

[†]The situation can, of course, be a bit more complicated. For example, if one considers the capacitance of the connecting leads, which may be large, C is increased and, hence, E_c is decreased. Then, the electron can tunnel with a charging energy that is lower than expected, based on (7.11), using C for the junction itself. In this case, it must be interpreted that the electron tunnels some distance down the leads, rather than simply to the other capacitor plate.

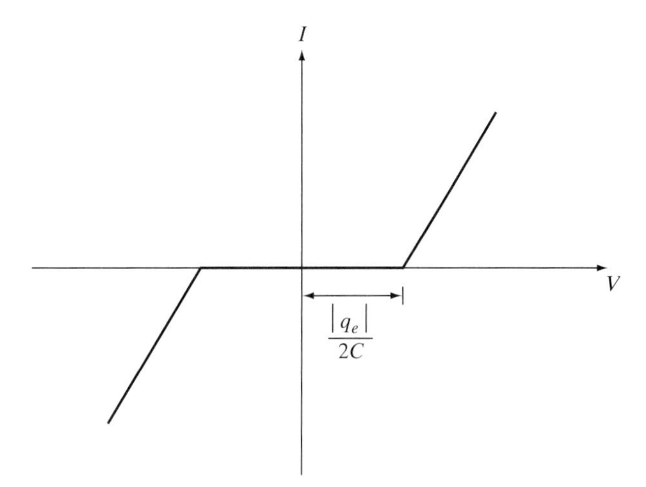

Figure 7.4 Coulomb blockade effect in a nano-scale capacitor.

$|V| > 1.4$ mV to have tunneling, and for the nanoscale capacitor with $C = 1.1 \times 10^{-19}$ F, $|V| > 0.73$ V is required. Thus, Coulomb blockade is not evident in macro-sized circuits because the charging energy is so low, but may be observed in nanometer scale circuits.

Temperature plays an important role in Coulomb blockade phenomena. For example, the preceding development is only strictly valid at $T = 0$ K. For higher temperatures, thermal energy is available such that the condition $\Delta E > 0$ (where E is based on (7.2)) can be relaxed. We typically need the charging energy to be greater than the thermal energy,

$$\frac{q_e^2}{2C} \gg \frac{1}{2} k_B T, \tag{7.12}$$

or

$$C \ll \frac{q_e^2}{k_B T}, \tag{7.13}$$

to observe Coulomb blockade, where k_B is Boltzmann's constant. Put simply, if (7.13) is not satisfied there is too much thermal energy available to the electron to be held back by the charging energy E_c of the capacitor. For the nanoscale capacitor considered previously ($C = 1.1 \times 10^{-19}$ F), $T \ll 16,911$ K, which is obviously the typical situation. Therefore, we should be able to observe Coulomb blockade for this capacitor at any practical temperature. However, for the macroscale capacitor ($C = 1.1 \times 10^{-13}$ F), $T \ll 0.017$ K, which is difficult to achieve. (This, of course, explains why Coulomb blockade is not seen in macroscale electronics at room temperature.) For the somewhat larger nano-scale capacitor ($C = 5.7 \times 10^{-17}$ F), $T \ll 32.6$ K.

To further consider the dimensions of the capacitor and the relation to temperature effects, consider that with (7.3), $C = \varepsilon A/d$, (7.13) becomes

$$\frac{A}{d} \ll \frac{q_e^2}{\varepsilon k_B T} = 7.17 \times 10^{-7} \tag{7.14}$$

for air separating the plates at $T = 293$ K. Assuming square capacitor plates of length L, and, for convenience, assuming that $d = L/10$, then to observe Coulomb blockade we must have

$$L \ll 70 \text{ nm}. \tag{7.15}$$

How far below 70 nm is necessary depends on the application to some extent, but usually a factor of between 10 (i.e., $L \sim 7$ nm) and 100 (i.e., $L \sim 0.7$ nm) would be sufficient. The main point is that if the junction capacitance is sufficiently small, room-temperature Coulomb blockade can be observed; otherwise, low temperatures must be used.

We can also gain an appreciation of the effect of thermal fluctuations by considering the following argument. If, rather than having a single-electron tunnel, we assume that n electrons tunnel as a group, i.e., that a charge of nq_e is the tunneling object, it is easily shown that

$$\frac{-nq_e}{2C} < V < \frac{nq_e}{2C}, \qquad E_c = \frac{(nq_e)^2}{2C}, \tag{7.16}$$

replace the expressions derived previously for the single charge case. Equating the charging energy to the thermal energy, we have

$$n = \sqrt{\frac{Ck_BT}{q_e^2}}, \tag{7.17}$$

which is, roughly, the number of electrons (or the uncertainty in the number of electrons) that could be transferred due to thermal energy. Assuming room temperature operation ($T = 293$ K), if $C = 1$ fF then $n = 13$ electrons can be transferred by thermal energy. However, if $C = 7$ aF, then $n = 1$. It is obvious that as C and/or T decrease, the uncertainty in the number of electrons transferred can be reduced to an acceptable value, at which point Coulomb blockade can be observed.

When is a Capacitor a Capacitor? Note that at the small-length scales considered for nanocapacitors, the concept of capacitance used earlier, the so-called "lumped" capacitance, is a bit problematic. First of all, from the viewpoint of classical electromagnetics, a parallel-plate capacitor will "act like" a capacitor when the capacitor dimensions and frequency of operation are such that $L \ll \lambda$, where L is the largest dimension of the capacitor and where λ is the usual electromagnetic wavelength. That is, the capacitor should be small compared to a wavelength; otherwise, a full electromagnetic analysis needs to be performed. If the condition $L \ll \lambda$ is not satisfied, the capacitor will also be shown to have inductive and radiative effects.†

For nanometer scale capacitors, L will be generally very small compared to an electromagnetic (photon) wavelength. Therefore, nanoscale capacitors can clearly be treated as lumped components in the (classical) electromagnetic sense. However, at the nanoscale, an

†Resistive effects, due to the capacitor being made from imperfect conductors, will also generally be present.

important wavelength is the de Broglie wavelength of the electron, $\lambda = h/p$, especially the value at the Fermi level, λ_F. This value typically ranges from a fraction of a nanometer in metals to tens of nanometers in semiconductors. Therefore, it may occur that the capacitor dimensions are not very small compared to this wavelength. In particular, if the separation between the conductors is small compared with λ_F, or on the order of λ_F , the electron is *delocalized*. That is, the electron is spread over the entire capacitor, and so it does not reside on either one of the conductor plates but is shared over both conductors.† This is discussed a bit further in the following, in relation to the tunneling resistance.

Nevertheless, the classical concept of capacitance yields results that are surprisingly good. In a more thorough quantum mechanical treatment, particles can actually tunnel through the Coulomb blockade, and quantum uncertainty in the junction charge should be taken into account.

Figure 7.5 shows an SEM image of source and drain leads separated by a 20 nm gap that contains arrays of sub-5 nm Au particles. For these structures, the $I-V$ curve for an 8×5 array of Au particles is shown in Fig. 7.6. (Curves have been offset for clarity.) Temperature dependence of the Coulomb blockade is clearly evident.

However, as described previously, Coulomb blockade phenomena can occur at room temperature (or higher) if the capacitance values are sufficiently small, corresponding to small particles. Fig. 7.7 shows the room temperature $I-V$ relationship for composite metallic particles (made up of several materials) having radius 2.1 nm.

7.1.2 Tunnel Junctions

The tunneling that occurs across the capacitor can be accounted for by considering the capacitor to be a leaky capacitor,‡ modeled by an ideal capacitance in parallel with a resistance $R_t = V/I$, where V is a d.c. voltage applied across the junction and I is the resulting current due to tunneling. This tunneling resistance is not an ordinary resistance, but conceptually allows electrons to cross the insulating junction as discrete events. The parallel combination of the capacitor and the tunneling resistance is called a *tunnel junction*, and is depicted in Fig. 7.8.

The tunneling effect accounted for by R_t should be weak enough (i.e., R_t should be sufficiently large) to prevent the charge of the tunneling electrons from becoming delocalized over (i.e., shared by) the capacitor plates. However, R_t must be finite, and not too large, so that tunneling can actually take place. In this case, the charge on the island is said to be "well quantized," and the leaky capacitor is considered to be a *tunnel junction*. Considering the discussion of tunneling in Section 6.1, we can appreciate that R_t is related to the potential energy barrier width, i.e., the separation between the capacitor plates.

In order to see Coulomb blockade, we need to limit tunneling to some degree, and we can get an estimate of this by considering the uncertainty relation between time and

†This is consistent with the idea of the "size" of an electron being its de Broglie (or Fermi) wavelength.

‡In macroscopic circuit theory a leaky capacitor is one where the dielectric separating the capacitor plates is an imperfect insulator. The result is that a (typically small) d.c. current will flow from one capacitor plate to the other when a d.c. voltage is applied across the capacitor. This current flow is modeled by an ordinary resistance R. The capacitor considered here is leaky due to tunneling, leading to a tunneling resistance R_t.

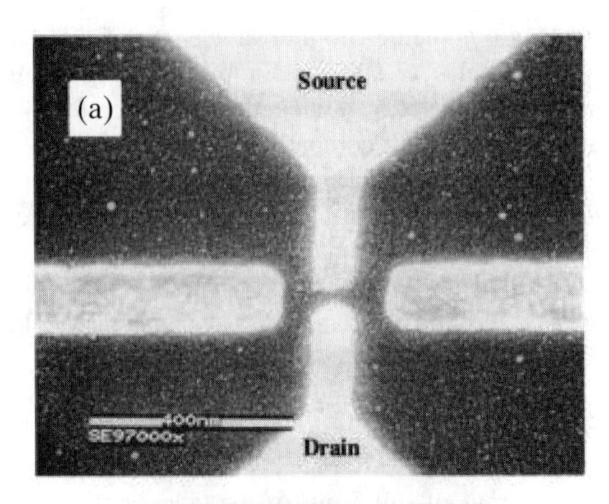

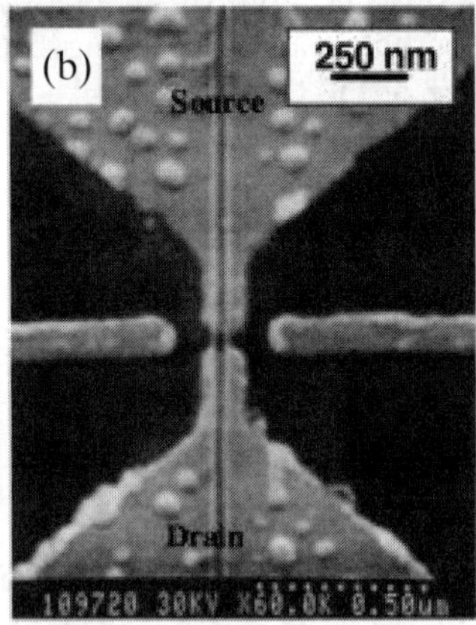

Figure 7.5 SEM images of tunnel junction devices made with 2D arrays of Au islands. (a) The Au islands cover the whole surface, and the width of the array is roughly given by the width of the source and drain electrodes (here, 80 nm; the interelectrode gap is 20 nm). (b) A patterned array where the islands are only kept under a narrow resist line between the contact electrodes. (From Pépin, A., et al., "Temperature Evolution of Multiple Tunnel Junction Devices Mode with Disordered Two-Dimensional Arrays of Metallic Islands," *Appl. Phys. Lett.* 74 (1999): 3047. © 1999, American Institute of Physics.)

energy (2.45),

$$\Delta E \Delta t \geq \hbar/2. \tag{7.18}$$

Given that the time constant of a parallel RC circuit is

$$\tau = RC, \tag{7.19}$$

then $\tau = R_t C$ is a characteristic time associated with tunneling events. This is not the time to tunnel through the junction, but, rather, the time between tunneling events; τ is considered

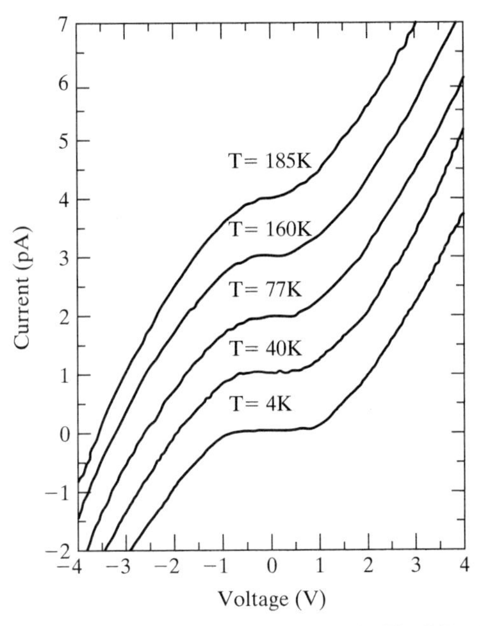

Figure 7.6 Typical $I-V$ characteristics of the devices shown in Fig 7.5, where an 8×5 array of Au islands occur in the interelectrode gap. (Curves have been offset for clarity.) (From Pépin, A., et al., "Temperature Evolution of Multiple Tunnel Junction Devices Mode with Disordered Two-Dimensional Arrays of Metallic Islands," *Appl. Phys. Lett.* 74 (1999): 3047. © 1999, American Institute of Physics.)

to be the approximate lifetime of the energy state of the electron on one side of the barrier. Thus, we then have an uncertainty in energy

$$\Delta E \geq \frac{\hbar}{2R_t C}. \tag{7.20}$$

To observe the Coulomb blockade effect, the charging energy must be much larger than this uncertainty, such that[†]

$$R_t \gg \frac{\hbar}{q_e^2} \simeq 4.1 \text{ k}\Omega. \tag{7.22}$$

Note that we really have two effects to consider, capacitance value, and the possibility of tunneling. First, in order to observe Coulomb blockade, we need very small values of

[†]Another, more careful analysis based on tunneling rates, called the *orthodox theory*, leads to

$$R_t \gg R_0 = \frac{h}{q_e^2} \simeq 25.8 \text{ k}\Omega, \tag{7.21}$$

where R_0 is called the *resistance quantum* (a name that is also used for the same quantity divided by two; see Section 10.2.3). As described previously, in order to observe Coulomb blockade, the electron should not be delocalized over the junction, which can occur unless the condition $R_t \gg R_0$ is met.

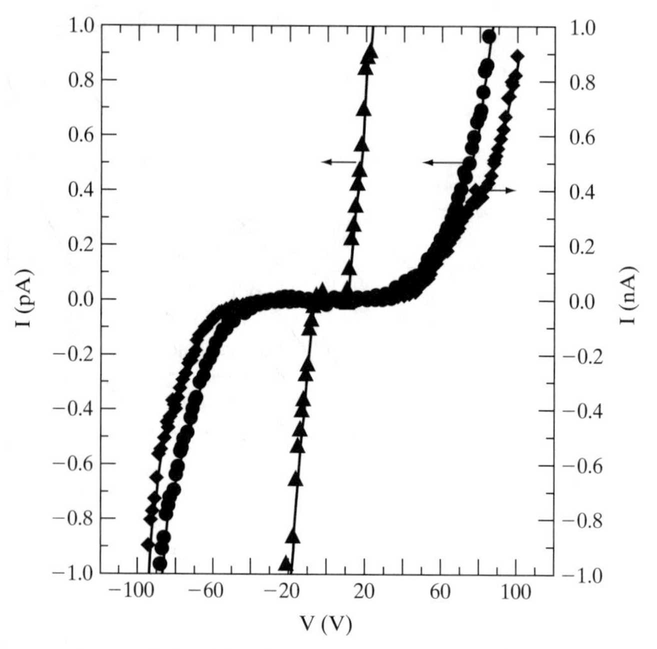

Figure 7.7 Current–voltage relationship of metal particles at 195 K (triangle) and 295 K (circle and diamond). (From Clarke, L., et al., "Room-Temperature Coulomb-Blockade Dominated Transport in Gold Nanocluster Structures," *Semicond. Sci. Technol.* 13 (1998): A111–A114. Courtesy Martin Wybourne.)

capacitance to obtain a sufficiently large charging energy. For the parallel-plate example (7.3),

$$C = \frac{\varepsilon A}{d}, \tag{7.23}$$

we see that small capacitance can be achieved from either a small plate area A, or a large plate separation d. From the charging energy standpoint, it makes no difference how the small value of C is obtained. However, we need an electron to be able to tunnel through the insulating region, and from (6.15), and, in particular, (6.27), the thickness of the insulating region must not be too great if tunneling is to occur. This rules out, in essence, creating a small capacitance by using large d, since this would lead to a too-large value of R_t.

7.1.3 Tunnel Junction Excited by a Current Source

An interesting circuit that leads to a surprisingly complex phenomenon consists of a tunnel junction fed by a constant current source I_s, as depicted in Fig. 7.9.

For a classical ideal capacitor (no tunneling allowed), the relation between voltage and current,

$$I = C\frac{dV}{dt}, \tag{7.24}$$

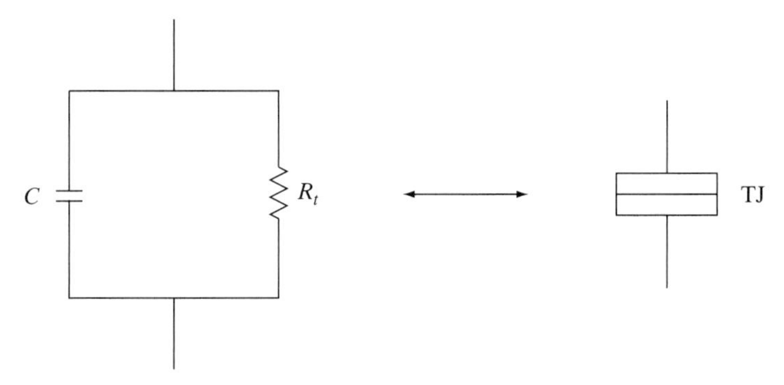

Figure 7.8 Left side: ideal capacitor in parallel with a tunnel resistance, the combination of which models an actual capacitor through which tunneling can occur (i.e., a tunnel junction). Right side: circuit symbol of a tunnel junction.

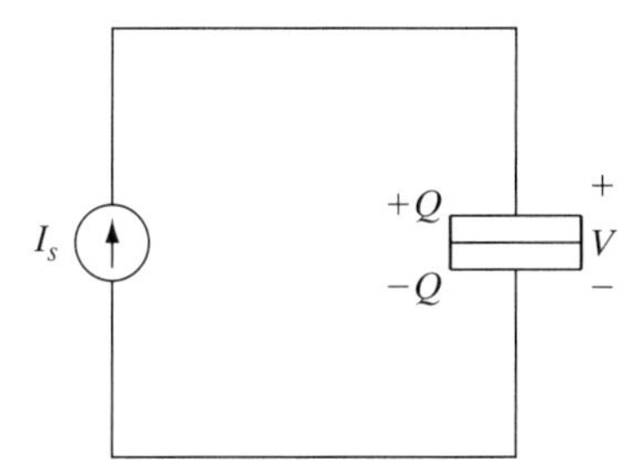

Figure 7.9 Tunnel junction excited by a constant current course.

leads to the conclusion that the voltage over a current-source-driven capacitor will increase linearly with respect to time,

$$V(t) = \frac{1}{C} \int_0^t I_s \, dt = \frac{I_s}{C} t, \tag{7.25}$$

as will the charge,

$$Q(t) = I_s \, t. \tag{7.26}$$

Even though no d.c. current can pass through the capacitor, one has the idea that the current source is pushing electrons onto the capacitor plate, where they are stored, building up the voltage and accumulated charge. For a real capacitor that admits tunneling, at the point when

$$Q > \frac{-q_e}{2}, \tag{7.27}$$

or, equivalently,

$$V > \frac{-q_e}{2C}, \tag{7.28}$$

tunneling from the lower to the upper capacitor plate can occur (since Coulomb blockade is then thwarted). This results in a decrease in the positive charge on the top plate,

$$Q \to Q + q_e = Q - |q_e|\,, \tag{7.29}$$

and an increase in the positive charge on the bottom plate,

$$-Q \to -Q - q_e = -Q + |q_e|\,. \tag{7.30}$$

The voltage then decreases by $|q_e|/C$ due to (7.9), and the process repeats itself, as shown in Fig. 7.10. The net result is that the voltage across the capacitor oscillates, which is known as *single-electron tunneling (SET) oscillations*.

The period of oscillation can be calculated using (7.25) as

$$T = \frac{|q_e|}{I_s} = \frac{e}{I_s}. \tag{7.31}$$

Therefore, the time between tunneling events is on the order of 10^{-10} s assuming nanoamp currents (i.e., $|q_e|/10^{-9} = 1.6 \times 10^{-10}$), which is much larger than the typical tunneling time of 10^{-14} s (discussed on page 215).

It turns out that the previously described oscillation is difficult to observe due to experimental factors, such as the generally high lead capacitance in experimental systems. To assess this issue, consider the world as seen by the tunnel junction. In a circuit model, the tunnel junction is connected to external bias and measurement circuitry by some lengths of electrical leads. The fact that there is no direct electrical contact between the island and the leads is modeled by the tunnel junction itself, via the tunnel resistance R_t and junction capacitance C. Using a Norton equivalent circuit for the external circuitry, we have the lumped element model shown in Fig. 7.11.

In this figure, I_s is an ideal current source, R_b is an equivalent bias resistance (taken such that $R_b \gg R_t$), C_b is the parasitic capacitance of the bias circuitry, and $Z_L = Z_L(\omega)$

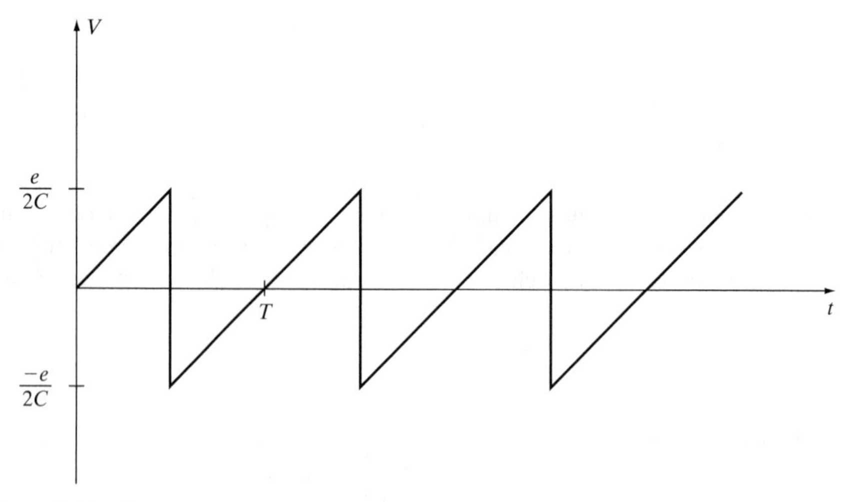

Figure 7.10 SET oscillations when a tunnel junction is driven by a constant current source.

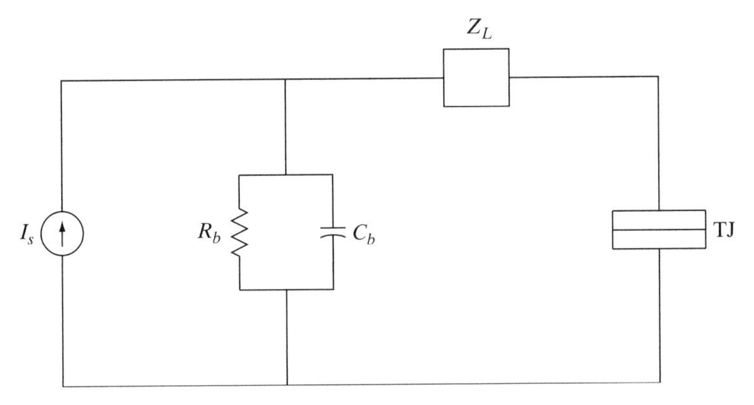

Figure 7.11 Current biased tunnel junction, including external circuitry.

is the impedance of the electrical leads closest to the junction. Since C_b is typically quite large, $C_b \gg C$, this capacitance remains charged as tunneling events occur (i.e., the time constant of the external circuitry is large compared with the time constant of the tunnel junction). Therefore, approximately, this capacitance acts like a constant voltage source V, resulting in the circuit depicted in Fig. 7.12.

At this point, we can identify two regimes of operation. First, for $|Z_L(\omega)|$ sufficiently small we have essentially a voltage source in parallel with the tunnel junction. As soon as charge tunnels through the junction, it is removed by the voltage source, to maintain V volts across the junction at all times. A careful treatment shows that this is, in fact, a purely quantum effect, and is valid even though the effective voltage source is located some distance from the junction. That is, the junction together with its environment behaves as a quantum mechanical system. In this regime there will be no Coulomb blockade.

This low impedance regime is typically characterized by

$$|Z_L(\omega)| \ll R_0 = \frac{h}{q_e^2} \simeq 25.8 \text{ k}\Omega. \tag{7.32}$$

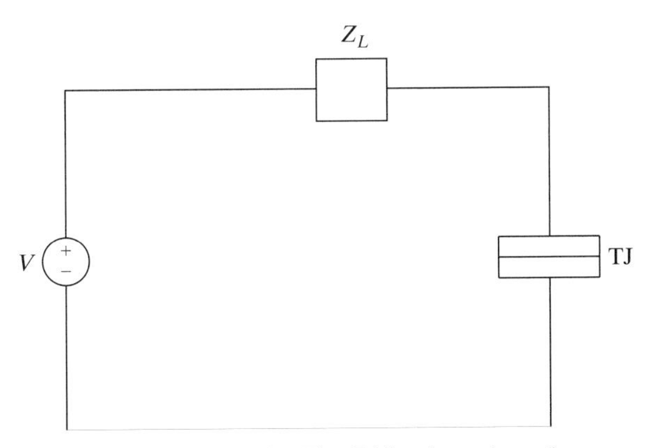

Figure 7.12 Simplified equivalent circuit for Fig. 7.11, where the voltage source accounts for the biasing of the large capacitance C_b in Fig. 7.11.

In particular, if we consider sufficiently high frequencies such that energy is radiated from the leads (like a classical antenna), it can be shown using ordinary transmission line analysis that $Z_L \simeq 377$ ohms, which is quite a low value of impedance.

The second regime of operation to consider is $|Z_L(\omega)|$ sufficiently large, usually taken as $|Z_L(\omega)| \gg R_K$. In this case, sufficient voltage is dropped over Z_L so that the junction can really act like a tunnel junction (i.e., the voltage over the junction is not fixed by the voltage source), and the oscillations depicted in Fig. 7.10 can occur. However, this case is difficult to achieve in practice.

7.1.4 Coulomb Blockade in a Quantum Dot Circuit

We are now in a position to consider a metallic quantum dot coupled to external leads, as shown previously in Fig. 7.1. We can model the system with the equivalent circuit shown in Fig. 7.13, where, to a first approximation, we can ignore the external environment. This is a valid approximation due to the presence of two tunnel junctions (and, hence, of an isolated charge island). This ensures some amount of decoupling from the environment, and also ensures that neither junction voltage will be held constant. In fact, characteristics similar to those described next would be obtained if one of the tunnel junctions was replaced by an ideal (no tunneling) capacitor.

The tunnel junctions represent the insulating regions isolating the island (dot) from the leads connected to an external source V_s. The tunnel junctions can be modeled by leaky capacitors, i.e., a capacitance in parallel with a resistance.† Therefore, for one junction, we have C_a, and R_{ta}, and for the other junction, C_b, and R_{tb}, where R_{ti} is the tunneling resistance of junction i. Associated with each tunnel junction, we have

$$Q_a = C_a V_a, \quad Q_b = C_b V_b, \tag{7.33}$$

such that the net charge on the island is the difference

$$Q = Q_b - Q_a. \tag{7.34}$$

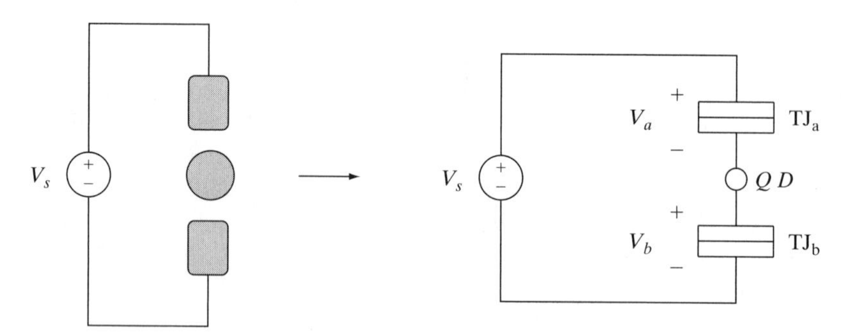

Figure 7.13 Equivalent circuit of a quantum dot connected to a voltage source via tunnel junctions.

†Since the dot is very small and may contain a relatively small number of electrons, the capacitance between the dot and a lead will actually be a function of the number of electrons on the dot, $C = C(n)$ (i.e., the dot does not act like a bulk metal), although we will ignore this complication here.

Electron tunneling allows a discrete number of electrons to accumulate on the island, and so

$$Q = nq_e, \tag{7.35}$$

where n is an integer. Note that in this semiclassical model, the discrete index n and the continuous charge Q are classical (not quantum) quantities, with well-defined values at all times.

The energy stored in the capacitors is

$$E_{se} = \frac{Q_a^2}{2C_a} + \frac{Q_b^2}{2C_b} = \frac{1}{2}\left(C_a V_a^2 + C_b V_b^2\right). \tag{7.36}$$

Kirchhoff's voltage law applied to the circuit yields

$$\begin{aligned} V_s &= V_a + V_b = V_a + \frac{Q_b}{C_b} \\ &= V_a + \frac{(Q + Q_a)}{C_b} = V_a + \frac{(Q + C_a V_a)}{C_b}, \end{aligned} \tag{7.37}$$

leading to, using $Q = nq_e$,

$$V_a = \frac{1}{C_s}\left(V_s C_b - nq_e\right), \tag{7.38}$$

where $C_s = C_a + C_b$. Similar manipulations lead to

$$V_b = \frac{1}{C_s}\left(V_s C_a + nq_e\right), \tag{7.39}$$

and we can write the stored energy (7.36) as

$$E_{se} = \frac{1}{2C_s}\left(C_a C_b V_s^2 + (nq_e)^2\right). \tag{7.40}$$

When charge tunnels across a junction, which is a current flow event, the power supply must do work transferring charge. This work is computed by integrating the power delivered to the junctions,

$$W = \int_{\Delta t} V_s I(t)\, dt = \int_{\Delta t} V_s \frac{dq}{dt} dt = \Delta Q V_s, \tag{7.41}$$

where ΔQ is the total charge transferred from the power supply during the charging period Δt. To determine this charge, we examine what happens when a tunneling event occurs.

Assume that initially there are n electrons on the island, and that the initial voltages across the junctions a and b are V_a^i and V_b^i, respectively, given by (7.38) and (7.39). The initial charges on the junctions are Q_a^i and Q_b^i. Let one electron tunnel onto the island through junction b. The voltage drops across junctions a and b become

$$V_a^f = \frac{1}{C_s}(V_s C_b - (n+1)\, q_e) = V_a^i - \frac{q_e}{C_s}, \tag{7.42}$$

$$V_b^f = \frac{1}{C_s}(V_s C_a + (n+1)\, q_e) = V_b^i + \frac{q_e}{C_s}, \tag{7.43}$$

such that $V_s = V_a + V_b$ is maintained, and the resulting charge stored by junction a is

$$Q_a^f = C_a V_a = C_a V_a^i - \frac{C_a q_e}{C_s} = Q_a^i - \frac{C_a q_e}{C_s}. \tag{7.44}$$

(Note that the charge supplied by the voltage source is not simply the one electron that tunnels across the junction.) The change in charge $\Delta Q_a = Q_a^i - Q_a$ must come from the power supply (the change in charge on junction b is associated with the tunneling event), and is associated with the supply doing work

$$W = V_s \frac{C_a q_e}{C_s}. \tag{7.45}$$

Upon the electron tunneling onto the island through junction b, the change in the total energy (the change in the stored energy minus the work done) is

$$\begin{aligned} \Delta E_t &= \Delta E_{se} - W \\ &= \frac{1}{2C_s}\left(C_a C_b V_s^2 + (n q_e)^2\right) - \frac{1}{2C_s}\left(C_a C_b V_s^2 + ((n+1)\, q_e)^2\right) - V_s \frac{C_a q_e}{C_s}. \end{aligned} \tag{7.46}$$

Requiring that this energy change be positive, i.e., that the tunneling event is energetically favorable, we have

$$\frac{-q_e}{C_s}\left(q_e\left(n + \frac{1}{2}\right) + V_s C_a\right) > 0, \tag{7.47}$$

such that

$$V_s > \frac{-q_e}{C_a}\left(n + \frac{1}{2}\right). \tag{7.48}$$

In a similar manner, when an electron tunnels off of the island through junction a, the voltages across the junctions change as

$$V_a^f = \frac{1}{C_s}(V_s C_b - (n-1)\, q_e) = V_a^i + \frac{q_e}{C_s}, \tag{7.49}$$

$$V_b^f = \frac{1}{C_s}(V_s C_a + (n-1)\, q_e) = V_b^i - \frac{q_e}{C_s}, \tag{7.50}$$

and the resulting charge stored in junction b is

$$Q_b^f = C_b V_b = C_b V_b^i - \frac{C_b q_e}{C_s} = Q_b^i - \frac{C_b q_e}{C_s},$$

associated with work

$$W = V_s \frac{C_b q_e}{C_s}. \tag{7.51}$$

The change in total energy is

$$\begin{aligned}\Delta E_t =& \Delta E_{se} - W \\ =& \frac{1}{2C_s}\left(C_a C_b V_s^2 + (nq_e)^2\right) - \frac{1}{2C_s}\left(C_a C_b V_s^2 + ((n-1)\,q_e)^2\right) - V_s \frac{C_b q_e}{C_s}.\end{aligned}$$

Requiring this to be positive (the system must evolve from a higher-to a lower-energy state, in the absence of any other energy input), we have

$$\frac{q_e}{C_s}\left(q_e\left(n - \frac{1}{2}\right) - V_s C_b\right) > 0, \tag{7.52}$$

which leads to

$$V_s > \frac{q_e}{C_b}\left(n - \frac{1}{2}\right). \tag{7.53}$$

As a special case, if we let $C_a = C_b = C$ and $n = 0$, then (7.48) and (7.53) both reduce to

$$V_s > \frac{-q_e}{2C}. \tag{7.54}$$

Now consider the opposite situation, where an electron tunnels onto the island through junction a, and off of the island through junction b. For $C_a = C_b = C$ and $n = 0$, we would find

$$V_s < \frac{q_e}{2C}. \tag{7.55}$$

Thus, under these conditions, we have

$$\frac{-q_e}{2C} < V_s < \frac{q_e}{2C}, \tag{7.56}$$

or

$$|V_s| > \frac{e}{2C}, \tag{7.57}$$

the same Coulomb blockade as encountered for the single capacitor, (7.10), and associated with the charging energy $E_c = q_e^2/2C$.

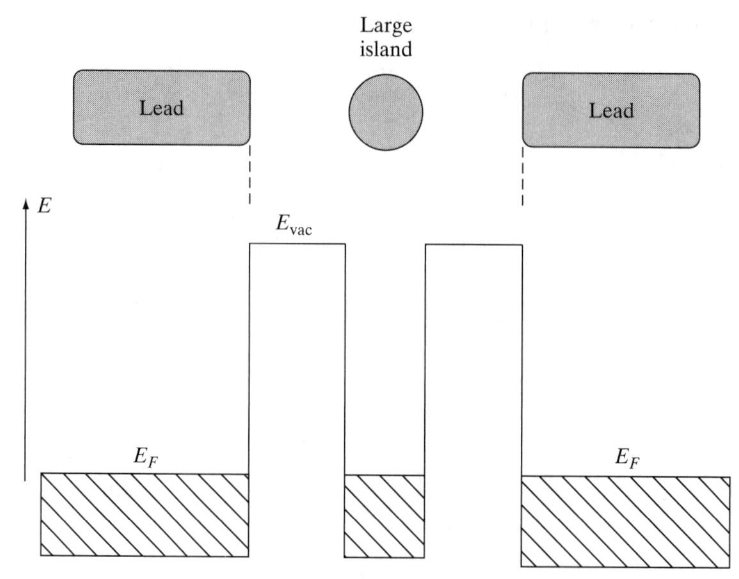

Figure 7.14 Large metallic island coupled to leads made of the same material at thermal equilibrium, in the absence of an applied bias. Fermi energy of the island correlates with the bulk value.

To obtain some further insight into this phenomenon, it is worthwhile to examine the energy band diagram for this structure. First, consider a junction with identical capacitances ($C_a = C_b = C$), and one where the island is very large. In this case, the charge on the island is not well discretized, and the Fermi energy on the island is approximately that of the bulk material. Assuming that the island and leads are made of the same material, and are at thermal equilibrium, we find that the energy band diagram is as shown in Fig. 7.14.

If a potential difference V_0 is applied across the structure, the energy levels will shift as shown in Fig. 7.15, allowing tunneling to occur onto or off of the island.

Now let the island be very small, such that the energy levels are well discretized and the charging energy is significant compared to the Fermi level.† In this case, for which the previous development applies, the charging energy is represented by an energy gap (a range of energies for which there are no states) of size $e^2/2C$, which appears symmetrically about the Fermi level if the junctions are symmetric, as shown in Fig. 7.16. On the island, empty states appear above the gap.

If a voltage is applied across the structure, then the band diagram shifts in a manner similar to Fig. 6.8 on page 194. If the applied voltage is slightly larger than $e^2/2C$, then the band diagram becomes as shown in Fig. 7.17, allowing an electron to tunnel onto the island.

However, after the electron tunnels onto the island, the energy on the island is increased by $e^2/2C$, readjusting the energy bands, and Coulomb blockade is re-established

†The presence of the leads will change somewhat the energy states on the island, compared with the case of an isolated island. However, this effect will not be considered here.

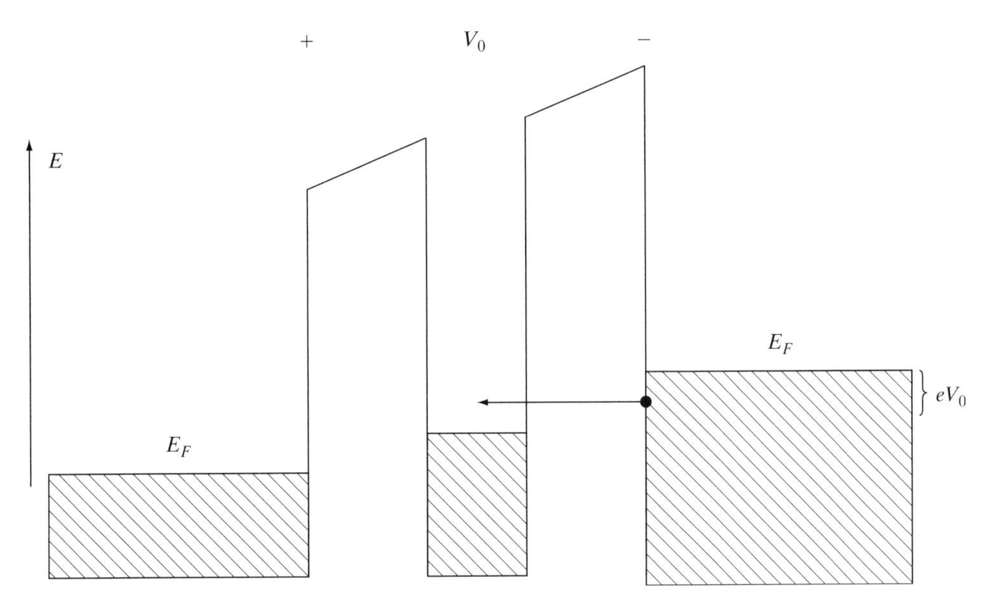

Figure 7.15 Large metallic island coupled to leads in the presence of an applied bias V_0.

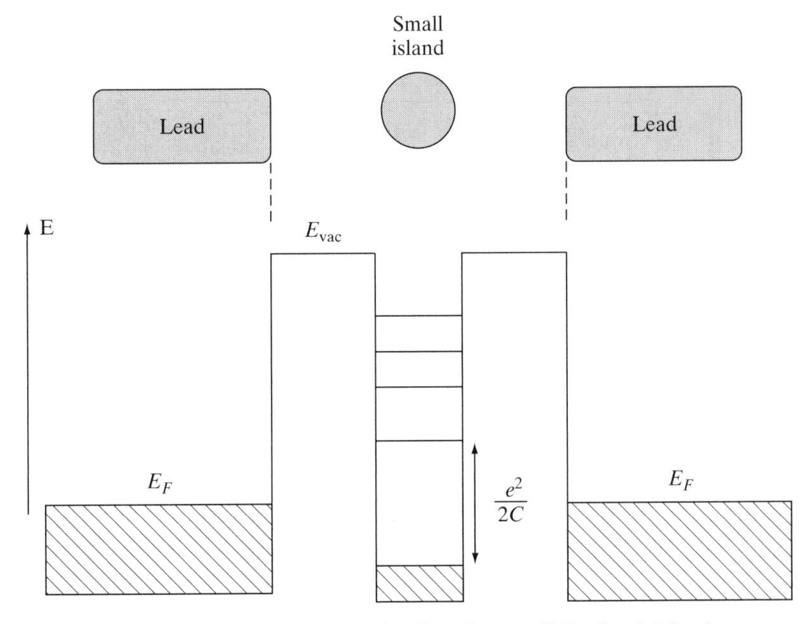

Figure 7.16 Energy band diagram for the double junction small Coulomb island structure. Discrete energy levels on the island are shown, and the energy gap $e^2/2C$ occurs due to the charging energy.

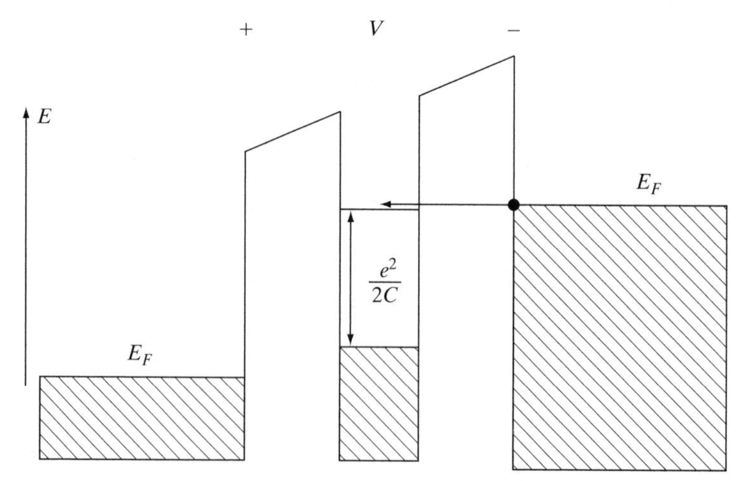

Figure 7.17 Same as Fig. 7.16, except that an applied potential V_0 slightly larger than the charging potential allows tunneling of an electron into the lowest state on the island, as shown.

until either $|V| > 3e/2C$ (as explained later) or the electron tunnels off of the island, reducing the island's energy by $e^2/2C$.

Now assume that one electron has already tunneled onto the island through junction b. If a second electron tunnels onto the dot through junction b, the change in energy is

$$\begin{aligned}\Delta E_t &= \Delta E_{se} - W \\ &= \frac{1}{2C_s}\left(C_a C_b V_s^2 + ((n+1)\,q_e)^2\right) - \frac{1}{2C_s}\left(C_a C_b V_s^2 + ((n+2)\,q_e)^2\right) - V_s \frac{C_a q_e}{C_s},\end{aligned} \tag{7.58}$$

and enforcing $\Delta E_t > 0$ leads to

$$\frac{q_e}{C_s}\left(-q_e\left(n + \frac{3}{2}\right) - V_s C_a\right) > 0, \tag{7.59}$$

such that

$$V_s > \frac{-q_e}{C_a}\left(n + \frac{3}{2}\right). \tag{7.60}$$

For the case $C_a = C_b = C$, and considering different tunneling directions, we obtain

$$|V_s| > \frac{3\,|q_e|}{2C}. \tag{7.61}$$

Considering a third tunneling event, we would obtain

$$|V_s| > \frac{5\,|q_e|}{2C}, \tag{7.62}$$

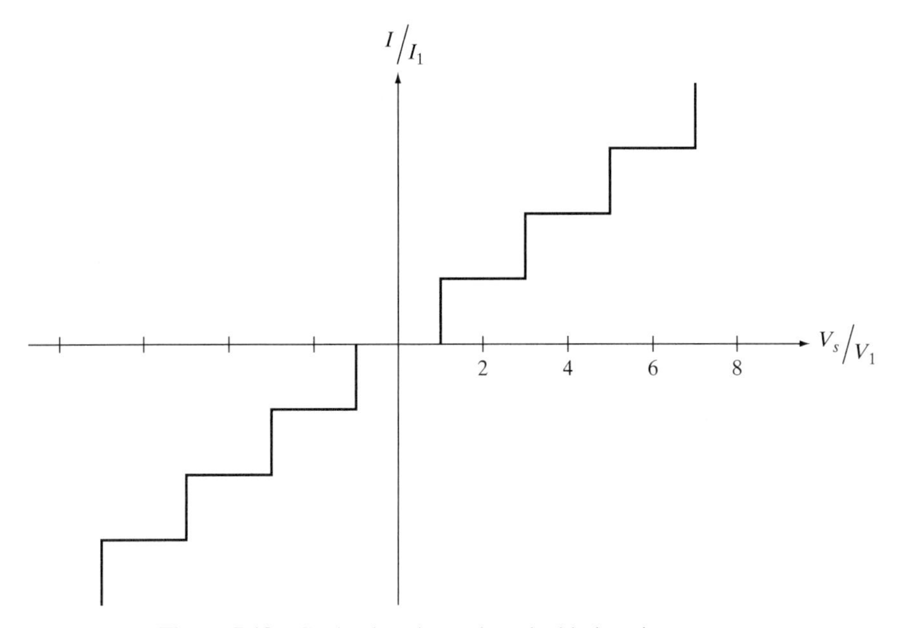

Figure 7.18 Coulomb staircase in a double junction system.

and, in general, we have

$$|V_s| > \frac{me}{2C}, \tag{7.63}$$

$m = 1, 3, 5, \ldots$, for $1, 2, 3, \ldots$ electron tunneling events, respectively. Thus, electron tunneling events, giving rise to currents, will occur at discrete voltage steps. The resulting I–V characteristic, known as a *Coulomb staircase*, is depicted in Fig. 7.18, where $V_1 = e/(2C)$, and $I_1 = e/(R_t C_s)$.

We usually need to have asymmetric junctions to observe the Coulomb staircase, and it is often convenient to have $C_a = C_b = C$, but $R_{ta} \gg R_{tb}$, or $R_{ta} \ll R_{tb}$. This can be achieved by having different-sized insulating gaps separating the leads from the island. In this way, the junctions have different tunneling rates. For example, assume that $R_{ta} \gg R_{tb}$. Then, tunneling is limited by junction a, so that as soon as one electron tunnels off of the island through junction a, it will be quickly replaced by a tunneling event through junction b. In this case, since junction a approximately controls tunneling, the resulting current will be $\Delta I \simeq \Delta V_a / R_{ta}$. From (7.49),

$$\Delta V_a = V_a^f - V_a^i = \frac{q_e}{C_s}, \tag{7.64}$$

and so the resulting current is

$$\Delta I \simeq \frac{\Delta V_a}{R_{ta}} = \frac{q_e}{R_{ta} C_s},$$

as shown in Fig. 7.18. (In the figure, I is normalized by $I_1 = |q_e| / (R_t C_s)$, where R_t is the larger of the tunnel resistances.) If the junctions are symmetric, the voltage offset about the origin is preserved, but the voltage steps tend to disappear. Thus, in practice, one can have $C_a \gg C_b$, or $R_{ta} \gg R_{tb}$, or both (or, of course, $C_a \ll C_b$, or $R_{ta} \ll R_{tb}$, or both) to achieve discrete voltage steps. However, if $C_a \gg C_b$ and $R_{ta} \ll R_{tb}$ such that the net effect is to have a somewhat symmetric junction (i.e., with the same RC time constants, $\tau = RC$), the voltage steps will tend to vanish.

As with the single capacitor, we need a sufficiently low temperature, or small capacitance, to observe Coulomb blockade. Analogous with (7.13), we require

$$C_s < \frac{q_e^2}{k_B T},$$

and sufficiently large tunneling resistance, $R_t \gg 4.1\ \text{k}\Omega$, where R_t is the smaller of the tunnel resistances of the junctions.

Experimental results showing the Coulomb staircase are shown in Fig. 7.19. In this experiment, the setup consisted of an In droplet separated from an Al ground plane by a tunneling oxide layer approximately 1 nm thick. An STM tip is brought into close proximity to the droplet, forming a metal–insulator–island–insulator–metal junction, as depicted in Fig. 7.20.

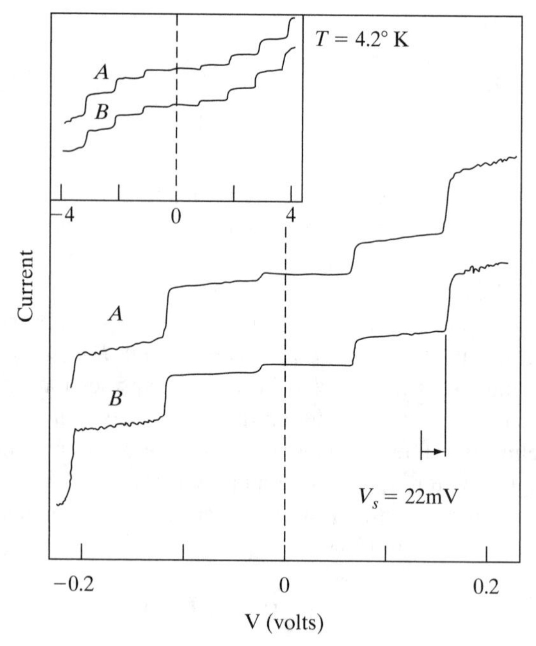

Figure 7.19 Curve A is the measured current–voltage characteristic for an In (indium) droplet Coulomb island (average droplet size is 30 nm). The peak current is 1.8 nA. Curve B is a theoretical fit to the data, with capacitance values in the aF range. (Based on a figure in Wilkins, R., E. Ben-Jacob, and R. C. Jaklevic, "Scanning-Tunneling-Microscope Observations of Coulomb Blockade and Oxide Polarization in Small Metal Droplets," *Phys. Rev. Lett.* 63 (1989): 801. © 1989, American Physical Society.)

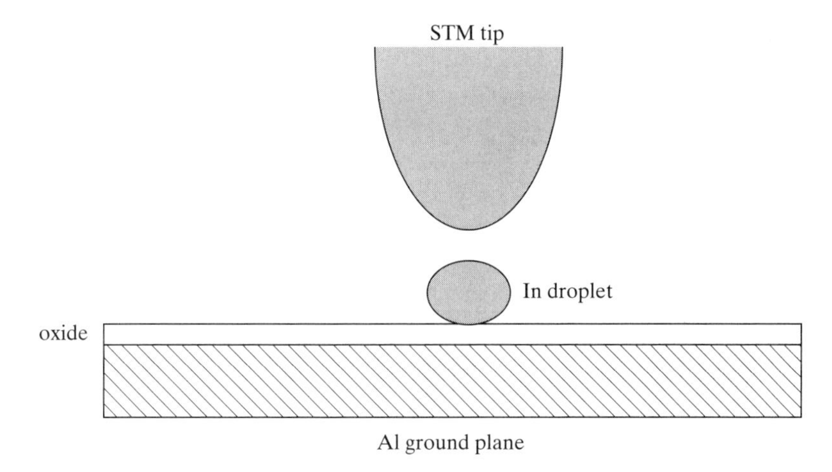

Figure 7.20 Experimental setup for generating the Coulomb staircase shown in Fig. 7.19.

For a metallic nanostructure, the energy levels on the island often form a quasi continuum, and for a semiconducting island, the energy levels show more clearly quantum confinement effects. As discussed previously, the difference between metallic and semiconducting dots is that in a metal, the Fermi wavelength of electrons is very small, typically on the order of a nanometer, compared to in a semiconductor (where the Fermi wavelength is typically on the order of tens of nanometers). Thus, the discreteness of energy levels for a metallic island will often not be evident unless the island is very small. In fact, since the energy levels in metal islands form a quasi continuum unless the dot is extremely small, the energy level spacing will typically be small compared with the thermal energy.

For semiconducting dots, the discreteness of energy levels is more readily apparent. In this way, they act more like artificial atoms (think of the hydrogen model), such that the properties of the dot are principally governed by electron confinement, as opposed to being due to the band behavior of the bulk material. This is discussed further in Section 9.3. In addition, the presence of magnetic fields affects the energy levels of the dot (spitting degeneracies in energy levels), and thus changing device behavior, although this effect won't be discussed here.

7.2 THE SINGLE-ELECTRON TRANSISTOR

We can gain additional control of the double-tunnel junction by adding a gate terminal, isolated from the island by an ideal (no tunneling) gate capacitance, and connected to a gate voltage, as depicted by the circuit shown in Fig. 7.21. This device is called a *single-electron tunneling transistor* (SET transistor), or simply a *single-electron transistor* (SET).[†]

The effect of the gate electrode can easily be understood from an energy band diagram similar to Fig. 7.16, as shown in Fig. 7.22. Application of a gate voltage $V_g > 0$ depresses the Fermi level on the island (compare with Fig. 7.16). Depending on how much gate voltage is applied, the top of the energy gap can be made to lie above, below, or even with

[†]The acronym SET is used for both single-electron tunneling and for the single-electron transistor.

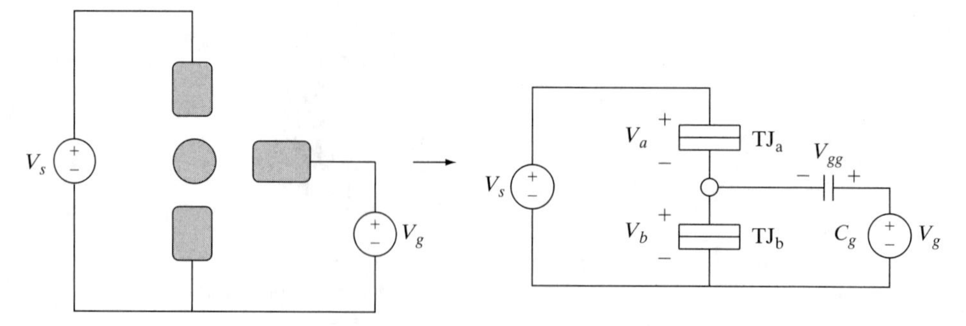

Figure 7.21 Circuit model of a single-electron transistor.

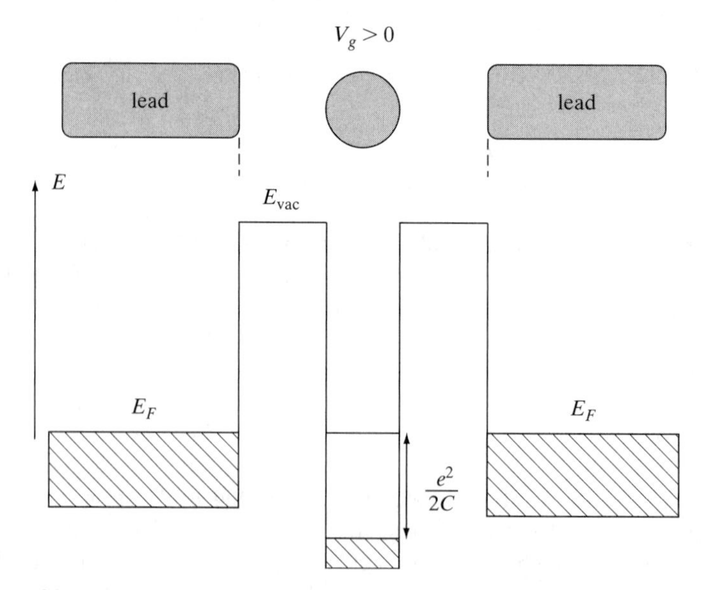

Figure 7.22 SET with positive gate voltage, depressing the Coulomb blockade gap.

the left and right Fermi levels. Thus, the strength of the applied source-drain voltage V_s required to drive a current through the device can be varied at the gate.

Following a development similar to that presented in the previous section, we have

$$Q_a = C_a V_a, \quad Q_b = C_b V_b, \tag{7.65}$$

$$Q_g = C_g \left(V_g - V_b \right), \tag{7.66}$$

such that the net charge on the island is the difference

$$Q = Q_b - Q_a - Q_g = nq_e. \tag{7.67}$$

The energy stored in the capacitors is

$$E_{se} = \frac{1}{2}\left(\frac{Q_a^2}{C_a} + \frac{Q_b^2}{C_b} + \frac{Q_g^2}{C_g}\right), \tag{7.68}$$

and Kirchhoff's voltage law applied to the circuit yields

$$V_s = V_a + V_b = \frac{Q_a}{C_a} + \frac{Q_b}{C_b}, \tag{7.69}$$

$$V_g = V_{gg} + V_b = \frac{Q_g}{C_g} + \frac{Q_b}{C_b}. \tag{7.70}$$

From (7.67) and (7.69)–(7.70), we obtain

$$Q_a = C_a \frac{V_s\left(C_b + C_g\right) - C_g V_g - Q}{C_s}, \tag{7.71}$$

$$Q_b = C_b \frac{C_a V_s + C_g V_g + Q}{C_s}, \tag{7.72}$$

$$Q_g = C_g \frac{V_g\left(C_b + C_a\right) - \left(Q + C_a V_s\right)}{C_s}, \tag{7.73}$$

where $C_s = C_a + C_b + C_g$, and the stored energy (7.68) is

$$E_{se} = \frac{1}{2C_s}\left(C_g C_a\left(V_s - V_g\right)^2 + C_a C_b V_s^2 + C_b C_g V_g^2 + Q^2\right). \tag{7.74}$$

When charge tunnels across a junction (i.e., a current flow event occurs), the power supplies must do work transferring charge. Assume that one electron tunnels onto the island through junction b. The change in stored energy is

$$\Delta E_{se} = \frac{\left(nq_e\right)^2 - \left(\left(n+1\right)q_e\right)^2}{2C_s} = -q_e^2 \frac{2n+1}{2C_s}, \tag{7.75}$$

and the change in charge on junction a and on the gate is

$$\Delta Q_a = C_a \frac{q_e}{C_s}, \tag{7.76}$$

$$\Delta Q_g = C_g \frac{q_e}{C_s}.$$

The work associated with supplying this change is

$$W_a = \Delta Q_a V_s = C_a \frac{q_e}{C_s} V_s, \tag{7.77}$$

$$W_g = \Delta Q_g V_g = C_g \frac{q_e}{C_s} V_g,$$

and, requiring that the tunneling event be energetically favorable,

$$\Delta E_t = \Delta E_{se} - \left(W_a + W_g\right) > 0, \tag{7.78}$$

we obtain

$$q_e\left(n + \frac{1}{2}\right) + \left(C_a V_s + C_g V_g\right) > 0. \tag{7.79}$$

In a similar manner, for an electron on the island to tunnel off of the island through junction a, we would find that

$$q_e\left(-n + \frac{1}{2}\right) + V_s\left(C_b + C_g\right) - C_g V_g > 0. \tag{7.80}$$

To consider current flow, assume that initially the island is charge neutral ($n = 0$), and that an electron tunnels onto the island through junction b. Then, for tunneling to occur, we need

$$V_s > \frac{1}{C_a}\left(\frac{-q_e}{2} - C_g V_g\right), \tag{7.81}$$

which is plotted in Fig. 7.23.

Now $n = 1$, and for the electron to tunnel off of the island through junction a, we need

$$V_s > \frac{1}{C_b + C_g}\left(\frac{q_e}{2} + C_g V_g\right), \tag{7.82}$$

which is depicted in Fig. 7.24.

For positive current to flow ($I > 0$; positive current flow is from top to bottom in Fig. 7.21), the electron must tunnel onto the island through junction b, then off of the island through junction a, and so both conditions need to be met, resulting in Fig. 7.25.

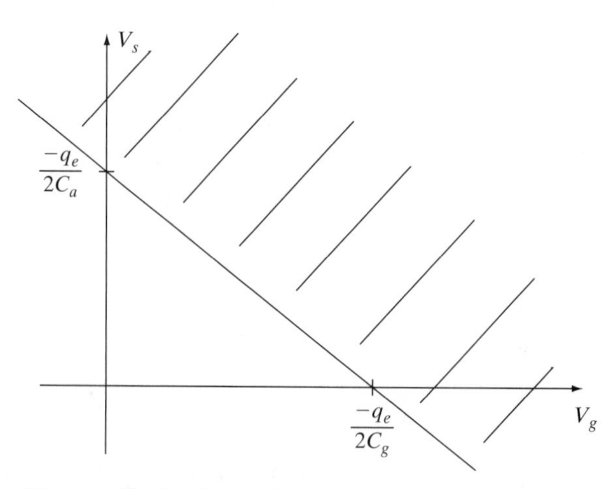

Figure 7.23 Region of the V_s–V_g plane given by (7.81).

Now if we assume that there is already one electron on the island, and another electron tunnels onto the island through junction b, then off of the island through junction a, we obtain

$$V_s > \frac{1}{C_a}\left(\frac{-3q_e}{2} - C_g V_g\right), \tag{7.83}$$

$$V_s > \frac{1}{C_b + C_g}\left(\frac{3q_e}{2} + C_g V_g\right). \tag{7.84}$$

Continuing in this manner, and considering the opposing tunneling events (that is, an electron tunneling onto the island through junction a, and then off of the island through junction b), we obtain the *Coulomb diamonds* shown in Fig. 7.26, where $V_1 = e/\left(2C_g\right)$. This is also called a *charge stability diagram*, since for combinations of V_s and V_g lying within the

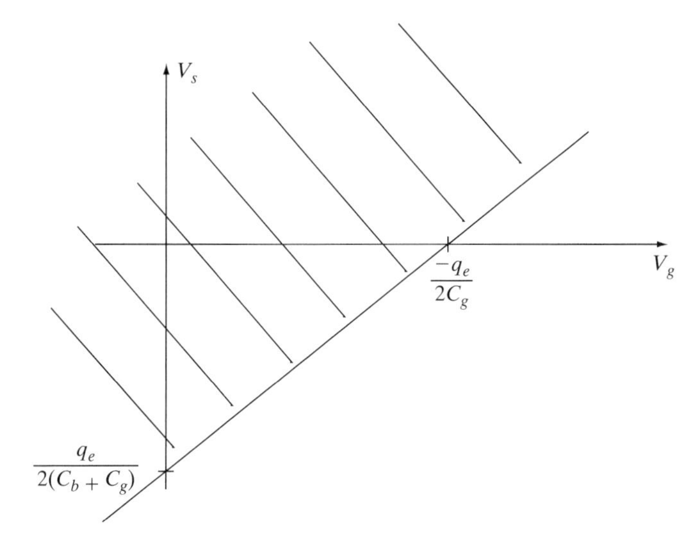

Figure 7.24 Region of the V_s–V_g plane given by (7.82).

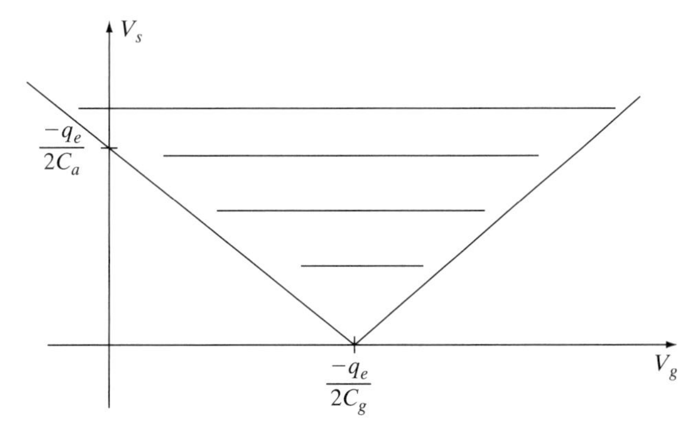

Figure 7.25 Region of the V_s–V_g plane given by (7.81) and (7.82).

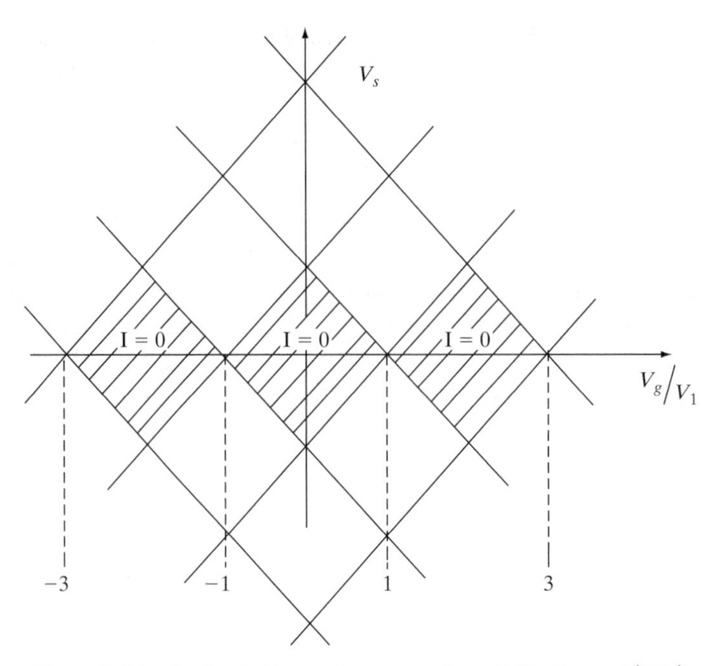

Figure 7.26 Coulomb diamond structure for a SET; $V_1 = e/\left(2C_g\right)$.

shaded regions, no tunneling is allowed ($I = 0$) and the charge on the island will remain stable. For the center shaded region, $n = 1$, and, moving outwards from the center, each shaded region corresponds to an integral change in the charge.

Note that by adjusting the gate voltage, one can tune the effective Coulomb gap seen by V_s. For $V_g = me/2C_g$, $m = \pm 1, \pm 3, \pm 5, \ldots$, there is no Coulomb gap, and current will flow for a small voltage V_s. However, for $m = 0, \pm 2, \pm 4, \pm 6, \ldots$, there is a maximum Coulomb gap that must be overcome in order for current to flow.

The Coulomb diamond has been observed in a variety of experiments where both V_g and V_s are varied. For example, in Fig. 7.27, the experimental conductance of a metal junction SET is shown, clearly exhibiting the Coulomb diamond structure.

Further, from Fig. 7.26, it is obvious that if V_s is held at a small constant value and V_g is varied, one should see strong fluctuations in the conductance through the island. This indeed occurs, and an example is shown in Fig. 7.28 for a semiconducting structure.

As with the other structures described previously, the observed effects are most easily observed at low temperatures, and tend to "wash out" as temperature increases. However, if a suitably small capacitance is used these effects can be observed at room temperature, as shown, for example, in Fig. 7.29.

Single-electron transistors offer the possibility of high-speed, high-density electronics, with low-power dissipation. However, they usually exhibit low gain, and suffer from random charge effects. To examine the random charge problem, consider a single charge impurity q_e trapped in the insulating oxide of a SET. The field of this impurity will polarize the island, in effect, creating an image charge on the island. This image charge will combine with the actual charge on the island, changing, for example, the condition for Coulomb blockade. In

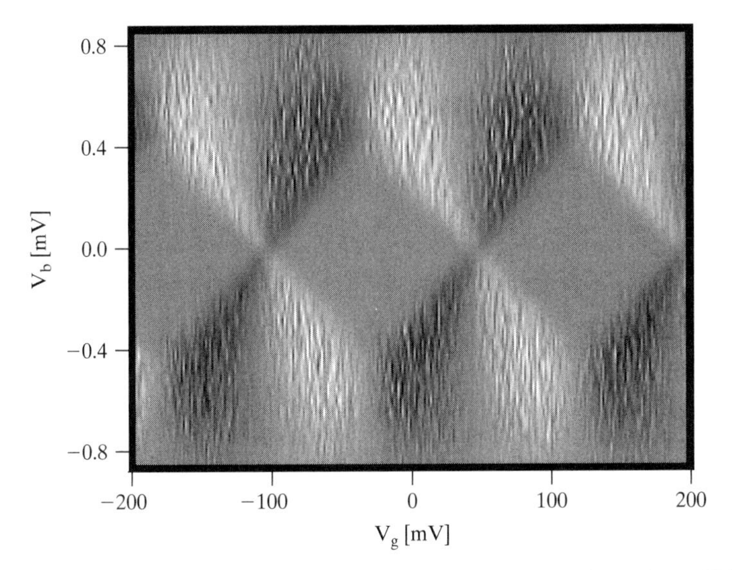

Figure 7.27 The conductance of a SET (dI/dV_g) plotted versus the bias voltage (V_s) and the gate voltage V_g. The gray diamonds are regions of Coulomb blockade (zero conduction). White is positive differential conductance and black is negative differential conductance; $T = 50$ mK. (From Hadley, P., et al., "Single-Electron Effects in Metals and Nanotubes for Nanoscale Circuits," in *Proceedings of the MIOP—The German Wireless Week, 11th Conference on Microwaves, Radio Communication and Electromagnetic Compatibility*, Stuttgart, Germany, 2001. 408–412. Courtesy of Leo Kouwenhoven, Kavli Institute of Nanoscience, Delft University of Technology.)

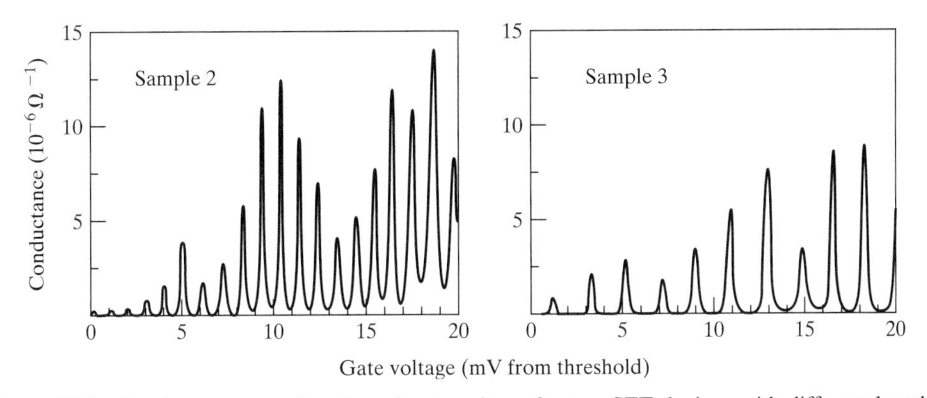

Figure 7.28 Conductance as a function of gate voltage for two SET devices with different lengths (sample 2 has $L = 0.8$ μm, and sample 3 has $L = 0.6$ μm). (Based on a figure from Kastner, M. A. "The Single-Electron Transistor," *Rev. Mod. Phys.* 64 (1992): 849. © 1992, American Physical Society.)

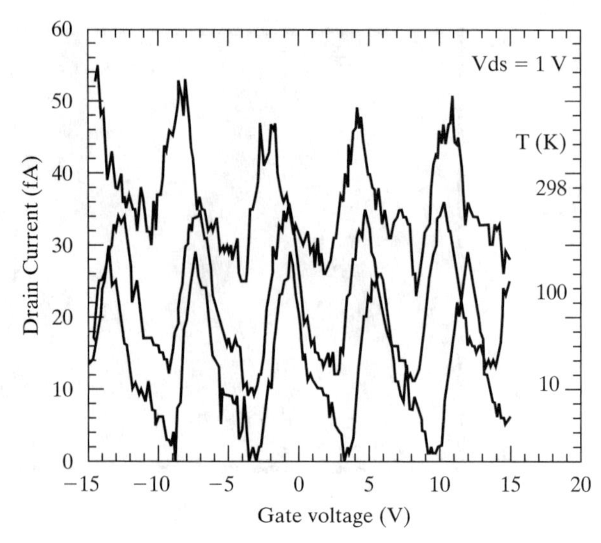

Figure 7.29 SET drain current versus gate voltage at several temperatures. Conductance fluctuations are present even at high temperatures due to the small C values ($C \simeq 6 \times 10^{-20}$ F). Source-drain voltage is 1 V. (Based on a figure from Shirakashi, J.-I., et al., "Single-Electron Changing Effects in Nb/Nb Oxide-Based Single-Electron Transistors at Room Temperature," *Appl. Phys. Lett.* 72 (1998): 1893. © 1998, American Institute of Physics.)

a macroscale circuit, the presence of a single stray charge would not normally be of much significance. For example, in a vacuum, the classical electric field due to a charge q_e is

$$\mathbf{E} = \widehat{r}\frac{q_e}{4\pi\varepsilon_0 r^2}, \tag{7.85}$$

where $\widehat{r}$ points radially away from the charge, and r is the radial distance from the charge. At a distance of 1,000 nm (1 μm), the field strength is $|\mathbf{E}| = 1.4$ V/m, or 1.4×10^{-9} V/nm. However, at a distance of 1 nm, the field strength is 1.4×10^9 V/m, or 1.4 V/nm. Thus, in spaces measured on the nm scale, even the field due to a single electron is relatively strong. Thus, whereas single-electron precision is a benefit in some respects, it has negative aspects as well. For example, in SET logic devices, charge impurities can lead to logical errors or unreliable operation. In recent years, schemes for SET memories have been proposed to alleviate this problem, and ways of constructing SET logic that are less sensitive to random charge effects have been considered.

The very near field of a single charge is actually so large that impurities subject to this field may migrate out of the oxide, thereby reducing the importance of oxide impurities. This is called the *self-cleaning effect*. However, this point is still a matter of debate. Moreover, impurity charge movement, if not to the point of self-cleaning, can lead to time-dependent circuit functionality, which is obviously intolerable.

In addition to the single-island, double-junction SET described previously, arrays of islands can be used as SETs. Such tunnel junction arrays often exhibit better performance (better isolation to the outside environment, higher temperature operation, etc.) than single-island SETs, although, of course, they are larger and more complicated.

In the following, we will discuss very briefly an application to SET logic, although at this point, SET memories may be more likely for near-term development. This is partly because of ameliorated random charge problems, and partly for other reasons. For example, SET devices have relatively poor current drive capabilities, which is often important in logic operations. In part because of this, it is likely that first-generation SET logic devices will involve some sort of combination with CMOS devices. Other than logic, memories, and other digital applications, SET analog applications have been proposed in the areas of electrometry, and current, resistance, and temperature standards.

7.2.1 Single-Electron Transistor Logic

The vast majority of current-generation integrated circuit chips use CMOS technology. As an example, a CMOS inverter is shown in Fig. 7.30, where V_{dd} represents a "high" logic level.

The circuit functions as an inverter in the following manner.† If V_{in} is "high" (typically, $V_{\text{in}} \simeq V_{dd}$), then $V_{gs} \simeq 0$ for the PMOS device. Since we would need $V_{gs} < -|V_t|$ to turn on the PMOS transistor, where the magnitude of the threshold voltage $|V_t|$ is usually on the order of a volt or two, and $|V_t| < V_{dd}$, this device is off. However, since $V_{gs} \simeq V_{dd} > V_t$ for the NMOS transistor, the NMOS device will be on, and therefore $V_{\text{out}} \simeq 0$, producing a "low" state.

If the input voltage is "low," i.e., $V_{\text{in}} \simeq 0$, then, conversely, $V_{gs} \simeq 0$ for the n-channel device, and so this transistor is turned off. However, $V_{gs} \simeq -V_{dd} < -|V_t|$ for the p-channel device, and so this transistor will be on, connecting the output to V_{dd}. Therefore, the output will be "high."

Single-electron transistors have been proposed for logic functions, such as inverters, since they are obviously very small and could lead to very high device densities on a chip. In particular, much attention has been given to constructing SET logic devices in a manner

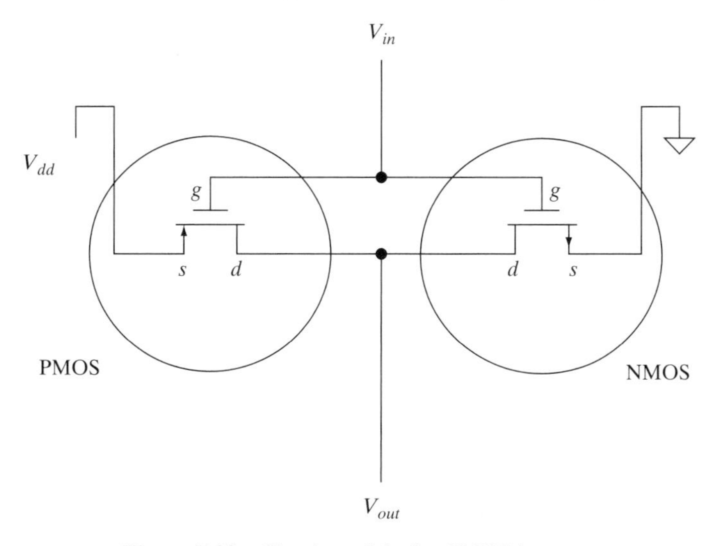

Figure 7.30 Circuit model of a CMOS inverter.

†A review of MOSFET operation is provided in Appendix C.

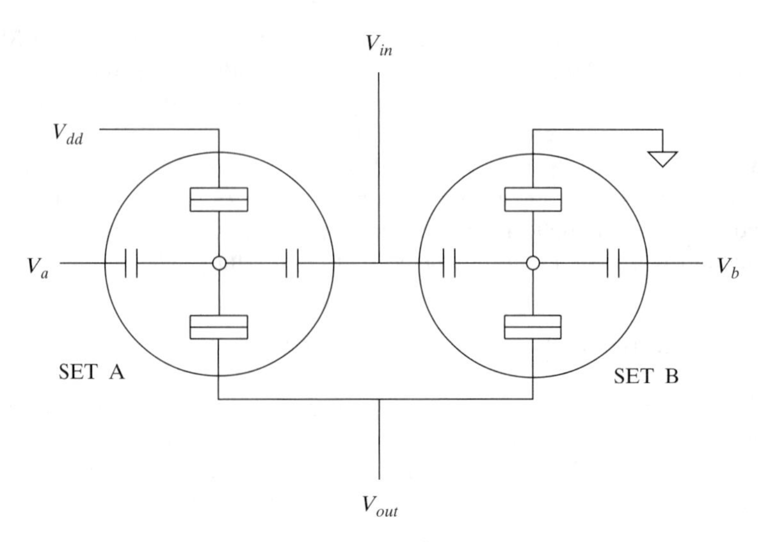

Figure 7.31 Circuit model of a SET inverter.

similar to CMOS topologies. These devices form what are called *voltage state logic devices*, since their characteristics, as viewed externally from the circuit, are based on voltage levels, as in current CMOS circuits. This is opposed to *charge state logic devices,* which operate by representing bits of information by the presence or absence of charge, which SET devices are also capable of doing.

As an example of a voltage state SET inverter, consider the circuit shown in Fig. 7.31. If SET A is "off" (i.e., if SET A is in the Coulomb blockade regime, such that there is, effectively, an open circuit between the top and bottom of the device) and SET B is "on" (i.e., electrons can cross the tunnel junctions, connecting the top and bottom leads of the device), then $V_{\text{out}} = 0$, forming a low logic level. If SET A is on and SET B is off, $V_{\text{out}} = V_{dd}$, forming a high logic level. By adjusting V_a and V_b, one can configure the SETs to be in the correct state (on or off), as shown in Figs. 7.32 and 7.33. Note, however, that multiple peaks occur in the transfer characteristics, so that there is no device saturation against the bias rails (as occurs in CMOS). Hence, this architecture will be sensitive to noise and bias shifts.

Voltage state SET logic is subject to static leakage currents leading to power dissipation for some examined structures on the order of 10^{-7} watts per transistor. Although this power level is, by itself, relatively low, for dense integration the total power dissipation can become quite large. For example, assume that each SET occupies 5 nm^2. Then, one could achieve 4.0×10^{12} transistors/cm^2, leading to 4×10^5 watts/cm^2, which is an enormous power density to be dissipated.

7.3 OTHER SET AND FET STRUCTURES

7.3.1 Carbon Nanotube Transistors (FETs and SETs)

In addition to forming SETs from two tunnel junctions as depicted in Fig. 7.21, a variety of other configurations have been shown to provide similar SET action. In particular, two promising structures for electronics use carbon nanotubes and molecules.

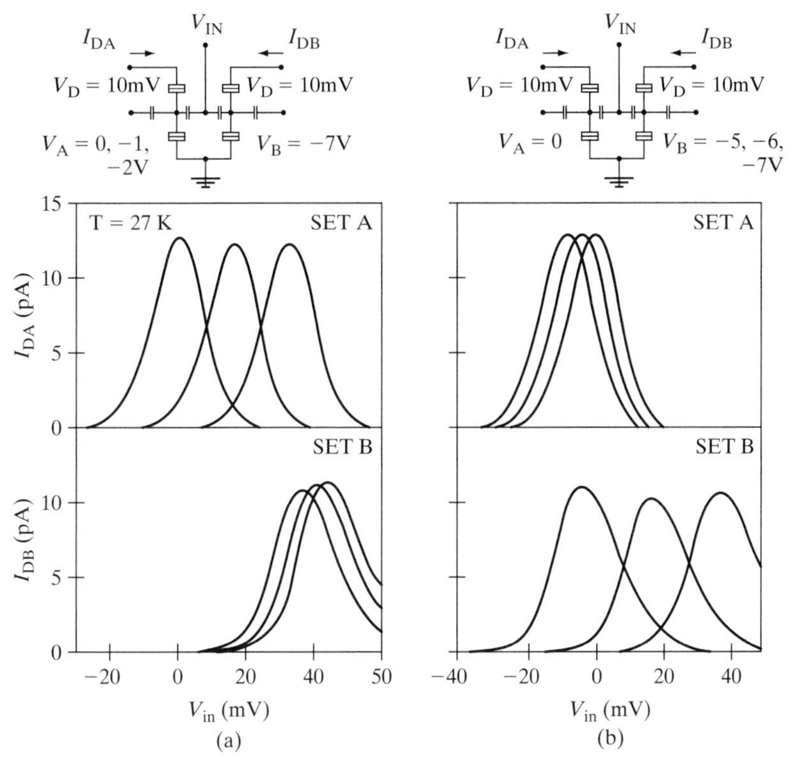

Figure 7.32 Measured drain currents I_{DA} and I_{DB} of the SETs as a function of the input voltage V_{in} applied to the top gate. In (a), the change in voltage V_A from 0 to -2 volts shifts the current curve of SET A to the positive V_{in} direction, while that of SET B is shifted very little. The situation in (b), where V_B is changed from -5 to -7 volts, is the converse of that in (a). The measurement configuration is shown at the top of each figure. (Based on a figure from Ono, Y., et al., "Si Complementary Single-Electron Inverter with Voltage Gain," *Appl. Phys. Lett.* 76 (2000): 3121. © 2000, American Institute of Physics.)

Carbon Nanotubes. The band structure of π electrons in carbon nanotubes was discussed in Section 5.5. Electronic applications of carbon nanotubes have included their use as transistors, gas sensors, field emission devices, and in a host of other devices. In particular;

- As discussed in Section 6.3.1, carbon nanotubes are attractive for field emission devices such as flat-panel displays, visible lights, and as X-ray sources. They are excellent emitters of electrons due to their high conductivity, and the fact that they naturally possess an extremely sharp tip, on the order of a nanometer, which concentrates the electric field to a very small region of space. This strong electric field enhances tunneling through the vacuum barrier.
- Carbon nanotubes are natural gas sensors. It has been found that the resistance of a semiconducting carbon nanotube is quite sensitive to the presence of nearby molecules. This change in resistance can be used to ascertain the presence or absence of certain chemicals in the nanotube's environment.

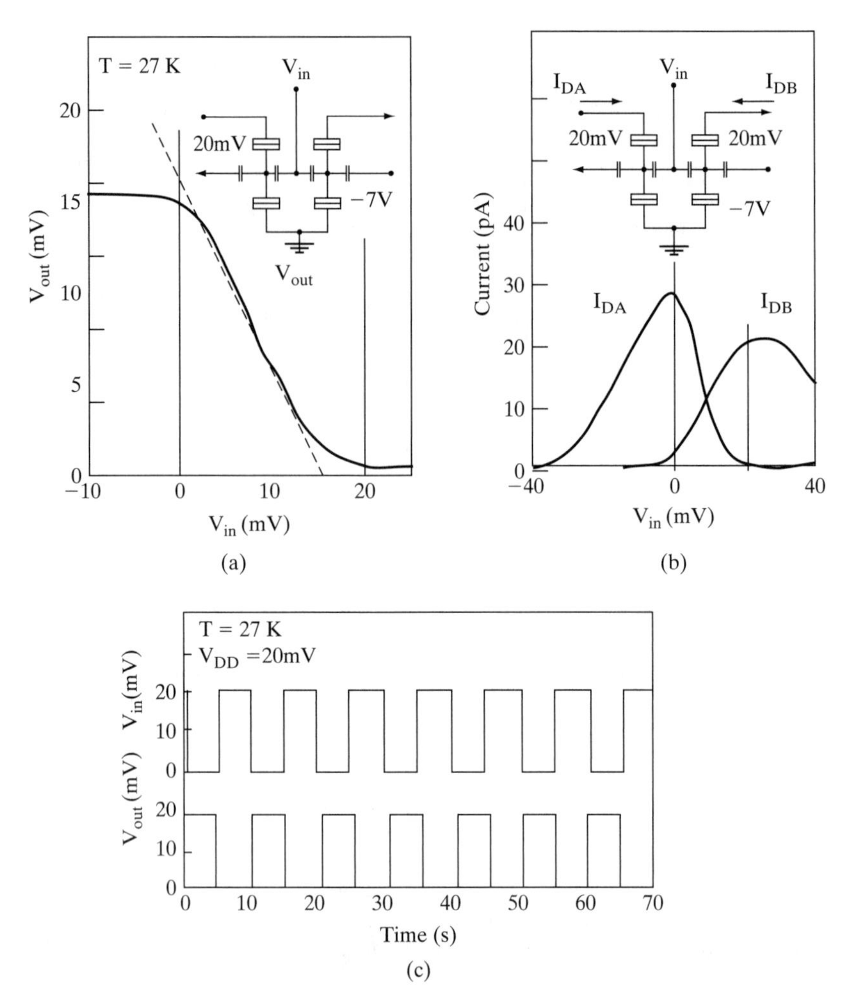

Figure 7.33 Transfer characteristics of the inverter for a power supply voltage of 20 mV, and (b), the corresponding current curves of the SETs in the inverter for a drain voltage of 20 mV. (c) The output voltage for a square-wave input with an amplitude of 20 mV. For all the figures, V_a and V_b were set to be 0 and −7 V, respectively. In (a) and (b), the insets show the measurement configuration. The vertical lines define the range of the input voltage used in the measurement shown in (c) (in (c) the power supply voltage is 20 mV). (Based on a figure from Ono, Y., et al., "Si Complementary Single-Electron Inverter with Voltage Gain," *Appl. Phys. Lett.* 76 (2000): 3121. © 2000, American Institute of Physics.)

- Carbon nanotubes are likely candidates for use as nanowires and nanoantennas and, in their semiconducting version, as channels in transistors.

Prototype devices in all of the preceding categories, and in other areas as well, have been made, but in many cases, technical difficulties still need to be overcome before commercial products become available. As with many nanotechnologies, inexpensive, high-throughput manufacturing techniques still largely await development. In particular, one would like to have a certain nanotube (say, semiconducting, of a specified radius) placed in some desired location in order to form a practical device. And, one would like to be able to do this reliably, millions of times, which cannot be accomplished as yet. Among other problems, establishing good contact between carbon nanotubes and electrodes that lead to "the outside world" can be difficult.

The use of carbon nanotubes as nanowires is discussed a bit more in Chapter 10, and here we consider carbon nanotube FETs and SETs. The use of a semiconducting carbon nanotube as a channel in a FET is depicted in Fig. 7.34.

An AFM image of an early CN FET is shown in Fig. 7.35, where the source, drain, and gate are implemented using Au. The gate is the middle gold contact, which is coplanar with the tube, whereas in the previous figure, the FET was backgated (gated from below).

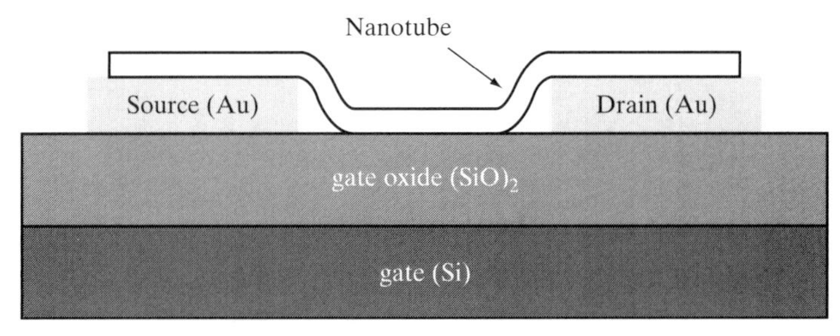

Figure 7.34 Depiction of a carbon nanotube FET. (Based on a figure from Avouris, P., et al., "Carbon Nanotube Electronics," *Proceedings of the IEEE*, 91 (2003): 1772–1784.)

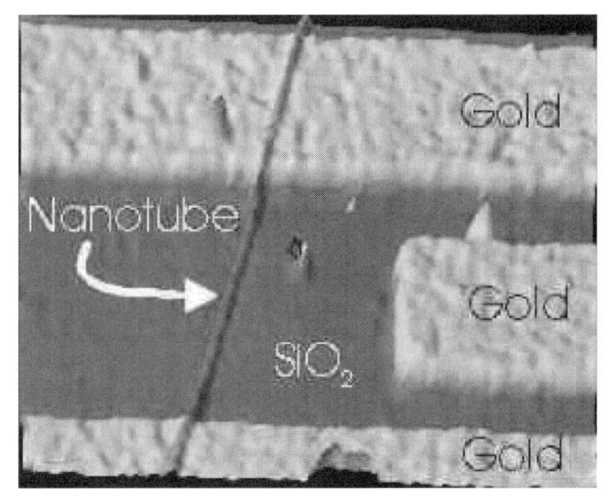

Figure 7.35 Carbon nanotube connecting source and drain, in the vicinity of a controlling gate. (From Avouris, P., et al., "Carbon Nanotube Electronics," *Proceedings of the IEEE*, 91 (2003): 1772–1784. Copyright 2003, IEEE.)

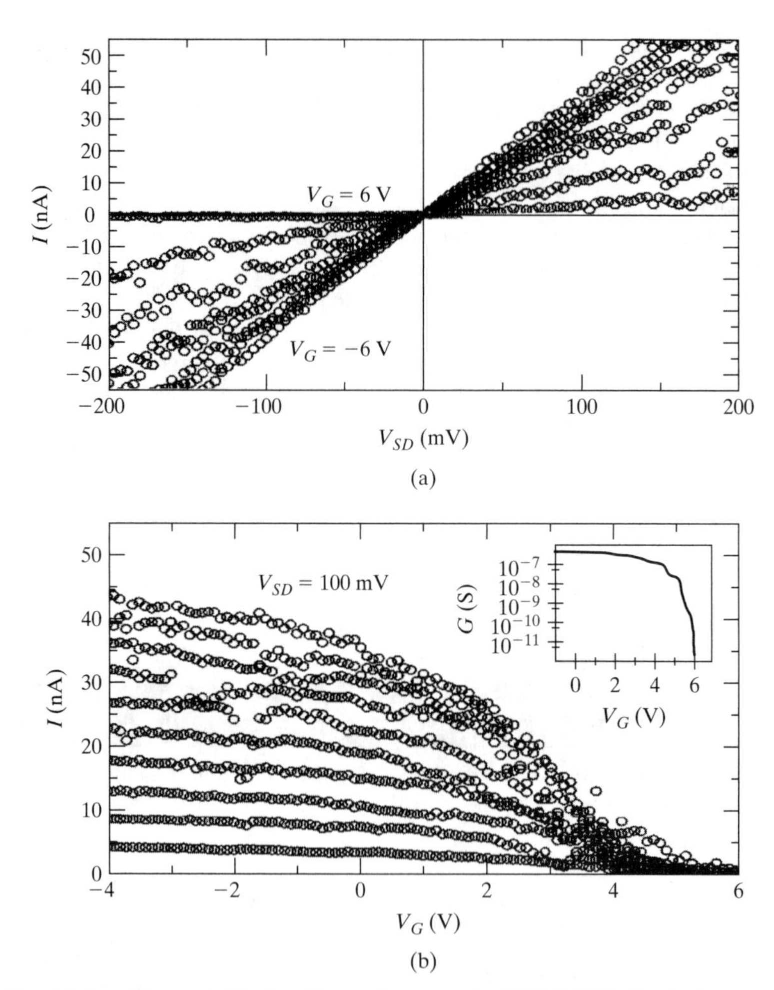

Figure 7.36 (a) I_{SD}–V_{SD} and (b) I_{SD}–V_{GATE} for an early SWNT FET. In the insert of (b), G is the low bias conductance of the tube. Tube diameter is 1.6 nm. (Based on a figure from R. Martel, et al., "Single- and Multi-Wall Carbon Nanotube Field-Effect Transistors." *Applied Physics Letters*, 73, 2447 (1998). © 1998, American Institute of Physics.)

The source-to-drain current (I_{SD}) flowing through the nanotube channel can be changed dramatically by changing the gate voltage (V_{GATE}), as can be seen in Fig. 7.36.

It was initially thought that the gate voltage modified the CN resistance, much as the gate voltage in a MOSFET varies the channel resistance, although investigations have shown that CN FETs function mainly based on the Schottky barriers (Section 6.2.1) present at the metal–CN junctions. For example, increasing the gate voltage lowers the Schottky barrier, resulting in increased current flow. Device physics depend somewhat on the method of fabrication and on the operating environment, since the Schottky barriers are sensitive to absorbed gases, although device characteristics essentially resemble those of ordinary FETs. Early CN FETs suffered from low drive currents and low transconductance, although recent

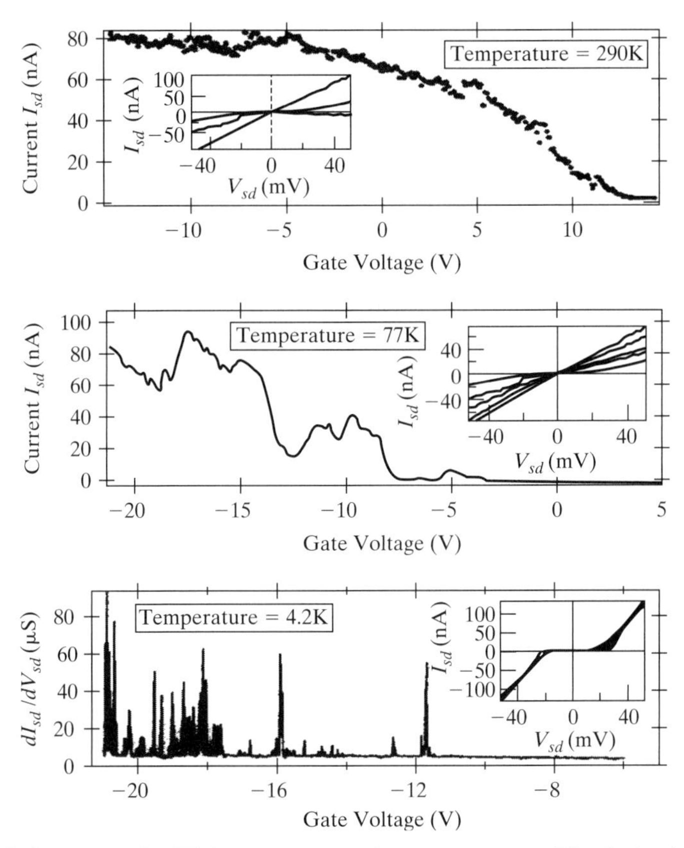

Figure 7.37 Carbon nanotube FET structure at various temperatures. The device behaves like an ordinary FET at 290 K, and like a SET at 4.2 K. (Based on a figure from H.R. Shea, R. Martel, T. Hertel, T. Schmidt and PH. Avouris, "Manipulation of Carbon Nanotubes and Properties of Nanotubes Field-Effect Transistors and Rings," *Microelectronic Engineering*, 46, 101–104 (1999). Used by permission from Elsevier.)

advances have shown that CN FETs may outperform conventional MOSFETs at similar size scales.

The preceding properties, where the device acts like an ordinary FET, were measured at room temperature. If the device is cooled to 4 K, it behaves like a SET, as shown in Fig. 7.37. Other room temperature carbon nanotube SETs have been made by introducing defects in the tube, such as sharp bends, which act like the insulating gaps in the tunneling island SET discussed previously.

At the low-temperature end, Fig. 7.37 can be compared with Fig. 7.28, which is, itself, explained from the Coulomb diamond description of SETs.

7.3.2 Semiconductor Nanowire FETs and SETs

In addition to using semiconducting carbon nanotubes as FET and SET channels, semiconducting nanowires can also serve a similar function. Semiconducting nanowires typically

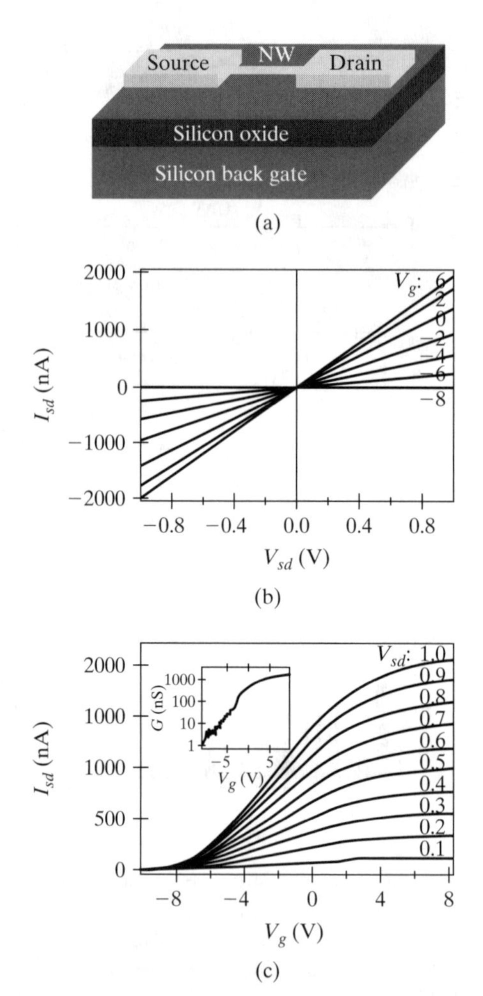

Figure 7.38 (a) Depiction of a nanowire FET. (b) and (c) show the source–drain current characteristics of a GaN nanowire FET, where in (c), the inset shows the conductance versus gate voltage. The diameter of the nanowire is 17.6 nm. (Based on a figure from Huang, Y., et al., "Gallium Nitride Nanowire Nanodevices," *Nanoletters*, 2 (2002): 101.)

have diameters in the range 10–100 nm, and controlled growth can result in nanowires that are quite long and straight, with few defects. Both p- and n-type nanowires can be fabricated, and a variety of devices have been demonstrated.

For example, Fig. 7.38 (a) depicts a nanowire FET, where it can be seen that the geometry is the same as for carbon nanotube FETs, with the nanowire replacing the tube in forming the channel. For a 17.6 nm diameter GaN nanowire, the source-drain current versus source-drain voltage is shown in Fig. 7.39(b), and the current versus gate voltage in Fig. 7.39(c). Results were obtained at room temperature, and typical FET characteristics are evident.

As with CN FETs, semiconducting nanowire FETs can be used to form SETs at very low temperatures. For example, the dashed curves in Fig. 7.39 show the current–voltage

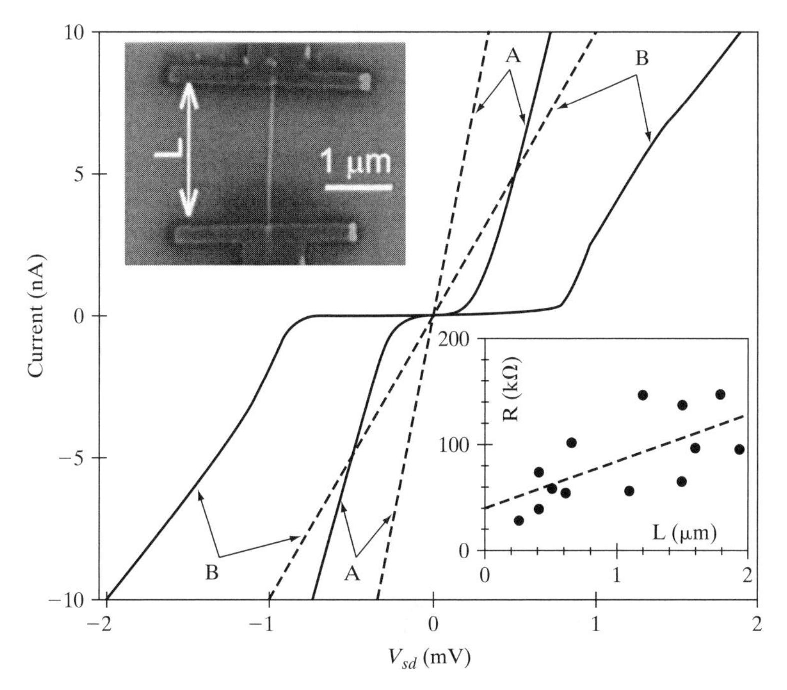

Figure 7.39 Current–voltage characteristics for an n-type InP nanowire FET. Dashed curves show room temperature behavior, and the solid curves were measured at 0.35 K (showing Coulomb blockade behavior). The designation A and B refers to two different devices, where device A has channel length $L = 0.20$ μm, and device B has $L = 1.95$ μm. The wire diameters were approximately 50 nm. The upper insert shows an SEM image of device B, and the lower insert shows the length dependence of the room temperature source-drain resistance. (From De Franceschi, S. and J. A. van Dam, "Single-Electron Tunneling in InP Nanowires," *Appl. Phys. Lett.*, 83 (2003): 344. © 2003, American Institute of Physics.)

characteristics for an n-type InP nanowire FET at room temperature. It is evident that the channel is ohmic. The solid curves were measured at 0.35 K, and clear Coulomb blockade behavior is observed. The low-temperature source-drain conductance G is shown in Fig. 7.40(a), where single-electron behavior is obviously present. (Compare with Fig. 7.28 on page 241, and with Fig. 7.37 in the previous section.) Coulomb diamond behavior is shown in Fig. 7.40(b).

7.3.3 Molecular SETs and Molecular Electronics

At this point it may be appreciated that a fundamental physical configuration of an electronic device is some element connecting two electrodes (the source and the drain), in the vicinity of a gate electrode that provides some control, as depicted in Fig. 7.41.

The object inside the box labeled "element" may be, for example, an n- or p-type silicon channel, as in an ordinary MOSFET, depicted in Fig. 6.14 on page 200. For a

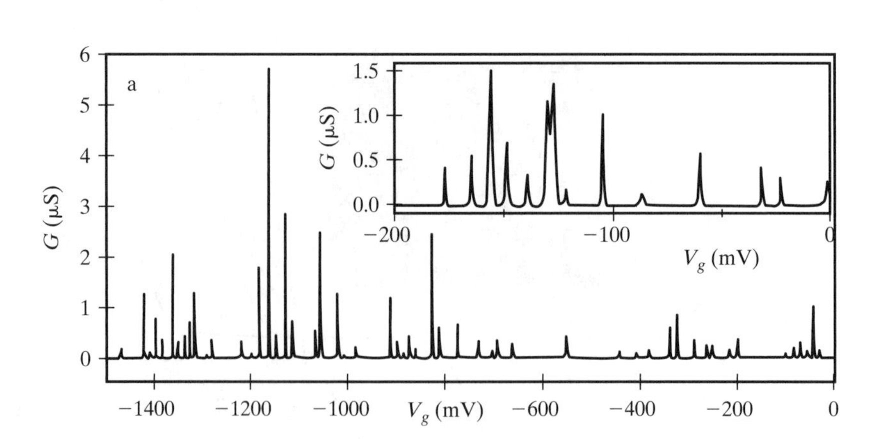

(a)

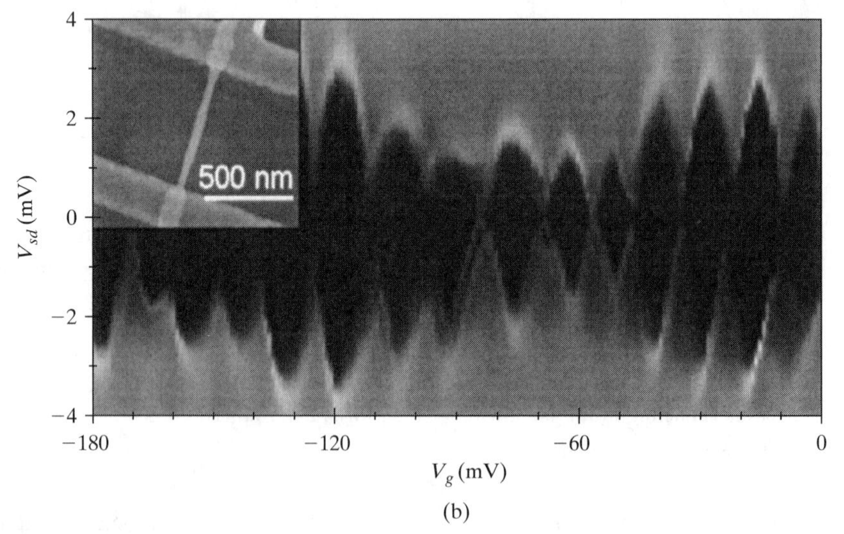

(b)

Figure 7.40 (a) Source-drain conductance for InP nanowire FETs having two different channel lengths. In the main image, $L = 0.65$ μm, and in the insert, $L = 1.6$ μm; note that both devices have channel lengths between those of devices A and B in the previous figure. Coulomb diamond behavior is shown in (b) for the $L = 0.65$ μm device. (From De Franceschi, S. and J. A. van Dam, "Single-Electron Tunneling in InP Nanowires," *Appl. Phys. Lett.*, 83 (2003): 344. © 2003, American Institute of Physics.)

nanoelectronic device, "element" may be a double-tunnel junction as shown in Fig. 7.21 on page 236, a carbon nanotube as shown in Fig. 7.34, a semiconducting nanowire as shown in the previous section, or something entirely different. "Element" should basically be capable of connecting the source and the drain such that the amount of current flowing is controlled by the voltage at the gate (and, of course, the drain-to-source voltage). For digital applications, the connection should be either "on" or "off," and for analog applications, the current I_{DS} should vary considerably with the gate voltage.

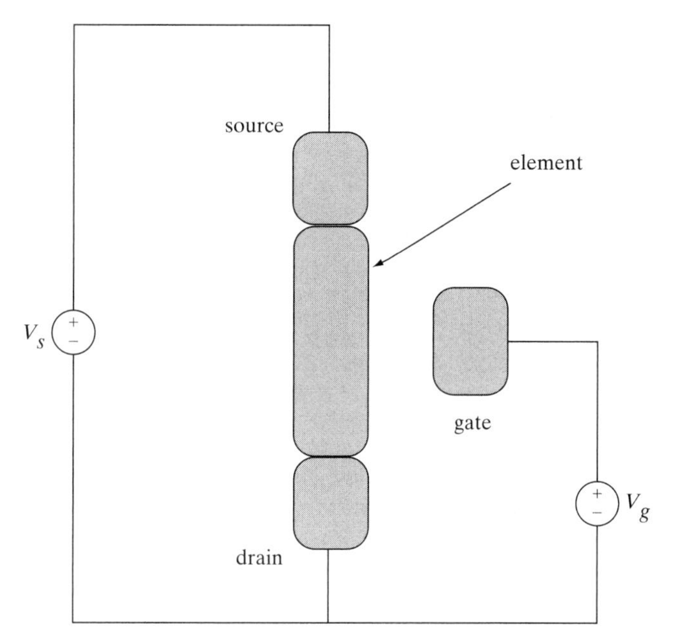

Figure 7.41 Depiction of a general electronic device consisting of an element connecting source and drain, in the presence of a control gate.

In addition to the preceding choices for "element," there has been considerable interest in using molecules, or chains of molecules (including DNA strands), to connect source and drain electrodes. Rather then possessing a band structure, as in a bulk crystal, energy levels in molecules are more similar to those in atoms. As described in Section 5.4.2, as N identical atoms are brought together to form a molecule, each energy level of the isolated atom splits into N levels, centered on the isolated atom level. If nonidentical atoms are brought together, energy levels also split, although in a more complicated manner. The resulting molecule has a discrete energy level structure that can be described in terms of molecular orbitals, with the highest-occupied/lowest-unoccupied orbitals being analogues to the highest-filled/lowest-unfilled bands in a bulk crystal.

Although the theory of electrical transport in molecular electronic devices is quite complicated, and incomplete, prototype devices have been experimentally demonstrated. For example, Fig. 7.42 shows a benzene molecule connecting source and drain, in the vicinity of a gate electrode having voltage V_G.

Fig. 7.43 shows the current through the device depicted in Fig. 7.42, as a function of the gate field E_g $(V/\mathring{A})$. Note the similarity with the resonant tunneling I–V curve shown in Fig. 6.25 on page 208; in general, for molecular devices, electron transfer can be described by resonant or nonresonant tunneling.

Molecular SETs have also been demonstrated experimentally. For example, Fig. 7.44 shows one such SET and the obtained I–V curves for different gate voltages (-0.4 to -1.0 volts, in steps of -0.15 volts).

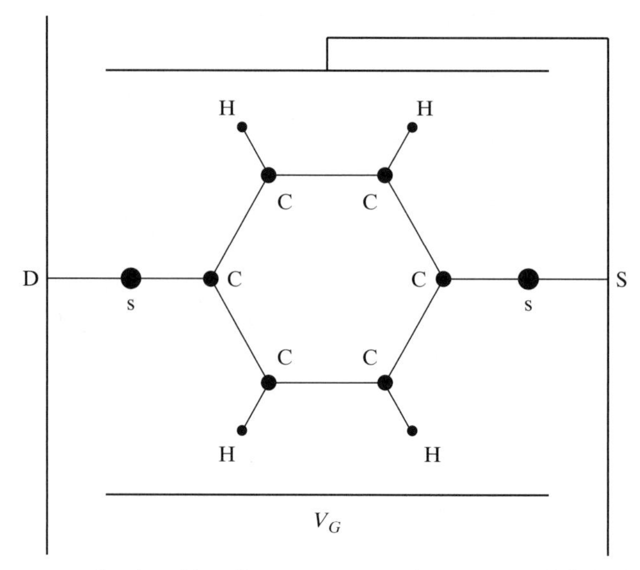

Figure 7.42 Benzene molecule with sulfur atoms connecting source and drain, in the vicinity of a gate electrode having voltage V_G.

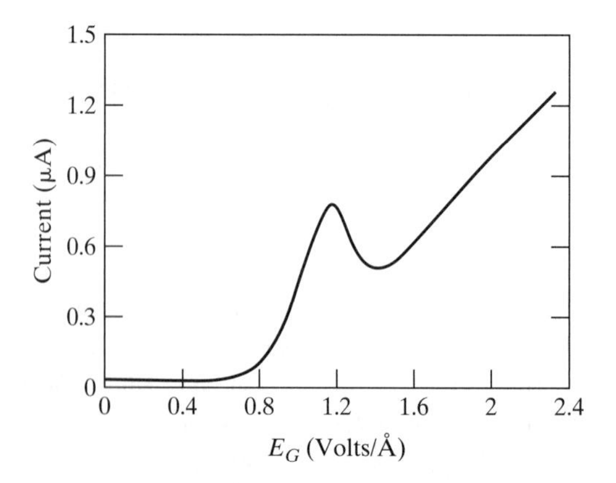

Figure 7.43 Calculated current through the device depicted in Fig. 7.42 as a function of the gate field E_g, in $V/\AA(V_{sd} = 10$ mV). (Based on a figure from Di Ventra, M., S. T. Pantelides, and N. D. Lang, "The Benzene Molecule as a Molecular Resonant-Tunneling Transistor," *Appl. Phys. Lett.*, 76 (2000): 3448. © 2000, American Institute of Physics.)

There is currently a lot of interest in developing molecular electronic devices. Advantages include an implicitly bottom-up approach (self-assembly based on chemistry), and extremely small device sizes. However, there are significant scientific and technological challenges to overcome. Complications include methods of connecting molecular devices to electrodes (the metal–molecule interface often significantly impacts device behavior), addressing of such small devices, the effect of chemical absorption on molecules' electrical

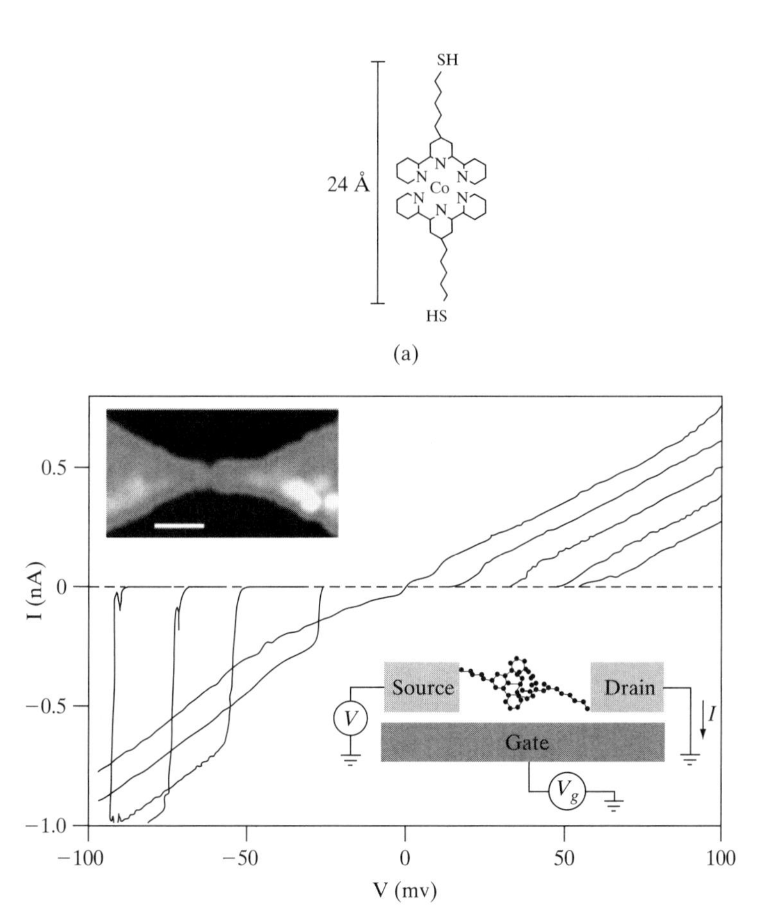

Figure 7.44 Molecular SET. (a) Molecular structure. (b) Family of d.c. I–V curves recorded for several values of the gate voltage. Left most curve is −0.4 V, and other curves are in increments of −0.15 V (so that the nearly straight line is for −1 V). Insert is an AFM image of the electrodes, and the scale bar is 100 nm. (From Park, J. et al., "Coulomb Blockade and the Kondo Effect in Single-Atom Transistors," *Nature*, 417 (2002): 722–725. © 2002, Macmillan Publishers, Ltd.)

behavior (this can significantly change device characteristics), and high temperature operation. At the current time, these issues largely remain to be solved.

7.4 MAIN POINTS

In this chapter, we have examined the Coulomb blockade effect in quantum dot circuits, and the single electron transistor. In particular, after studying this chapter, you should understand

- the Coulomb blockade effect;
- current-biased tunnel junctions;

- the influence of capacitance and energy on tunneling and Coulomb blockade, and basic Coulomb blockade structures;
- tunnel junction models, tunneling in quantum dot circuits, and the Coulomb staircase;
- the SET and its applications;
- electronic channels based on carbon nanotubes and semiconducting nanowires;
- the basic idea of molecular electronics.

7.5 PROBLEMS

1. For a tunnel junction with $C = 0.5$ aF and $R_t = 100$ kΩ, what is the RC time constant? What does this value mean for the tunnel junction circuit?
2. For a tunnel junction having $C = 0.5$ aF and $R_t = 100$ kΩ, what is the maximum temperature at which you would expect to find Coulomb blockade? Repeat if $C =$ 1.2 pF.
3. For the capacitor depicted in Fig. 7.2 on page 215, for an electron to tunnel from the negative terminal to the positive terminal we determined the condition

$$V > \frac{-q_e}{2C}. \tag{7.86}$$

(See (7.8).) Show, including all details, that for an electron to tunnel from the positive terminal to the negative terminal we need

$$V < \frac{q_e}{2C}. \tag{7.87}$$

4. The energy stored in a capacitor is

$$E = \frac{Q^2}{2C}, \tag{7.88}$$

where $\pm Q$ is the charge on the capacitor plates ($+Q$ on one plate, and $-Q$ on the other plate). Using this equation, *derive* the condition on charge Q and voltage V for N electrons to tunnel across the junction at the same time, in the same direction.

5. There is always a capacitance between conductors separated by an insulating region. For the case of conductors associated with different circuits (i.e., circuits that should operate independently from one another), this is called *parasitic capacitance*, and, generally, $C \propto 1/d$, where d is some measure of the distance between the conductors. For two independent circuits, d is often fairly large, and, thus, very small parasitic capacitance values are generally present.
   (a) Using the formula for the impedance of a capacitor, show that even for very small values of C, at sufficiently high frequencies the impedance of the capacitor can be small, leading to significant unintentional electromagnetic coupling between the circuits.
   (b) Describe why these small values of parasitic capacitance, which can easily be aF or less, do not lead to Coulomb blockade phenomena.

6. For the SET oscillator shown in Fig. 7.10 on page 224, derive the oscillation period given by (7.31).

7. As another way of seeing that Coulomb blockade is difficult to observe in the current-biased junction shown in Fig. 7.9 on page 223, consider the equivalent circuit shown in Fig. 7.11 on page 225. If R_b is sufficiently large so that it can be ignored, and if $|Z_L|$ is sufficiently small, construct an argument, from a total capacitance standpoint, for why the amplitude of the SET oscillations will tend towards zero.

8. For the quantum dot circuit depicted in Fig. 7.13 on page 226, assume that initially there are $n = 100$ electrons on the dot. If $C_a = C_b = 1.2$ aF, what is the condition on V_s for an electron to tunnel onto the dot through junction b?

9. For the quantum dot circuit depicted in Fig. 7.13 on page 226, we found that for an electron to tunnel onto the dot through junction b, and then off of the dot through junction a, we need

$$V_s > \frac{-q_e}{2C} \tag{7.89}$$

if initially there were no electrons on the dot ($n = 0$), where $C_a = C_b = C$. It was then stated that for the opposite situation, where an electron tunnels onto the dot through junction a, and then off of the dot through junction b, we would need

$$V_s < \frac{q_e}{2C}. \tag{7.90}$$

Prove this result, showing all details.

10. Consider the double-junction Coulomb island system depicted in Fig. 7.13 on page 226, and refer to Fig. 7.17 on page 232. Assume that an electron tunnels onto the island subsequent to applying a voltage $V = e^2/2C$. Draw the resulting energy band diagram, showing the readjusted energy bands, and the re-establishment of Coulomb blockade.

11. For the SET shown in Fig. 7.21 on page 236, assume that $C_a = C_b = 10$ aF, $C_g = 1.4 \times 10^{-16}$ F, and $V_g = 0.1$ V. If initially there are 175 electrons on the island, then what is the condition on V_s for an electron to tunnel across junction b and onto the island?

12. For the SET, we considered electrons tunneling onto the island through junction b, then off of the island through junction a, resulting in positive current flow (top to bottom). Explicit details were provided for the generation of Coulomb diamonds for $I > 0$, and the results merely stated for $I < 0$. Fill in the details of the derivation predicting Coulomb diamonds for $I < 0$.

13. Draw the energy band diagram for a SET

(a) under zero bias ($V_s = V_g = 0$)
(b) when $V_s > 0$ and $V_g = 0$
(c) when $V_s > 0$ and $V_g = |q_e| / (2C_g)$

14. Consider an electron having kinetic energy 5 eV.

(a) Calculate the de Broglie wavelength of the electron.

(b) If the electron is confined to a quantum dot of size $L \times L \times L$, discuss how big the dot should be for the electron's energy levels to be well quantized.
(c) To observe Coulomb blockade in a quantum dot circuit, is it necessary to have energy levels on the dot quantized? Why or why not?

15. Assume that a charge impurity $q = 2q_e$ resides in an insulating region. Determine the force on an electron 10 nm away from the impurity. Repeat for the case when the electron is 10 μm away from the impurity.

16. Universal conductance fluctuations (UCF) occur in Coulomb blockade devices due to the interference of electrons transversing a material by a number of paths. Using other references, write one-half to one page on UCFs, describing the role of magnetic fields and applied biases.

17. Summarize some of the technological hurdles that must be overcome for molecular electronic devices to be commercially viable.

18. There is currently a lot of interest in spintronics, which rely on the spin of an electron to carry information. (See Section 10.4.) In one-half to one page, summarize how spin can be used to provide transistor action.

Part III

MANY ELECTRON PHENOMENA

In the previous part of the text, we considered nanoelectronic devices that are sensitive to the charge of a single electron. In general, these devices operate using a very small number of electrons. In this part, we still consider nanoscale devices and structures, but ones which operate, in some sense, using relatively larger numbers of electrons. This distinction is actually somewhat artificial, since we will discuss, for example, quantum dots and quantum wires, where we may still be interested in a relatively small number of electrons. The real distinction between the structures and associated analysis methods considered in this section and those of previous sections is that here we will use the concepts of *density of states* and *quantum statistics*. These concepts are often used in the physics of solids and gases, where an extremely large number of particles are typically of interest. For example, a cube of copper having side length 1 cm can be modeled as a free electron gas consisting of approximately 10^{22} electrons.

However, the concepts of density of states and quantum statistics are still useful at the nanoscale for several reasons. First, although nanoscale regions of space do not hold such a large number of electrons, they still may be made up of many hundreds to hundred of thousands of atoms. Although the properties of these structures differ from large, macroscopic bulk regions, there are still enough particles such that density of states and quantum statistical concepts are useful. Second, in the case of, for example, ballistic transport on quantum wires, these wires may be connected to macroscopic leads, which serve as large reservoirs of electrons. Again, density of states concepts and quantum statistical principles have been found to be quite useful in modeling such structures.

In Chapter 8, quantum statistics are described for large collections of particles, leading to the knowledge of which energy states or bands are occupied (depending on the temperature, the Fermi level, etc.) The important concept of the density of states is developed, and related material for semiconductor materials is presented. Then, in Chapter 9, we revisit quantum well, quantum wire, and quantum dot structures, and discuss implementations using semiconductor materials. Important applications of semiconducting quantum dots are described, especially relating to their optical properties.

Chapter 10 describes the movement of charge over nanoscopic (and sometimes larger) length scales. Of course, the flow of electrical current in wires and in macroscopic circuits is well known, and the concept of resistance is integral to any discussion of current flow. Modeling a material as having a certain resistivity is a standard technique in electrical technology, with the classical concept of resistivity being related to electron collisions with the material lattice, or with impurities.[†] The length an electron travels between collisions depends on the material and on the temperature, but it is generally on the order of tens of nanometers, or more. However, when device length scales on the order of a few nanometers are of interest, such that, on average, few or no collisions will take place, obviously a collision-based model will not suffice. This is the regime of what is called *ballistic transport*, and in Chapter 10, the general concepts of ballistic transport are described. In this chapter as well, transport in carbon nanotubes and nanowires is discussed, as is the transport of spin.

[†]However, recall that in the quantum model described in Chapter 5, collisions are actually necessary to frustrate Bloch oscillations and allow current to flow.

Chapter

8

PARTICLE STATISTICS AND DENSITY OF STATES

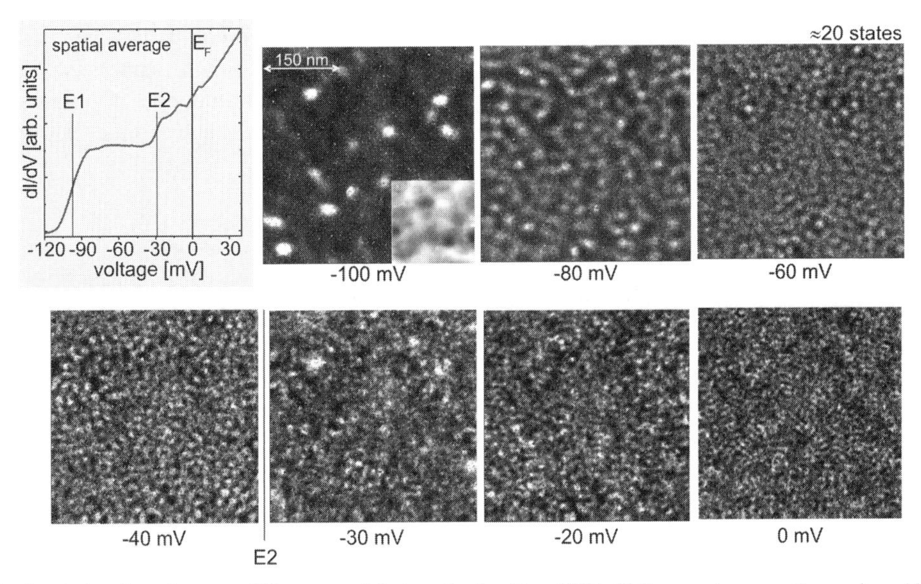

The local density of states of Fe-covered InAs, obtained by STM. Voltage values are the various bias voltages used. (Courtesy of Prof. Dr. Roland Wiesendanger, Executive Director of the Institute of Applied Physics (IAP) and Interdisciplinary Nanoscience Center Hamburg (INCH).)

In this chapter, the concepts of density of states and particle statistics are introduced. In short, density of states relates to how many electronic states are available in a certain structure at a certain energy, and particle statistics give the probability that a certain energy state will be occupied. Recall that in previous quantum well problems, the *allowed* energy states were obtained (e.g., (4.53)), but there was no way to say which states would actually be "filled" by electrons, except if we had a certain number of electrons in the ground state. With the concepts developed in this chapter, we can examine how many allowed states are near an energy of interest, and the probability that those states will actually be filled with electrons.

Therefore, density of states and particle statistics concepts are indispensable in the study of bulk materials, and, as previously discussed, have utility when considering relatively small material samples as well.

8.1 DENSITY OF STATES

Considering again the case of an electron in a three-dimensional bounded region of space (Section 4.3.2), we often find it useful to know how many quantum states lie within a particular energy range, say, between E and $E - \Delta E$. From the equation for energy, (4.54), the number of states below a certain energy E_n is equal to the number of states inside a sphere of radius

$$n = \sqrt{n_x^2 + n_y^2 + n_z^2} = \sqrt{\frac{E_n}{E_1}}, \tag{8.1}$$

as shown in Fig. 8.1.

We will consider the hard-wall case with nonperiodic boundary conditions ($\alpha = 1/2$ in (4.60); the results using periodic boundary conditions are the same). We only count the first octant of the aforementioned sphere, since sign changes don't lead to additional states. Then, the total number of states N_T having energy less than some value (but with $E \gg E_1$) is approximately the volume of the octant,

$$N_T = \frac{1}{8}\frac{4}{3}\pi n^3 = \frac{\pi}{6}\left(\frac{E}{E_1}\right)^{3/2}. \tag{8.2}$$

The total number of states having energy in the range $(E, E - \Delta E)$ is

$$\Delta N_T = \frac{\pi}{6E_1^{3/2}}\left((E)^{3/2} - (E - \Delta E)^{3/2}\right) \tag{8.3}$$

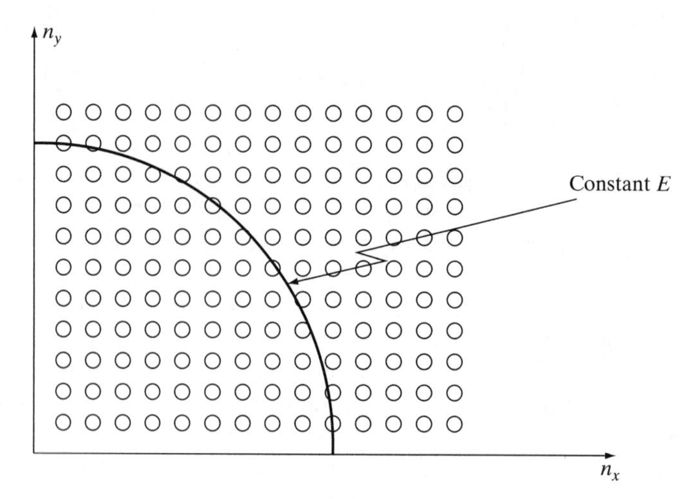

Figure 8.1 Electronic states in the n_y–n_x plane. The states below a certain energy E_n lie inside a circle of radius E_n/E_1.

$$= \frac{\pi}{6E_1^{3/2}}\left((E)^{3/2} - E^{3/2}\left(1 - \frac{\Delta E}{E}\right)^{3/2}\right)$$

$$\simeq \frac{\pi E^{3/2}}{6E_1^{3/2}}\left(1 - \left(1 - \frac{3}{2}\frac{\Delta E}{E}\right)\right)$$

$$= \frac{\pi E^{1/2}}{4E_1^{3/2}}(\Delta E),$$

where we used $(1 - x)^p \simeq 1 - px$ for $x \ll 1$.

The *density of states* (DOS), $N(E)$, is defined as the number of states per unit volume per unit energy around an energy E. The total number of states in a unit volume in an energy interval dE around an energy E is (replacing ΔN_T, ΔE with dN_T, dE, respectively)

$$dN_T = N(E)\,dE = \frac{\pi E^{1/2}}{4E_1^{3/2}}dE, \tag{8.4}$$

so that

$$N(E) = \frac{\pi E^{1/2}}{4E_1^{3/2}} = \frac{2^{3/2} m_e^{*3/2} E^{1/2}}{4\hbar^3\pi^2}$$

(since the density of states is per unit volume, we set $L^3 = 1$ in the expression for E_1). Accounting for spin, we multiply by 2, such that

$$N(E) = \frac{2^{1/2} m_e^{*3/2} E^{1/2}}{\hbar^3\pi^2}, \tag{8.5}$$

and if the electron has potential energy V_0, the density of states becomes

$$N(E) = \frac{2^{1/2} m_e^{*3/2} (E - V_0)^{1/2}}{\hbar^3\pi^2}, \tag{8.6}$$

where $E > V_0$. The density of states is shown in Fig. 8.2.

Note that (8.6) was derived assuming the energy relationship (4.54), which holds for an electron in an ideal quantum well. More generally, the parabolic E–k relationship usually holds for electrons in real materials near the bandedges, which are generally the most important energies. However, energy relationships different from (4.54) will result in different density of states. For example, electrons at energies removed from the bandedges generally do not have a parabolic E–k relationship. Phonons also have a different density of states, since their energy equation is different from (4.54).

To gain an appreciation of (8.6), assuming $m_e^* = m_e$ we have

$$N(E) = 1.06 \times 10^{56}\,(E_{\text{Joules}} - V_0)^{1/2} \qquad \text{J}^{-1}\text{m}^{-3}, \tag{8.7}$$

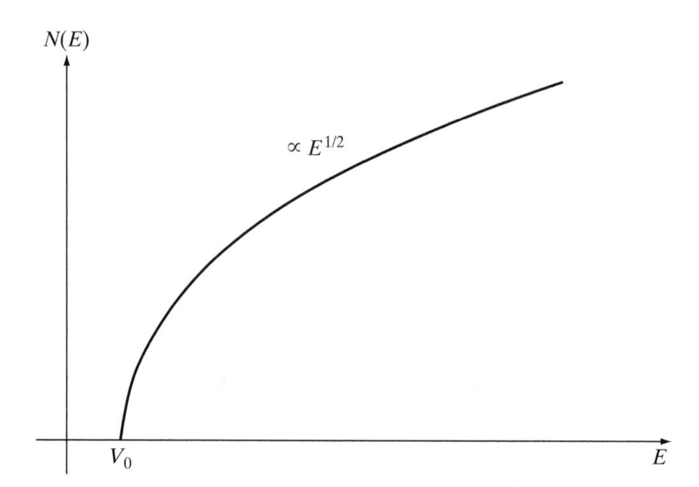

Figure 8.2 Density of states for a three-dimensional system.

where E_{Joules} indicates that the energy is in units of joules. Often the DOS is expressed in units of eV^{-1}cm^{-3}, where $E_{\text{Joules}} = E_{\text{eV}} \times e$. Therefore,

$$\begin{aligned} N(E) &= 1.06 \times 10^{56} \left(E_{\text{Joules}} - V_0\right)^{1/2} \frac{1}{\text{J}}\frac{1}{\text{m}^3} \times \frac{\text{J}}{\frac{1}{e}\,\text{eV}} \frac{\text{m}^3}{(100\ \text{cm})^3} \\ &= 1.7 \times 10^{31} \left(E_{\text{Joules}} - V_0\right)^{1/2} \qquad \text{eV}^{-1}\text{cm}^{-3} \\ &= 6.8 \times 10^{21} \left(E_{\text{eV}} - V_{0,\text{eV}}\right)^{1/2} \qquad \text{eV}^{-1}\text{cm}^{-3}. \end{aligned} \tag{8.8}$$

For example, if $E = 0.1$ eV and $V_0 = 0$, then $N(E) = 2.15 \times 10^{21}$ eV^{-1} cm^{-3}.

Thinking physically, if there are enough electrons to fill the various states (or, at least fill the states of interest), then the density of states $N(E)$ is the density of electrons having energy E. Furthermore, often one is interested in the *local density of states* (LDOS), which is the density of states as a function of position in a material. This can be obtained from STM images by measuring the differential conductance dI/dV, since the tunneling current is proportional to the local electron density. Such an image is shown on page 261.

8.1.1 Density of States in Lower Dimensions

In discussing electron transport in nanosystems we will often need the density of states in sub-three-dimensional systems. In one dimension, such as for a quantum wire, the density of states is defined as the number of available states per unit length per unit energy around an energy E. To be specific, for an electron confined to a line segment of length L, from (4.35),

$$E_n = \frac{\hbar^2 \pi^2}{2m_e^* L^2} n^2, \tag{8.9}$$

and the total number of states below a certain energy is equal to the number of states inside an interval of length

$$N_T = n = \left(\frac{E}{E_1}\right)^{1/2}. \tag{8.10}$$

The total number of states having energy in the range $(E, E - \Delta E)$ is

$$\begin{aligned}\Delta N_T &= \frac{1}{E_1^{1/2}}\left(E^{1/2} - (E - \Delta E)^{1/2}\right) = \frac{1}{E_1^{1/2}}\left(E^{1/2} - E^{1/2}\left(1 - \frac{\Delta E}{E}\right)^{1/2}\right) \\ &\simeq \frac{1}{E_1^{1/2}}\left(E^{1/2} - E^{1/2}\left(1 - \frac{\Delta E}{2E}\right)\right) = \frac{1}{E_1^{1/2}E^{1/2}}\left(\frac{\Delta E}{2}\right). \end{aligned} \tag{8.11}$$

The total number of states in a unit length in an energy interval dE around an energy E is (again, replacing ΔN_T, ΔE with dN_T, dE, respectively),

$$dN_T = N(E)\,dE = \frac{1}{2}\frac{1}{E_1^{1/2}E^{1/2}}dE, \tag{8.12}$$

so that

$$N(E) = \frac{1}{2}\frac{1}{E_1^{1/2}E^{1/2}} = \frac{\sqrt{2m_e^*}}{2\hbar\pi}E^{-1/2}. \tag{8.13}$$

(Since the density of states is per unit length, we set $L = 1$.) Accounting for spin, we multiply by 2 to obtain

$$N(E) = \frac{\sqrt{2m_e^*}}{\pi\hbar}E^{-1/2}, \tag{8.14}$$

and if the electron has potential energy V_0, we have

$$N(E) = \frac{\sqrt{2m_e^*}}{\pi\hbar}(E - V_0)^{-1/2}, \tag{8.15}$$

where $E > V_0$. The density of states for an electron confined to a one-dimensional line segment is shown in Fig. 8.3.

Using similar ideas, we find that the density of states for an electron confined to a two-dimensional region of space (such as a quantum well) is

$$N(E) = \frac{m_e^*}{\pi\hbar^2}, \tag{8.16}$$

for $E > V_0$, which is shown in Fig. 8.4. For example, if an electron is confined to two dimensions, with $m_e^* = m_e$,

$$N(E) = \frac{m_e}{\pi\hbar^2} = 2.6 \times 10^{37} \quad \text{J}^{-1}\text{m}^{-2} = 4.17 \times 10^{14} \quad \text{eV}^{-1}\text{cm}^{-2}. \tag{8.17}$$

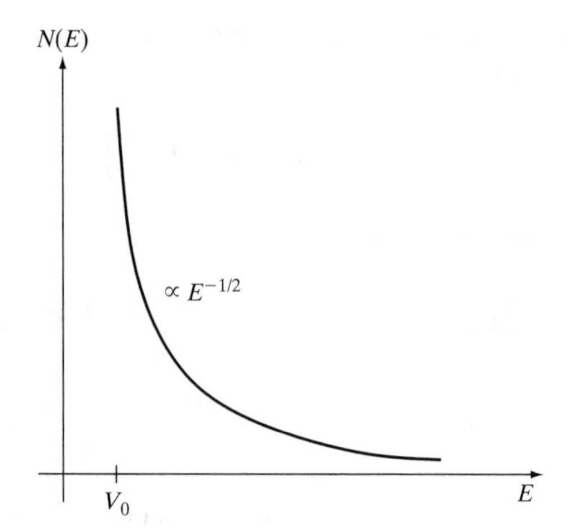

Figure 8.3 Density of states for a one-dimensional system.

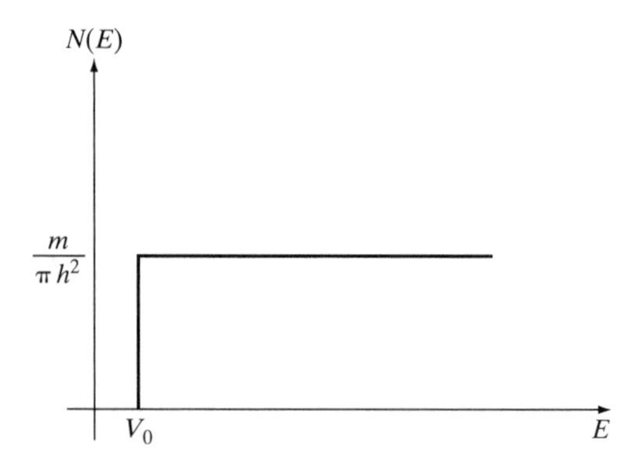

Figure 8.4 Density of states for a two-dimensional system.

The previous derivations made use of the fact that, because the structure was large in three, two, or one dimension, the energy states E_n formed a quasi continuum. That is, we start with the discrete states E_n, but then assume that the energy levels are closely spaced in obtaining the density of states. However, in a zero-dimensional system (such as a quantum dot), the states are truly discrete (i.e., they don't form a quasi continuum). So in this case, the density of states is merely a delta function,

$$N(E) = 2 \sum_n \delta(E - E_n), \tag{8.18}$$

where the factor of two accounts for spin. In real quantum dots, the infinitely narrow delta functions are broadened by electron collisions in the material comprising the dot.

All physically realizable structures are actually three dimensional, and so it may seem strange to consider low-dimensional systems. However, as discussed previously, three-dimensional structures that can appropriately confine electrons can implement an effectively two-, one-, and even zero-dimensional system. The density of states in quantum structures will be further discussed in Chapter 9.

8.1.2 Density of States in a Semiconductor

We found previously that the three-dimensional density of states for electrons in a constant potential V_0 (i.e., in an empty box) is (8.6),

$$N(E) = \frac{2^{1/2} m_e^{*3/2} (E - V_0)^{1/2}}{\hbar^3 \pi^2}, \tag{8.19}$$

where $E > V_0$. In a crystalline solid, this expression also holds because the effective mass can account for the crystal structure, and the potential can account for energy bandedges. For example, the three-dimensional density of states in a semiconductor is given by

$$N_c(E) = \frac{2^{1/2} m_c^{*3/2} (E - E_c)^{1/2}}{\hbar^3 \pi^2}, \tag{8.20}$$

$$N_v(E) = \frac{2^{1/2} m_v^{*3/2} (E_v - E)^{1/2}}{\hbar^3 \pi^2}, \tag{8.21}$$

where the subscript indicates either the conduction or valence band, $m_{c,v}^*$ is the effective mass appropriate to the band,† $E > E_c$ for N_c, and $E < E_v$ for N_v. For semiconductors where the effective mass differs along different directions, the most appropriate effective mass to use is $m^* = \left(m_1^* m_2^* m_3^*\right)^{1/3}$, where m_i^* is the effective mass in the ith direction. For silicon the effective mass varies in two directions, resulting in the longitudinal and transverse effective mass, m_l^* and m_t^*, respectively. Therefore, in the conduction band of silicon, accounting for the six equivalent conduction band valleys, we find that an appropriate density of states is

$$N_c(E) = 6 \frac{2^{1/2} m_c^{*3/2} (E - E_c)^{1/2}}{\hbar^3 \pi^2}, \tag{8.22}$$

where

$$m_c^* = \left(m_l^* m_t^{*2}\right)^{1/3}. \tag{8.23}$$

8.2 CLASSICAL AND QUANTUM STATISTICS

In earlier chapters, we have found solutions to Schrödinger's equation for free electrons, and for electrons confined to certain regions of space. These solutions represented *possible*

†Note that the density of states is higher for larger effective mass values. Thus, the heavy holes tend to denominate the properties of the valence band, although both heavy and light holes are important.

states of the corresponding quantum system. In the previous section, we obtained the density of states, which describes the number of these states in the vicinity of a certain energy. What remains to be found is the number of occupied states, i.e., which states, and how many states, are actually filled.† For this we need to know the probability of a state being occupied at an energy E. This probability is given by the distribution $f(E)$, and next we consider several common distributions.

1. **Classical or Boltzmann Distribution**
   This distribution applies when particles are distinguishable from each other (such as classical particles), with no other constraints on their behavior. The Boltzmann distribution arises from classical statistical mechanics, and is given by

$$f^B(E) = f^B(E, \mu, T) = e^{-\frac{E-\mu}{k_B T}}. \tag{8.24}$$

   The quantity μ is the chemical potential, usually simply replaced by the Fermi energy E_F, and k_B is Boltzmann's constant ($k_B = 1.38 \times 10^{-23}$ J/K).

2. **Fermi–Dirac Distribution**
   The Fermi–Dirac distribution applies when the particles in question are indistinguishable, and when only one particle can occupy a particular state (e.g., electrons). Therefore, the Fermi–Dirac distribution applies to quantum particles that obey the exclusion principle, and is given by

$$f(E) = f(E, \mu, T) = \frac{1}{e^{\frac{E-\mu}{k_B T}} + 1}, \tag{8.25}$$

   where again we usually make the replacement $\mu = E_F$. When $E - E_F \gg k_B T$ (known as the *Boltzmann approximation*; $k_B T \simeq 0.025$ eV at room temperature), the first term in the denominator is large compared to the second term, and the Fermi–Dirac distribution becomes the classical Boltzmann distribution (8.24). In the limit $T = 0$, the Fermi–Dirac distribution becomes a step function,

$$f(E, E_F, T = 0) = \begin{cases} 0, & E > E_F, \\ 1, & E < E_F, \end{cases} \tag{8.26}$$

   as shown in Fig. 8.5. Thus, at $T = 0$ K, all states below E_F are completely filled and all states above E_F are empty. Although the step function is not well defined at $E = E_F$, since for all $T > 0$ the Fermi–Dirac distribution crosses through the point $f(E) = 1/2$ when $E = E_F$, then $f(E = E_F, E_F, 0)$ can be defined to be $1/2$. The Fermi–Dirac distribution for $T \gg 0$ is also shown in Fig. 8.5, where it can be seen that only states within a few $k_B T$ of the Fermi energy are likely to be excited. In many materials (especially metals) at even quite high temperatures, $k_B T$ is very small

†As previously remarked, by a "filled state" we mean that an electron is in that state, i.e., that an electron is represented by the corresponding state function. For a one-electron system, one state would be filled; the electron is in a certain state. For N electrons, N states would be filled.

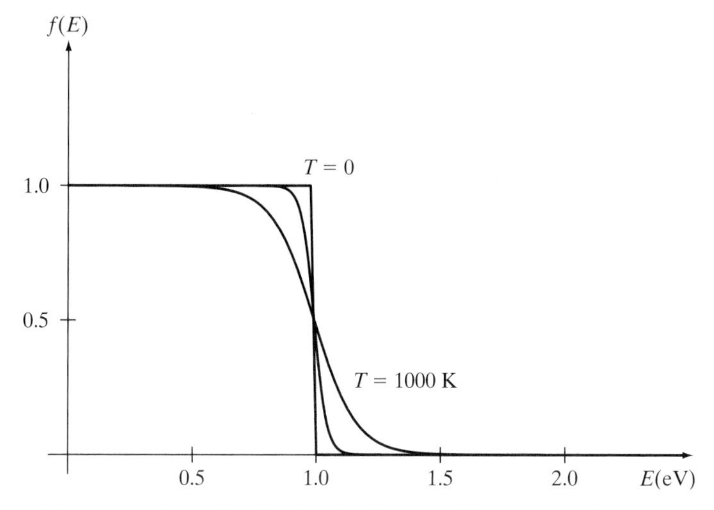

Figure 8.5 Fermi–Dirac distribution for $E_F = 1$ eV at $T = 0, 300$, and 1,000 K.

compared to the Fermi energy, and, thus, electrons at the Fermi energy are of principal importance. In fact, to use the Fermi sea notion described in Section 4.4, since k_BT is often small compared to E_F, the excited states above E_F are said to be merely "ripples on the Fermi sea."

3. **Bose–Einstein Distribution**
   The Bose–Einstein distribution applies when the particles are indistinguishable, but can occupy the same available states. That is, it applies to quantum particles that do not obey the exclusion principle—for example, photons and phonons. The Bose–Einstein distribution is

$$f^{BE}(E) = f^{BE}(E, \mu, T) = \frac{1}{e^{\frac{E-\mu}{k_BT}} - 1}. \tag{8.27}$$

We can see that at large values of energy, all three distributions are equal. This is because of the de Broglie wavelength (2.15)—as energy increases, the de Broglie wavelength decreases, and for very high energies, the de Broglie wavelength is vanishingly small, such that the particle becomes like a classical particle. Since we are interested in electrons, we will be concerned primarily with the Fermi–Dirac distribution.

Using the Fermi–Dirac distribution, we find that since the probability of a level being occupied by an electron is $f(E)$, then the probability of a level not being occupied by an electron (i.e., being occupied by a hole), is

$$1 - f(E) = \frac{1}{e^{\frac{E_F-E}{k_BT}} + 1}, \tag{8.28}$$

since†

$$\frac{1}{e^{\frac{E-E_F}{k_B T}}+1}+\frac{1}{e^{\frac{E_F-E}{k_B T}}+1}=1. \tag{8.30}$$

8.2.1 Carrier Concentration in Materials

With the concept of the density of states and of the Fermi–Dirac distribution, we can now find the total number of filled electronic states per unit volume, and also obtain a useful formula for the Fermi level. In general, we can determine the electron concentration n (m^{-3}) by multiplying the electronic density of states, $N(E)$, and the probability that a state is occupied, $f(E)$, and integrating over energy,

$$n=\int N(E)\, f(E, E_F, T)\, dE. \tag{8.31}$$

The limits of integration would depend on the specific circumstances, such as the energy range of interest, etc. The number of filled states per unit volume at $T=0$ is

$$N_f=\int_0^\infty N(E)\, f(E, E_F, T=0)\, dE \tag{8.32}$$

$$=\int_0^{E_F} N(E)\, dE. \tag{8.33}$$

For the electron confined to a three-dimensional space, using (8.6) we have‡

$$N_f=\int_0^{E_F}\frac{2^{1/2} m_e^{*3/2} E^{1/2}}{\hbar^3\pi^2}\, dE=\frac{2^{1/2} m_e^{*3/2}}{\hbar^3\pi^2}\frac{2}{3}E_F^{3/2}, \tag{8.34}$$

which must equal the total number of electrons per unit volume, N. Then, if we know the total number of electrons in a system, we can find the Fermi level as

$$E_F=\left(3N\frac{\hbar^3\pi^2}{\left(2m_e^*\right)^{3/2}}\right)^{2/3} \tag{8.35}$$

†To see this, let $u=e^{\frac{E-E_F}{k_B T}}$, such that $1/u=e^{\frac{E_F-E}{k_B T}}$. Then,

$$\frac{1}{e^{\frac{E-E_F}{k_B T}}+1}+\frac{1}{e^{\frac{E_F-E}{k_B T}}+1}=\frac{1}{u+1}+\frac{1}{\frac{1}{u}+1} \tag{8.29}$$

$$=\frac{1}{u+1}+\frac{u}{1+u}=1.$$

‡Integrating over the specified limits and using (8.6) implies that this calculation is valid for an electron gas model. This is applicable to most metals, and serves as an approximation for semiconductors that are doped at a sufficiently high level (so that we can consider the material to be a particle gas of the dopant carriers). An analogous calculation for semiconductors not in the highly doped limit is given in the next section.

$$= \left(\frac{\hbar^2}{2m_e^*}\right)\left(3N\pi^2\right)^{2/3}. \tag{8.36}$$

In a typical metal, E_F is in the range of a few eV. For example, assuming one electron per 0.1 nm^3 (10^{22} electrons/cm^3), which is a typical order of magnitude for many materials, we have

$$E_F = 1.67 \text{ eV}. \tag{8.37}$$

In two dimensions, using (8.14) we have

$$N = N_f = \int_0^{E_F} N(E)\, dE = \int_0^{E_F} \frac{m_e^*}{\pi\hbar^2} dE = \frac{m_e^*}{\pi\hbar^2} E_F, \tag{8.38}$$

leading to

$$E_F^{(2d)} = \frac{\pi\hbar^2}{m_e^*} N. \tag{8.39}$$

In one dimension,

$$E_F^{(1d)} = \left(\frac{\hbar^2}{2m_e^*}\right)\left(\frac{\pi N}{2}\right)^2. \tag{8.40}$$

Later we will need a few other useful quantities related to the Fermi level. The *Fermi wavevector* is the wavevector at the Fermi energy. If the energy–wavevector relationship is[†]

$$E = \frac{\hbar^2 k^2}{2m_e^*}, \tag{8.41}$$

then the Fermi wavevector is

$$k_F = \frac{\left(2m_e^* E_F\right)^{1/2}}{\hbar} \tag{8.42}$$

$$= \left(3N\pi^2\right)^{1/3} \quad \text{in three dimensions,} \tag{8.43}$$

$$= (2\pi N)^{1/2} \quad \text{in two dimensions,} \tag{8.44}$$

$$= \frac{\pi N}{2} \quad \text{in one dimension.} \tag{8.45}$$

Of course, N is the appropriate number density (m^{-3}, m^{-2}, m^{-1} in three dimensions, two dimensions, and one dimension, respectively). The relationship between Fermi energy and Fermi wavevector is depicted in Fig. 8.6.

[†]This quadratic dependence between E and k occurs for free electrons (Section 4.1), spatially confined electrons in otherwise empty space (Section 4.3), and for electrons in crystalline materials near energy bandedges (Section 5.4.4). Recall that for semiconductors, the bandedges are the most important sections of the energy band.

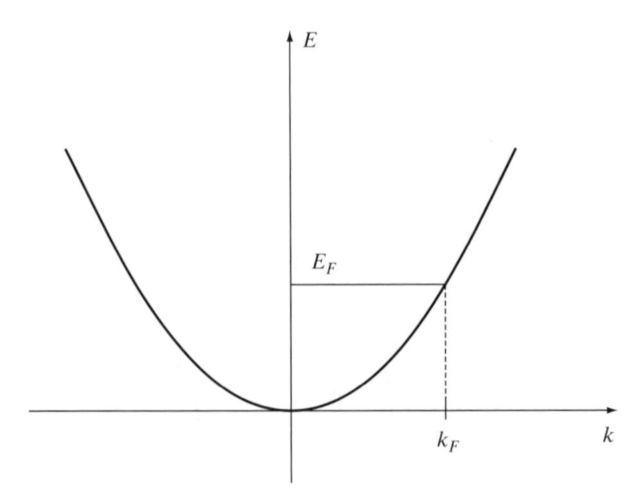

Figure 8.6 Energy versus wavevector showing the Fermi level and Fermi wavevector.

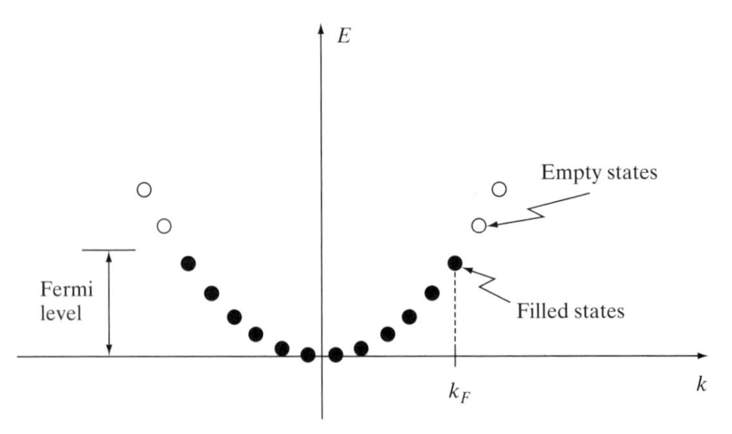

Figure 8.7 Energy versus wavenumber for discrete energy levels. Filled circles denote filled states, and empty circles denote empty states. Only values of energy and wavenumber represented by circles are allowed.

We are usually interested in confined regions of space, leading to discrete states, such as considered in Section 4.3. In this case, the wavenumber k will be discrete along some (perhaps multiple) coordinate(s), and the energy-wavevector diagram will look like Fig. 8.7. In this figure, note that the Fermi level is shown halfway between the highest filled state and the next higher empty state. This may occur, or it may lie at the level of the highest filled state, depending on the number of electrons, per the definition of chemical potential given in Section 4.4. This is discussed in Section 8.2.3, where the Fermi level is considered to be the energy state that has a probability of $\frac{1}{2}$ of being occupied by an electron at $T = 0$ K.

If we consider confinement in two dimensions, say, along the x- and y-coordinates, then we can view the occupation of states as being those states within the Fermi circle, as

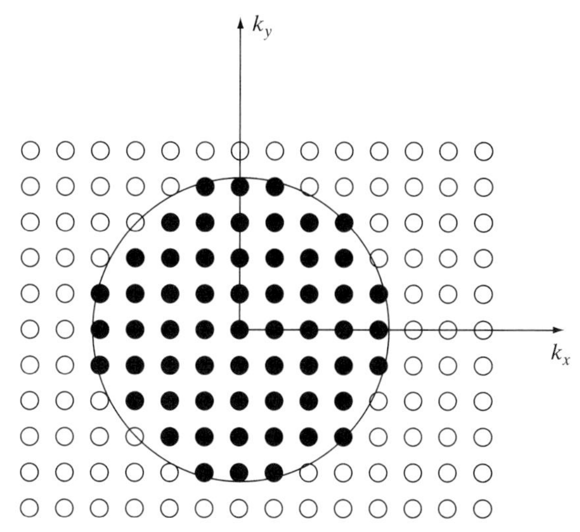

Figure 8.8 Electronic states in the k_x–k_y plane. The states within the Fermi circle shown are occupied.

shown in Fig. 8.8. For confinement in three dimensions, we obtain a Fermi sphere, rather than a Fermi circle.

The Fermi wavelength is

$$\lambda_F = \frac{2\pi}{k_F}, \tag{8.46}$$

and the *Fermi velocity* is, from the momentum–wavevector relationship (2.15) ($m_e^* v = p = \hbar k$),

$$v_F = \frac{\hbar k_F}{m_e^*}. \tag{8.47}$$

Note that the relationship between the Fermi wavelength, Fermi wavenumber, and Fermi velocity are completely general, unlike the parabolic relationship between Fermi energy and Fermi wavenumber (8.41).

Table 8.1 shows the Fermi energy, velocity, and wavelength for a few common elements in three dimensions. More values are provided in Table II in Appendix B.

TABLE 8.1 FERMI ENERGY, FERMI VELOCITY, AND FERMI WAVELENGTH FOR SEVERAL COMMON ELEMENTS.

Element	E_F (eV)	v_F ($\times 10^6$ m/s)	λ_F (nm)
Li	4.74	1.29	0.56
Na	3.24	1.07	0.68
K	2.12	0.86	0.85
Cu	7.00	1.57	0.46
Ag	5.49	1.39	0.52

8.2.2 The Importance of the Fermi Electrons

Before leaving this section, it is worthwhile to examine some ideas related to the Fermi level. For example, it is a fact that many electrons contribute to conduction in metals, and yet we have said previously that only electrons having energy near the Fermi level contribute to conduction (i.e., that electrons near the Fermi level are the "important" electrons). However, there is no contradiction; there are many electrons that have energy near to the Fermi level, and these are the many electrons that contribute to conduction. Considering copper, for example, we find that using $E_F = 7.0$ eV, (8.34) leads to the density of electrons as $N(E_F) = 1.8 \times 10^{22}$ cm^{-3}. Thus, there are a large number of states around the Fermi level.

In addition, at room temperature (around 293 K), $k_B T = 0.025$ eV. This energy is very small compared with the Fermi level in most metals, as shown in Table 8.1. That is, since Boltzmann's constant is so small, the thermal energy gained by electrons, even at quite high temperatures, is small compared to the Fermi energy. Therefore, in conductors, electrons in the absence of an applied field can often be considered to have energy E_F (i.e., thermal energy can often be ignored). For example, consider again copper, where $E_F = 7$ eV. Let's assume that in order for thermal effects to be important, we need $k_B T = 3.5$ eV, which is one-half the Fermi energy. The corresponding temperature is $T \sim 40,600$ K, and even at this temperature $k_B T$ is only one-half of the Fermi level! Therefore, filled states deep within the Fermi sphere will not play much role in electronic interactions, and we are most interested in electrons and states near the Fermi surface.

However, in semiconductors, thermal energy can liberate a large number of electrons, especially from donors or acceptors (creating free electrons and holes), changing significantly the material's electronic properties. In this case, electron energies are typically above the Fermi level, as described in Section 5.4.

Aside from thermal effects, it turns out that even in the event of other energy inputs, in metals the electron's energy is still approximately the Fermi energy. For example, consider what happens to the electron's energy if we apply an electric field. The resulting drift velocity (the velocity of the net current) is very low, typically on the order of mm/s (see Section 10.1), which is many orders of magnitude smaller than the Fermi velocity. Even assuming that we apply a large field of 100 V/m to copper, and that the field accelerates an electron over the mean free path of 40 nm, the energy gained from the field is

$$\begin{aligned} E = Fd &= e\,(100)\left(40 \times 10^{-9}\right) = 6.4 \times 10^{-25}\ \text{J} \\ &= 4 \times 10^{-6}\ \text{eV}, \end{aligned} \tag{8.48}$$

which is miniscule compared to the Fermi energy. So again, electrons in metals act approximately as if they have energy E_F. This is true even at very high temperatures, even near the melting temperature of a metal.

8.2.3 Equilibrium Carrier Concentration and the Fermi Level in Semiconductors

In semiconductors, in particular, we are often interested in determining the carrier concentration available for electrical conduction. We can determine the carrier concentration in the conduction band by multiplying the electron density of states in the conduction band, $N_e(E)$, and the probability that a state is occupied, $f(E, E_F, T)$, and summing over all energies in the band,

$$n = \int_{E_c}^{\infty} N_e(E) f(E, E_F, T) \, dE, \tag{8.49}$$

where $f(E, E_F, T)$ is the Fermi–Dirac function. This leads to

$$\begin{aligned} n &= \int_{E_c}^{\infty} \left(\frac{\sqrt{2} m_e^{*3/2} (E - E_c)^{1/2}}{\pi^2 \hbar^3} \right) \left(\frac{1}{e^{\left(\frac{E-E_F}{k_B T}\right)} + 1} \right) dE \\ &= \frac{1}{2\pi^2} \left(\frac{2m_e^*}{\hbar^2} \right)^{3/2} \int_{E_c}^{\infty} \left(\frac{(E - E_c)^{1/2}}{e^{\left(\frac{E-E_F}{k_B T}\right)} + 1} \right) dE. \end{aligned} \tag{8.50}$$

If $(E - E_F) / (k_B T) \gg 1$ (i.e., the Boltzmann approximation), then

$$n = \frac{1}{2\pi^2} \left(\frac{2m_e^*}{\hbar^2} \right)^{3/2} e^{\frac{E_F}{k_B T}} \int_{E_c}^{\infty} (E - E_c)^{1/2} e^{-\frac{E}{k_B T}} dE. \tag{8.51}$$

With

$$\begin{aligned} \int_{E_c}^{\infty} \left((E - E_c)^{1/2} e^{-\frac{E}{k_B T}} \right) dE &= e^{-\frac{E_c}{k_B T}} \int_{E_c}^{\infty} (E - E_c)^{1/2} e^{-\frac{E-E_c}{k_B T}} dE \\ &= e^{-\frac{E_c}{k_B T}} (k_B T)^{1/2} (k_B T) \int_0^{\infty} u^{1/2} e^{-u} du \\ &= e^{-\frac{E_c}{k_B T}} (k_B T)^{3/2} \frac{1}{2} \sqrt{\pi}, \end{aligned} \tag{8.52}$$

using the change of variables $u = (E - E_c) / (k_B T)$, $du = dE / (k_B T)$, and

$$\int_0^{\infty} u^{1/2} e^{-u} du = \frac{1}{2} \sqrt{\pi}, \tag{8.53}$$

then

$$n = 2 \left(\frac{m_e^* k_B T}{2\pi \hbar^2} \right)^{3/2} e^{\frac{E_F - E_c}{k_B T}} = N_c e^{-\frac{(E_c - E_F)}{k_B T}}. \tag{8.54}$$

N_c is known as the *effective density of states at the conduction bandedge*. For example, for silicon, $N_c \simeq 2.8 \times 10^{19}$ cm^{-3} at room temperature.

Taking the natural logarithm of the previous equation, we have an expression for the Fermi level,

$$E_F = E_c + k_B T \ln \frac{n}{N_c}. \tag{8.55}$$

A similar calculation for the holes leads to

$$p = 2\left(\frac{m_h^* k_B T}{2\pi\hbar^2}\right)^{3/2} e^{\frac{E_v - E_F}{k_B T}} = N_v e^{-\frac{(E_F - E_v)}{k_B T}}, \tag{8.56}$$

where N_v is called the *effective density of states at the valence bandedge.*

In summary, we have

$$n = N_c e^{-\frac{(E_c - E_F)}{k_B T}}, \tag{8.57}$$

$$p = N_v e^{-\frac{(E_F - E_v)}{k_B T}}. \tag{8.58}$$

Note that we haven't indicated whether (8.57) and (8.58) refer to intrinsic or extrinsic semiconductors. In fact, they typically apply to both in thermal equilibrium (assuming the Boltzmann approximation is valid), where the presence or absence of dopant atoms is reflected in the position of the Fermi level. For intrinsic semiconductors, the condition $n_i = p_i$ leads to

$$2\left(\frac{m_e^* k_B T}{2\pi\hbar^2}\right)^{3/2} e^{\frac{E_F - E_c}{k_B T}} = 2\left(\frac{m_h^* k_B T}{2\pi\hbar^2}\right)^{3/2} e^{\frac{E_v - E_F}{k_B T}} \tag{8.59}$$

$$\Rightarrow E_{F_i} = \frac{E_c + E_v}{2} + \frac{3k_B T}{4} \ln \frac{m_h^*}{m_e^*}.$$

Therefore, the intrinsic Fermi level is near the middle of the bandgap. (If the effective electron and hole masses are equal, the intrinsic Fermi level is precisely in the middle of the bandgap.)† The top of the valence band is usually set to $E = 0$, and in this case, $E_{F_i} = E_g/2$. Furthermore, for intrinsic semiconductors, obviously $n_i p_i = n_i^2$, so that

$$n_i = \sqrt{N_c N_v} e^{-\frac{E_g}{2k_B T}}. \tag{8.60}$$

From (8.60) we see that carrier concentration increases exponentially as bandgap decreases. For silicon at room temperature, $n_i \simeq 1.5 \times 10^{10}$ cm^{-3}.

For extrinsic n-type semiconductors, if we assume that the dopant concentration N_d is much higher that the intrinsic carrier concentration n_i (typical doping levels in conventional devices are on the order of 10^{15} atoms/cm^3 or more), and that all dopant atoms are ionized (the usual situation), then for an n-type material,

$$n \simeq N_d. \tag{8.61}$$

†It is convenient to view the Fermi level as the energy state that has a probability of $\frac{1}{2}$ of being occupied by an electron. Since at $T = 0$ K all states below E_v are filled, and all states above E_c are empty, then the equal probability point lies in the middle of the bandgap, even though no allowed states are located in the gap. This is consistent with the definition of the chemical potential in Section 4.4.

Substituting this into (8.57), we have

$$E_F = E_c - k_B T \ln\left(\frac{N_c}{N_d}\right). \tag{8.62}$$

Of course, with this Fermi level, (8.57) yields $n = N_d$.

Note that if the doping level is very high such that $N_d \simeq N_c$, then E_F approaches E_c and the semiconductor is said to be *degenerate*. However, in this case, the Boltzmann approximation breaks down, and a more careful analysis must be used.

In a similar manner, for a p-type semiconductor with an acceptor doping density $N_a \gg p_i$,

$$E_F = E_v + k_B T \ln\left(\frac{N_v}{N_a}\right),$$

which approaches the value $E_F = E_v$ for $N_a \simeq N_v$ (e.g., $N_v \simeq 10^{19}$ cm^3 for silicon at room temperature).

Since the product

$$np = N_c N_v e^{-\frac{E_g}{k_B T}} = n_i^2 \tag{8.63}$$

is independent of the Fermi level, it should hold for doped (extrinsic) semiconductors as well; (8.63) is known as the *mass-action law*. It holds in the extrinsic case since an increase in one carrier type (electrons or holes) tends to diminish, through recombination, the other carrier type.

8.3 MAIN POINTS

In this chapter, we have considered the idea of particle statistics and the density of states, and implications to the Fermi level and carrier concentrations. In particular, after studying this chapter you should understand

- the concept of density of states in various spatial dimensions, and the significance of the density of states;
- how the density of states can be measured;
- quantum and classical statistics for collections of large numbers of particles, including the Boltzmann, Fermi–Dirac, and Bose–Einstein distributions;
- the role of density of states and quantum statistics in determining the Fermi level;
- applications of density of states and quantum statistics to determine carrier concentration in materials, including in doped semiconductors.

8.4 PROBLEMS

1. Energy levels for a particle in a three-dimensional cubic space of side L with hard walls (boundary conditions (4.50)) were found to be (4.54)

$$E_n = \frac{\hbar^2 \pi^2}{2m_e^* L^2}\left(n_x^2 + n_y^2 + n_z^2\right), \tag{8.64}$$

$n_{x,y,z} = 1, 2, 3, \ldots$, which leads to the density of states (8.6). Using periodic boundary conditions (4.55), we found energy levels to be (4.59)

$$E_n = \frac{2\hbar^2\pi^2}{m_e^* L^2}\left(n_x^2 + n_y^2 + n_z^2\right), \tag{8.65}$$

$n_{x,y,z} = 0, \pm 1, \pm 2, \ldots$. Following a derivation similar to the one shown for (8.6), show that the same density of states arises from (8.65).

2. Derive (8.16), the density of states in two dimensions.

3. The density of states in a one-dimensional system is given by (8.15),

$$N(E) = \frac{\sqrt{2m_e^*}}{\pi\hbar} E^{-1/2}, \tag{8.66}$$

assuming zero potential energy.

(a) Use this formula to show that the Fermi energy in terms of the total number of filled states at $T = 0$ K, N_f, is (8.40),

$$E_F = \frac{\hbar^2}{2m_e^*}\left(\frac{\pi N_f}{2}\right)^2. \tag{8.67}$$

(b) If there are N electrons in a one-dimensional box of length L, show that the energy level of the highest energy electron is E_F, given by (8.67). Use the fact that the energy levels in a one-dimensional box are (4.35),

$$E_n = \frac{\hbar^2}{2m_e^*}\left(\frac{n\pi}{L}\right)^2, \quad n = 1, 2, 3, \ldots \tag{8.68}$$

4. Assume that the density of states in a one-dimensional system is given by

$$N(E) = \frac{\sqrt{2m}}{\pi\hbar} E^{-1/3} \tag{8.69}$$

at zero potential energy.

(a) Use this formula to obtain the Fermi energy.

(b) What is the relationship between the de Broglie wavelength and the Fermi wavelength?

5. The electron carrier concentration in the conduction band can be determined by multiplying the electron density of states in the conduction band, $N_e(E)$, and the probability that a state is occupied, $f(E, E_F, T)$, and then summing over all energies to yield (8.54) on page 275. Perform the analogous calculations for determining the hole carrier concentration in the valence band, leading to (8.56).

6. The Fermi wavelength in three-dimensional copper is $\lambda_F = 0.46$ nm. Determine the Fermi wavelength in two-dimensional copper.

7. Determine the Fermi wavelength of electrons in three-dimensional aluminum and zinc.

8. What is the electron concentration in an n-type semiconductor at room temperature if the material is doped with 10^{14} cm^{-3} donor atoms? How would one determine the hole concentration?

9. The concept of the Fermi energy can be used to give some confidence of electron–electron screening in conductors (i.e., the ability to ignore interactions among electrons), previously described in Sections 3.5 and 4.2. To see this, approximate the electron's kinetic energy by the Fermi energy. Then, for an electron density N m^{-3}, assuming that the average distance between electrons is $N^{-1/3}$, show that the ratio of Coulomb potential energy to kinetic energy goes to zero as N goes to infinity, showing that the electrons essentially screen themselves.

10. Constructive and destructive interference of electromagnetic waves (light, radio-frequency signals, etc.) is one of the most commonly exploited phenomena in classical high-frequency devices. For example, resonance effects result from wave interference, and are used to form filters, impedance transformers, absorbers, antennas, etc. In contrast to these electromagnetic (i.e., photon) devices, which obey Bose–Einstein statistics, electron waves are fermions, and obey Fermi–Dirac statistics and the exclusion principle. Describe how it is possible for a large collection of bosons to form a sharp interference pattern, yet this will not be observed by a large collection of electrons in a solid. Does this make sense considering Fig. 2.6 on page 29, where sharp electron interference was, in fact, observed?

Chapter

9

MODELS OF SEMICONDUCTOR QUANTUM WELLS, QUANTUM WIRES, AND QUANTUM DOTS

GaSb-InAs-AlSb heterostructure. (From Harper, J., et al., "Microstructure of GaSb-on-InAs Heterojunction Examined with Cross-Sectional Scanning Tunneling Microscopy," *Appl. Phys. Lett.* 73 (1998): 2805.)

In Section 4.7, the idea of a quantum well, a quantum wire, and a quantum dot was introduced using a simple particle-in-a-box model. The main differentiating characteristic among the three structures was the size of the structure in each coordinate, with respect to a particle's de Broglie wavelength at the Fermi energy (i.e., with respect to λ_F). Assuming the electron E–k relation

$$E = \frac{\hbar k^2}{2m_e^*}, \tag{9.1}$$

which, for instance, typically holds near the bottom of the conduction band in a semiconductor, we have

$$\lambda_F = \frac{h}{\sqrt{2m_e^* E_F}}. \tag{9.2}$$

If the electron's environment is large compared to λ_F, then the electron will behave approximately as if it is free. If the electron's wavelength is on the order of, or is large compared with, its environment, then it will behave in a confined fashion.[†]

To appreciate the size scales involved, consider that in three dimensions the Fermi wavelength is, using (8.36),

$$\lambda_F = \frac{2\pi}{\left(3N_e\pi^2\right)^{\frac{1}{3}}}. \tag{9.3}$$

For copper, $N_e \simeq 8.45 \times 10^{28}$ /m^3, such that $\lambda_F = 0.46$ nm ($\lambda_F = 0.52$ nm for gold), whereas for GaAs[‡] assuming a doping level such that $N_e \simeq 10^{22}$ m^{-3}, $\lambda_F = 94$ nm. More generally, in typical metals,

$$\lambda_F \sim 0.5 - 1 \text{ nm}, \tag{9.4}$$

and in typical semiconductors,

$$\lambda_F \sim 10 - 100 \text{ nm}, \tag{9.5}$$

although this value depends on the doping level. Thus, confinement effects tend to become important in semiconductors at much larger dimensions than for conductors. This can also be considered from the effective mass point of view, due to the small effective mass in semiconductors.

To summarize, the three structures are shown in Fig. 9.1, where

- if $\lambda_F \ll L_x, L_y, L_z$, then we have an effectively three-dimensional system—the system, in all directions, is large compared to the size scale of an electron;
- if $L_x \leq \lambda_F \ll L_y, L_z$, then we have an effectively two-dimensional system—a two-dimensional electron gas or quantum well;
- if $L_x, L_y \leq \lambda_F \ll L_z$, then we have an effectively one-dimensional system—a quantum wire;
- if $L_x, L_y, L_z \leq \lambda_F$, then we have an effectively zero-dimensional system—a quantum dot.

In the previous discussion, we simply considered the structures to be empty space bounded by hard walls. However, many important structures are made from semiconducting materials, and, in any event, it is often crucial to take into account the material properties of the object. In this chapter, we reconsider quantum wells, wires, and dots, taking into account material properties of the structure.

[†]Coherence length, which will be described in Section 10.2, is another factor in determining system dimensionality. Furthermore, for quantum dots, often the exciton radius is taken as the measure of quantum confinement, as described in Section 9.3.1.

[‡]For semiconductors, the use of (8.36) in (9.2) corresponds to approximating the material as a bulk electron gas having electron density equivalent to the doping density. See also Problem 9.2 for another method, as well as the discussion in Section 9.3.1.

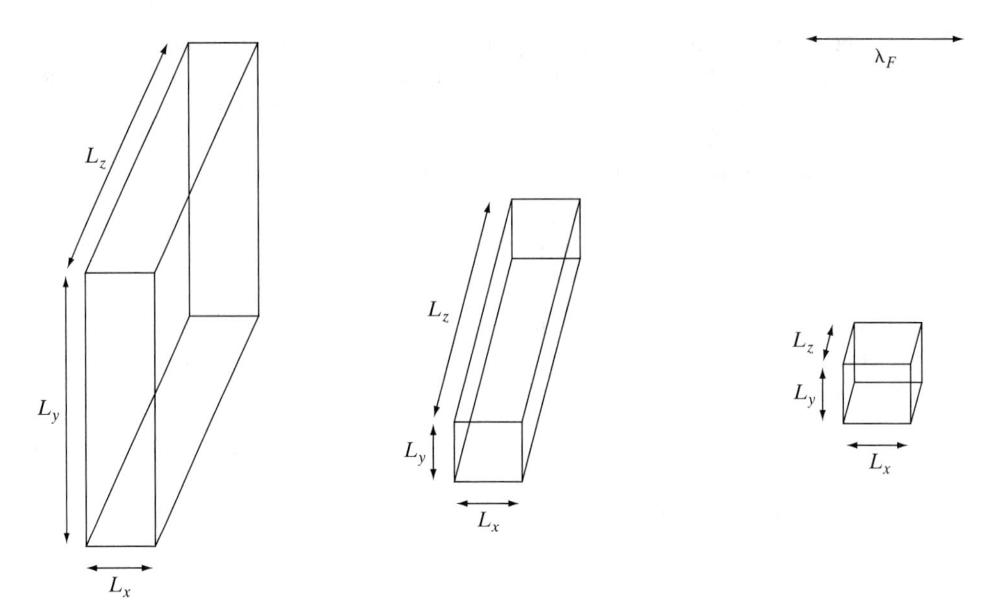

Figure 9.1 Effectively two-, one-, and zero-dimensional regions of space, where it is assumed that a spatial dimension can be neglected when $L \leq \lambda_F$. The size of λ_F is depicted in the upper-right corner.

9.1 SEMICONDUCTOR HETEROSTRUCTURES AND QUANTUM WELLS

Crystal growth techniques can produce atomically abrupt interfaces between two materials (see, for example, the image at the start of this chapter). especially if the lattice types and lattice constants of the two materials are similar. For example, we could sandwich a small bandgap material, such as GaAs (perhaps tens of angstroms thick) between thick layers of a large bandgap material, such as AlGaAs, as shown in Fig. 9.2. Note that in this model, the AlGaAs is a semi-infinite half space, and the GaAs is an infinite planar slab having thickness L_x.

This sandwich is called a *semiconductor heterostructure*. Quantum wells are formed in both the conduction and valence bands, as shown in the real-space energy band diagram in Fig. 9.3.

To a good degree of accuracy, semiconductor heterostructures composed of crystalline materials can be analyzed by replacing the actual crystal structure with a potential energy profile, and using the value of effective mass appropriate to the material. This approximation

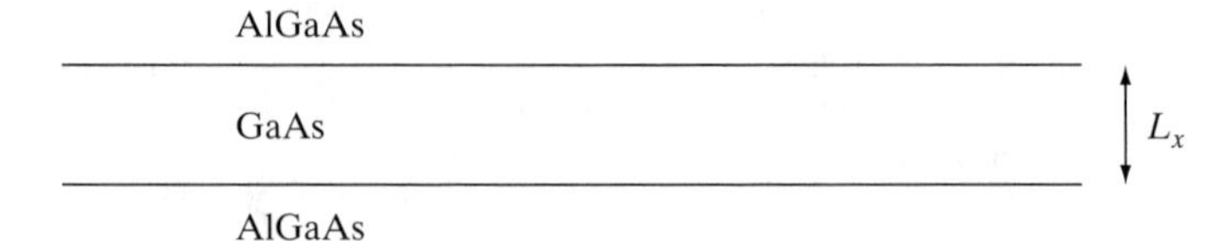

Figure 9.2 Semiconductor heterostructure composed of AlGaAs (large bandgap material) and GaAs (small bandgap material).

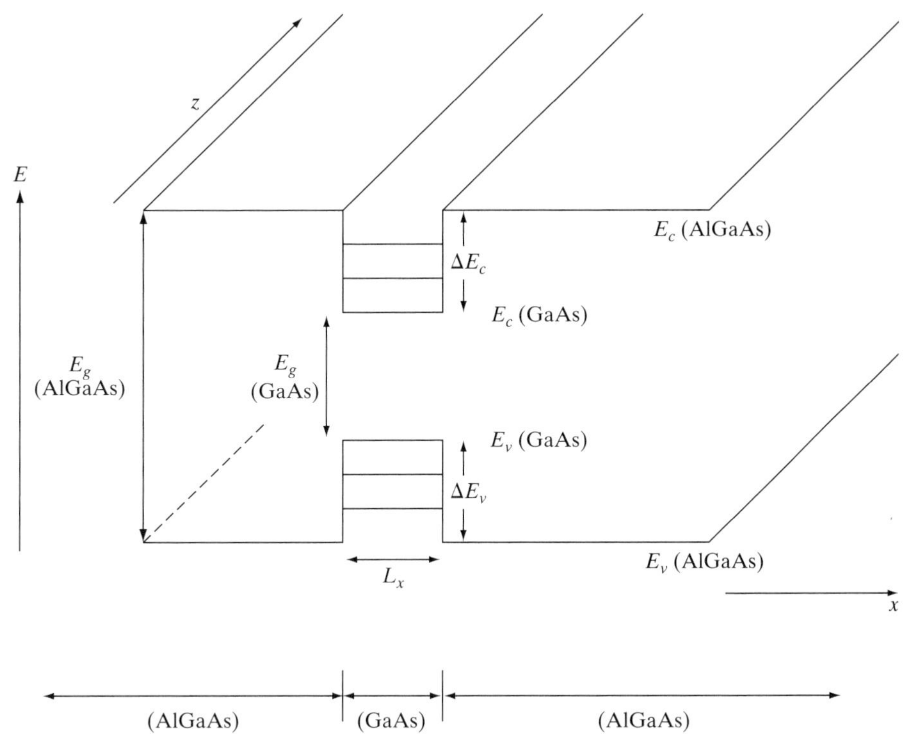

Figure 9.3 Semiconductor heterostructure made from AlGaAs and GaAs. Quantum wells form in both the conduction and valence bands, resulting in subband energy levels.

presupposes that the Bloch functions (5.7) in each region are the same, or similar. This is often a reasonable approximation for typical III-V heterostructures, i.e., those constructed from elements in groups III and V of the periodic table, and II-VI heterostructures are also of interest.

We use the effective mass Schrödinger's equation (5.41),

$$\left(-\frac{\hbar^2}{2m^*}\nabla^2 + V(x)\right)\psi(\mathbf{r}) = E\psi(\mathbf{r}), \tag{9.6}$$

where m^* is the effective mass of the particle at a given point in the heterostructure, so that $m^* = m^*(x)$. Since the heterostructure is comprised of piece-wise constant regions, we can solve (9.6) in the usual way. That is, we solve (9.6) in each region, and match the solutions across each interface using the continuity conditions of Schrödinger's equation (3.143).

The effective mass Schrödinger's equation actually does not strictly hold for alloys such as AlGaAs, which do not have pure periodic (crystalline) structure. However, it often suffices to treat the alloy as having a periodic nature. For instance, AlGaAs is modeled as having properties between AlAs and GaAs. This is called the *virtual-crystal approximation.* The bandgap of the alloy is approximately the arithmetic average of the bandgaps of the two materials. For AlAs, $E_g \simeq 2.2$ eV, and for GaAs, $E_g \simeq 1.5$ eV, and so AlGaAs has a bandgap on the order of $E_g \simeq 1.85$ eV. More specifically, AlGaAs is an alloy, written as $Al_\alpha Ga_{1-\alpha}As$, where α is the mole fraction of Al. The band gap can be empirically approximated as

$$E_g \simeq 1.426 + 1.247\alpha \tag{9.7}$$

for $\alpha < 0.45$. (See Table VI in Appendix B for properties of AlGaAs). For this range of α, $Al_\alpha Ga_{1-\alpha}As$ is a direct bandgap material, whereas for $0.45 < \alpha < 1.0$, it is an indirect bandgap material ($\alpha = 0.3$ is a common value, resulting in $E_g \simeq 1.80$). Furthermore, AlAs and GaAs have similar lattice constants (Table V in Appendix B), and thus can be combined without significant lattice strain.

Using the separation of variables technique presented in Section 3.2, we solve (9.6) by writing

$$\psi(x, y, z) = \psi_x(x)\,\psi_y(y)\,\psi_z(z), \tag{9.8}$$

which results in

$$\left(\frac{1}{\psi_x}\frac{\partial^2}{\partial x^2}\psi_x + \frac{1}{\psi_y}\frac{\partial^2}{\partial y^2}\psi_y + \frac{1}{\psi_z}\frac{\partial^2}{\partial z^2}\psi_z + \frac{2m^*}{\hbar^2}(E - V(z))\right) = 0. \tag{9.9}$$

Using the separation argument, we obtain

$$\frac{1}{\psi_y}\frac{\partial^2}{\partial y^2}\psi_y = -k_y^2 \rightarrow \psi_y(y) = Ce^{ik_y y} + De^{-ik_y y}, \tag{9.10}$$

$$\frac{1}{\psi_z}\frac{\partial^2}{\partial z^2}\psi_z = -k_z^2 \rightarrow \psi_z(z) = Ae^{ik_z z} + Be^{-ik_z z}, \tag{9.11}$$

leading to

$$\left(-k_y^2 - k_z^2 + \frac{1}{\psi_x}\frac{\partial^2}{\partial x^2}\psi_x + \frac{2m^*}{\hbar^2}(E - V(x))\right) = 0. \tag{9.12}$$

Therefore, in the unconstrained directions (y and z), the wavefunction is represented by plane waves. Since k_y and k_z can be allowed to take on positive or negative values, for our purposes it is sufficient to take

$$\psi_y(y)\,\psi_z(z) = Ae^{ik_y y}e^{ik_z z}. \tag{9.13}$$

It remains to solve†

$$\left(-\frac{\hbar^2}{2m^*}\frac{\partial^2}{\partial x^2} + \frac{\hbar^2}{2m^*}\left(k_y^2 + k_z^2\right) + V(x)\right)\psi_x(x) = E\psi_x(x). \tag{9.14}$$

†Another approach is to assume plane waves in the unconfined coordinates, in accordance with our experiences with free (unconfined) electrons, such that,

$$\psi(x, y, z) = e^{ik_y y}e^{ik_z z}\psi_x(x),$$

which, when substituted into (9.6), leads to the same result.

In what follows, we will want to make use of the various one-dimensional problems considered in Sections 4.1, 4.3, and 4.5. However, there we solved the one-dimensional Schrödinger's equation (3.140),

$$\left(-\frac{\hbar^2}{2m}\frac{d^2}{dx^2}+V(x)\right)\psi(x)=E\psi(x). \tag{9.15}$$

This equation and (9.14) will have the same form, and, thus, the solutions of (9.15) can be used as the solutions of (9.14), if we define an effective potential

$$V_e(x,k)=\frac{\hbar^2}{2m^*(x)}\left(k_y^2+k_z^2\right)+V(z), \tag{9.16}$$

such that (9.14) becomes

$$\left(-\frac{\hbar^2}{2m^*}\frac{\partial^2}{\partial x^2}+V_e(x,k)\right)\psi_x(x)=E\psi_x(x). \tag{9.17}$$

The effective potential depends on the confining potential $V(x)$, and also on the longitudinal wavenumber $k_l=\sqrt{k_y^2+k_z^2}$. Furthermore, the effective mass will depend on position x as well.

This is as far as we can go without considering a specific model of the heterostructure, which we will do in the next section. In summary, we have

$$\psi(x,y,z)=Ae^{ik_y y}e^{ik_z z}\psi_x(x), \tag{9.18}$$

where k_y and k_z are continuous variables (as in the free-electron case), and ψ_x (and k_x) will be determined by solving (9.17) subject to boundary conditions in the x-coordinate.

9.1.1 Confinement Models and Two-Dimensional Electron Gas

To continue with our analysis of the heterostructure we need to determine the wavefunction ψ_x, and wavenumber k_x. The effective potential V_e depends on the longitudinal wavenumber k_l, and serves to modify the potential seen by the electron from $V(x)$ to $V_e(x,k)$.

The simplest approximation is to assume that the well (here referring to the well in either the conduction or valence band) is infinitely deep, that is,

$$\begin{aligned} V&=0, \quad 0\le x\le L_x,\\ V&=\infty, \quad x<0,\ x>L_x, \end{aligned} \tag{9.19}$$

which is the hard-wall model considered in Section 4.3.2. Schrödinger's equation inside the well is

$$\left(-\frac{\hbar^2}{2m^*}\frac{\partial^2}{\partial x^2}+\frac{\hbar^2k_l^2}{2m^*}\right)\psi_x(x)=E\psi_x(x), \tag{9.20}$$

with k_l being the longitudinal wavenumber and m^* the effective mass in the well region (i.e., the GaAs region in Figs. 9.2 and 9.3). The solution of (9.20) is

$$\psi_x(x) = Fe^{ik_x x} + Ge^{-ik_x x}, \tag{9.21}$$

where

$$k_x^2 = \frac{2m^*}{\hbar^2}E - k_l^2. \tag{9.22}$$

Applying the boundary conditions

$$\psi_x(0) = \psi_x(L_x) = 0 \tag{9.23}$$

to (9.21), we obtain, as found in Section 4.3.1,

$$k_x = k_{x,n} = \frac{n\pi}{L_x}, \tag{9.24}$$

$n = 1, 2, 3, ...,$ so that

$$\psi(x) = \left(\frac{2}{L_x}\right)^{1/2} \sin\frac{n\pi}{L_x}x. \tag{9.25}$$

From (9.18),

$$\psi(x, y, z) = A\left(\frac{2}{L_x}\right)^{1/2} \sin\left(\frac{n\pi}{L_x}x\right) e^{ik_y y}e^{ik_z z}. \tag{9.26}$$

Energy is obtained from (9.22) as

$$E = \frac{\hbar^2}{2m^*}\left(\left(\frac{n\pi}{L_x}\right)^2 + k_y^2 + k_z^2\right) \tag{9.27}$$

$$= E_n + \frac{\hbar^2}{2m^*}\left(k_y^2 + k_z^2\right), \tag{9.28}$$

where

$$E_n = \frac{\hbar^2}{2m^*}\left(\frac{n\pi}{L_x}\right)^2. \tag{9.29}$$

The given energy levels are the subbands previously obtained in Section 4.7 (i.e., E_n is the bottom of the nth subband), and depicted as the quantized energy levels in Fig. 9.3. The resulting collections of electrons confined to the well form a two-dimensional electron gas; the structure is effectively two-dimensional in the sense that electrons can only move freely in two dimensions (the x- and y-directions). Two-dimensional electron gases formed at semiconductor heterojunctions typically have thickness values on the scale of a few nanometers, resulting in relatively large, well-separated subband energies. The energy–wavevector dispersion diagram is shown in Fig. 9.4 (a).

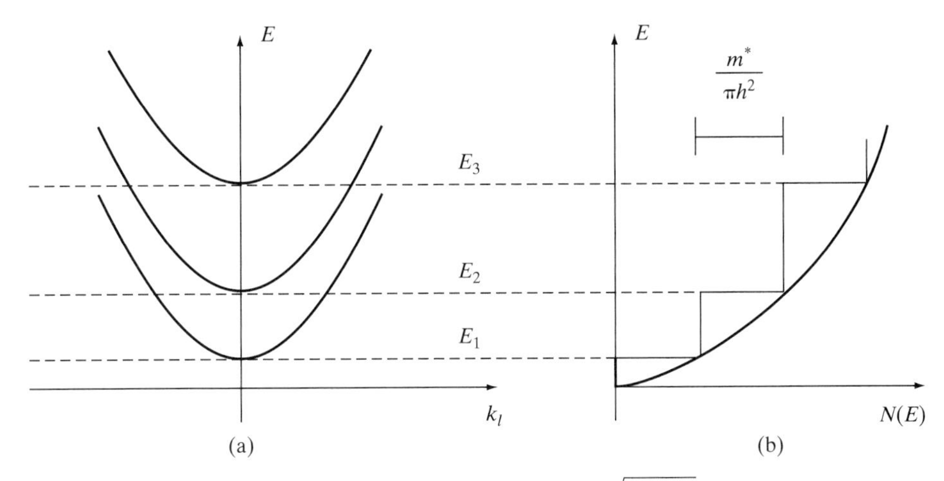

Figure 9.4 (a) Energy versus continuous wavevector $k_l = \sqrt{k_y^2 + k_z^2}$, showing three subbands. (b) Energy versus two-dimensional density of states. For comparison, the continuous parabolic curve is the density of states for three-dimensional free electrons.

Earlier, we found that the density of states in two dimensions is constant with respect to energy,

$$N(E) = \frac{m^*}{\pi\hbar^2}. \tag{9.30}$$

In the present case, however, associated with each subband is a density of states. The total density of states for any energy is the sum of all subband density of states at or below that energy, leading to the density of states depicted in Fig. 9.4(b), and given by

$$N(E) = \frac{m^*}{\pi\hbar^2} \sum_j H(E - E_j), \tag{9.31}$$

where H is the Heaviside function,

$$H(E) = \begin{cases} 1, & E > 0, \\ 0, & E < 0. \end{cases} \tag{9.32}$$

Absorption in an $Al_\alpha Ga_{1-\alpha}As/GaAs$ quantum well structure is shown in Fig. 9.5, where the effects of the step-like two-dimensional density of states is evident.

If the Fermi energy is a bit greater than the jth subband energy, $E_F > E_j$, then approximately j two-dimensional subbands will be filled. More specifically, from (8.49), the density of electrons in the jth two-dimensional subband will be given by

$$n_j = \int_{E_j}^{\infty} N(E)\, f(E, E_F, T)\, dE \tag{9.33}$$

$$= \frac{m^*}{\pi\hbar^2} \int_{E_j}^{\infty} f(E, E_F, T)\, dE = \frac{m^* k_B T}{\pi\hbar^2} \ln\left(1 + e^{\left(\frac{E_F - E_j}{k_B T}\right)}\right), \tag{9.34}$$

and the total density of electrons per unit area is $n = \sum_j n_j$.

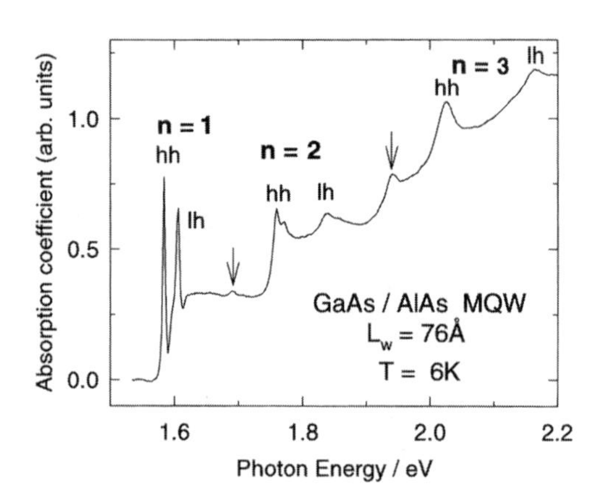

Figure 9.5 Absorption in an AlGaAs/GaAs multiple quantum well structure (40 wells, each of width 7.6 nm), showing clearly the effects of the step-like two-dimensional density of states. The relatively sharp peaks at the onset of absorption are due to excitons (hh and lh indicate heavy holes and light holes, respectively). (Based on a figure from N. Mayhew, D. Phil. thesis, Oxford University, 1993. (Published in A.M. Fox, "Optoelectronics in Quantum Well Structures," *Contemporary Physics* 37, 111–125 (1996) (http://www.tandf.co.uk/journals)). Reprinted with permission from A.M. Fox.)

It should be noted that for the subband model to be appropriate, the thermal energy $k_B T$ should be much less than the difference between energy subbands; otherwise, the discrete nature of the subbands will be obscured. Therefore, we must have

$$\frac{\hbar^2}{2m^*}\left(\frac{\pi}{L_x}\right)^2 \gg k_B T. \tag{9.35}$$

Considering GaAs with $m^* = 0.067m_e$ for electrons at room temperature, we find that (9.35) yields $L_x \ll 15$ nm. At $T = 4$ K, $L_x \ll 128$ nm.

9.1.2 Energy Band Transitions in Quantum Wells

As with a bulk semiconductor, energy band transitions can occur between states in a quantum heterostructure. These transitions tend to govern the optical properties of the device, and can be engineered for desired properties. Since quantum wells are formed in both the valence and conduction bands, various transitions can be envisioned as described next.

Interband Transitions. Interband transitions are from the nth state in the valence band well to the mth state in the conduction band well (or vice versa). From Fig. 9.3, it is evident that these can occur for incident photon energies†

$$\hbar\omega = E_g + E_n + E_m \tag{9.36}$$

†Here we assume bandedge-to-bandedge direct transitions (i.e., at $k = 0$). See Section 5.4.5 for more details on other types of transitions.

$$= E_g + \frac{\hbar^2\pi^2}{2m_h^* L_x^2} n^2 + \frac{\hbar^2\pi^2}{2m_e^* L_x^2} m^2, \tag{9.37}$$

where E_g is the gap energy in the well, and E_n and E_m are the energies in the valance and conduction well subbands, respectively, measured from the band edges.

The shift (from the bulk value E_g) in the required photon energy for absorption is characteristic of quantum-confined structures, and the possible tuning of the shift by adjustment of the well thickness and well material is one advantage of confined structures over bulk materials. For transitions among the lowest states ($n = m = 1$) we have

$$\hbar\omega = E_g + \frac{\hbar^2}{2}\left(\frac{\pi}{L_x}\right)^2 \left(\frac{1}{m_h^*} + \frac{1}{m_h^*}\right) \tag{9.38}$$

$$= E_g + \frac{\hbar^2\pi^2}{2m_r^* L_x^2}, \tag{9.39}$$

where m_r^* is the reduced mass, $m_r^{-1} = m_e^{*-1} + m_h^{*-1}$.

It should be noted that not all transitions are possible, and only certain pairs of n and m lead to permissible transitions. This is governed by Fermi's golden rule,† and leads to the idea of *selection rules* for deciding which transitions are permissible. The allowed transitions are dependent on the type of incident energy; for example, it turns out that for incident light polarized in the plane of the well (the z–y plane in Fig. 9.3), only transitions $n = m$ are allowed, although this rule is only strictly applicable to the infinite-wall model. These transitions are depicted in Fig. 9.6 for a $Al_\alpha Ga_{1-\alpha}As$/GaAs system.

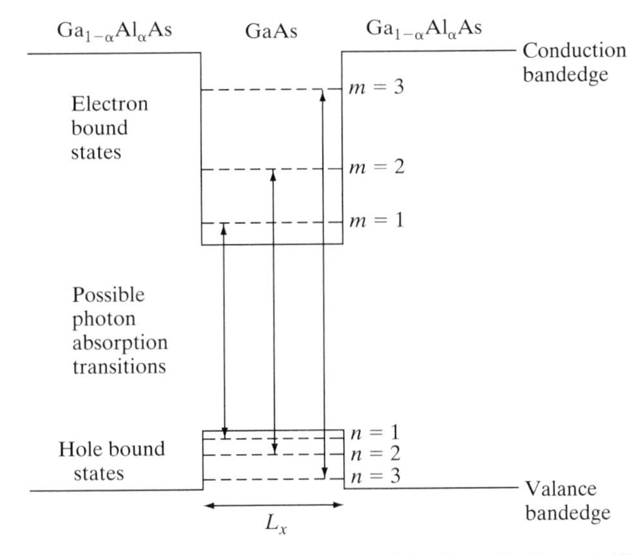

Figure 9.6 Allowed transitions in an infinite-well model of an $Al_\alpha Ga_{1-\alpha}As$/GaAs quantum well structure for incident light polarized in the plane of the well.

†Fermi's golden rule is discussed in many solid state and quantum mechanics texts; a good introduction can be found in [17].

Excitonic Effects. Excitonic effects are more prominent in quantum well structures then in bulk materials. As described in Section 5.4.5, for bulk semiconductors, excitons (bound electron–hole pairs) are typically only important at very low temperatures. Room temperature thermal energy can easily overcome the exciton binding energy (5.55), which is on the order of a few meV. However, the situation changes significantly in quantum-confined structures. In quantum wells, the binding energy is enhanced since electrons and holes are forced to be closer together. This is obvious for a suitably small well (obviously the electron and hole cannot be separated by 14 nm, the bulk GaAs exciton radius as given in Section 5.4.5, in a 5 nm well!), although the effect of the finite size of the structure is felt by the exciton even if the well size is bigger than the exciton radius. In fact, as described later for quantum dots, the relative size between the exciton radius and the well width can be used to gauge the importance of confinement effects. The closer spacing between the electron and hole in a confined exciton results in quantum well excitons being generally stable at room temperatures. Excitonic effects are seen in Fig. 9.5 as the relatively sharp peaks at the onset of absorption.

Intersubband Transitions. Aside from transitions from the valence band well to the conduction band well, transitions between subbands in each of these wells can occur. These are known as *intersubband transitions*, and obviously involve much lower energies than interband transitions since the transition energies are simply

$$\Delta E = E_\alpha - E_\beta = \frac{\hbar^2}{2m^*}\left(\frac{\pi}{L_x}\right)^2 (\alpha^2 - \beta^2) \tag{9.40}$$

for the subbands indexed by α and β.

Intersubband transitions in the conduction band well are depicted in Fig. 9.7.

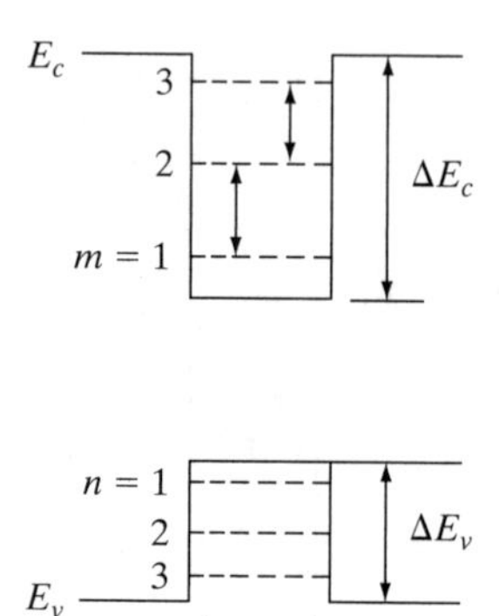

Figure 9.7 Depiction of intersubband transitions in the conduction band of a quantum well.

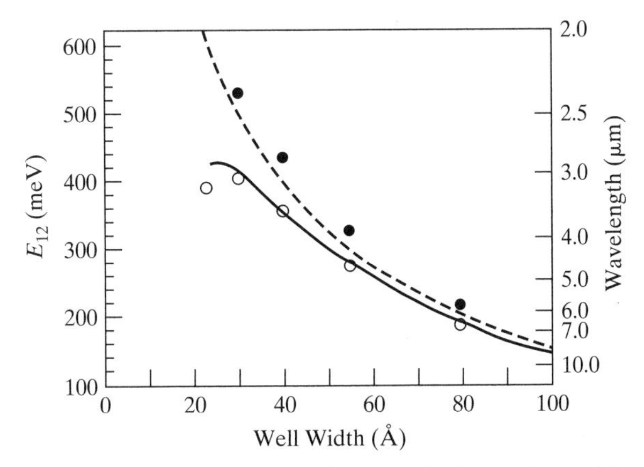

Figure 9.8 Measured quantum well width dependence on the lowest intersubband transition ($\alpha = 2$, $\beta = 1$) peak energies for $In_{0.5}Ga_{0.5}As/Al_{0.45}Ga_{0.55}As$ (hollow circles) and $In_{0.5}Ga_{0.5}As/AlAs$ (solid circles) quantum wells. Solid and dashed lines are calculated results for $Al_{0.45}Ga_{0.55}As$ (solid line) and AlAs (dashed line) quantum wells using an effective mass model. (Based on a figure from Chui, H. C., et al., "Short Wavelength Intersubband Transitions in In GaAs/AlGaAs Quantum Wells Grown on GaAs," *Appl. Phy. Lett.* 64 (1994): 736. © 1994, American Institute of Physics.)

Modeling the well as being infinitely high and considering light polarized in the direction of confinement† (the x-direction in Fig. 9.3), we find that selection rules dictate that only transitions corresponding to $(\alpha - \beta)$ being an odd number are allowed. Therefore, the lowest transition occurs for

$$\Delta E = E_2 - E_1 = \frac{3\hbar^2}{2m^*}\left(\frac{\pi}{L_x}\right)^2, \tag{9.41}$$

which for a 10 nm GaAs conduction band well is $\Delta E = 0.168$ eV. This corresponds to $f = 40.62$ THz, or $\lambda = 7.39$ μm, and, thus, intersubband transitions can be used for infrared detectors and emitters.

As an example, the width dependence of the peak energies of a quantum well for the lowest intersubband transition ($\alpha = 2$, $\beta = 1$) is shown in Fig. 9.8. The structures are $In_{0.5}Ga_{0.5}As/Al_{0.45}Ga_{0.55}As$ (hollow circles) and $In_{0.5}Ga_{0.5}As/AlAs$ (solid circles) quantum wells. Circles are measured results, and solid and dashed lines are calculated results for $Al_{0.45}Ga_{0.55}As$ (solid line) and AlAs (dashed line) quantum wells using an effective mass model.

Comments on the Model. The simple infinite well model is a fairly gross approximation to a realistic heterostructure, since the difference in energy bands in the different

†In quantum well structures such as $Al_{\alpha}Ga_{1-\alpha}As/GaAs$, intersubband transitions will essentially only be induced by light polarized in the direction of confinement, which necessitates features such as gratings to couple light into the well.

materials making up the heterostructure can be as small as a few tenths of an eV. For example, the bandgap for GaAs is 1.5 eV, and for AlGaAs E_g is on the order of 1.8 eV. Modeling this difference as being infinitely large is then not a very good approximation (although the infinite-well model gives an order-of-magnitude approximation; see Problem 4.17 and Problem 9.8). Furthermore, it can be appreciated that the result (9.28),

$$E = E_n + \frac{\hbar^2}{2m^*} k_l^2, \tag{9.42}$$

and, in particular, the existence of energy subbands, is quite general, and is independent of the specific confining potential in the x coordinate. For example, we could have solved (9.17) subject to a finite potential well as considered in Section 4.5, a parabolic well as considered in Section 4.5.2, or a triangular well as considered in Section 4.5.3. From an energy standpoint, the main difference is that in those cases, E_n would not be given by (9.29), but by the solution of (4.83) or (4.85) for the finite well case, by (4.98) for the parabolic well case, or (4.100) for the triangular well case. Other potential profiles can be considered, obviously. Thus, qualitatively, we expect our solution to be reasonable, but for good quantitative results, we would need to account for the specific confining potential.

However, if we assume a more realistic finite well model, a complication arises, since an electron's effective mass will be different in each different region (the well itself, and the exterior regions). We can take this into account by using the appropriate effective mass in the boundary conditions for the derivative of the wavefunction, as given in (3.143). This aspect was ignored in our previous analysis of finite wells in Chapter 4, although it is considered in Problems 9.7 and 9.8 in this chapter. For example, as discussed in Problem 9.7, if we assume a finite square well profile, accounting for effective mass results in equations for the energy values that are a bit more complicated than (4.83) and (4.85), but which can still be easily solved numerically in the same manner as the constant mass equations.

Yet another complication for the finite potential well model is that we need to solve (9.17) involving the effective potential,

$$V_e(x, k) = \frac{\hbar^2 k_l^2}{2m^*} + V(x). \tag{9.43}$$

For example, consider the two adjacent regions at either side of a finite square well. On one side, we have the large bandgap material, which we'll call the barrier, and on the other side, we have the smaller bandgap material of the well. We may be interested in either the conduction band, the valence band, or both bands. In the barrier region, we take $V(x) = E^b$, where E^b is the bandedge in either the conduction or valence band (i.e., E^b is either E_c^b or E_v^b), and in the well, the corresponding bandedge is $V(x) = E^w$. As shown in Fig. 9.3, the potential difference between the two materials is simply the difference in bandedges. (The absolute reference of each band is not important.) Therefore, in the barrier, Schrödinger's equation (9.17) is

$$\left(-\frac{\hbar^2}{2m_b^*} \frac{\partial^2}{\partial x^2} + \frac{\hbar^2 k_l^2}{2m_b^*} + E^b \right) \psi_x(x) = E \psi_x(x), \tag{9.44}$$

and in the well,

$$\left(-\frac{\hbar^2}{2m_w^*}\frac{\partial^2}{\partial x^2}+\frac{\hbar^2 k_l^2}{2m_w^*}+E^w\right)\psi_x(x)=E\psi_x(x), \tag{9.45}$$

where $m_{b,w}^*$ is the appropriate effective mass in each region. Because $m_b^* \neq m_w^*$, in general, the actual potential difference seen by a particle is not simply $E^b - E^w$ (although not described at that time, the potential V_0 used in Section 4.5 represents this difference in bandedges), but is

$$V_0 = E^b - E^w + \frac{\hbar^2 k_l^2}{2}\left(\frac{1}{m_b^*}-\frac{1}{m_w^*}\right).$$

Therefore, the solution actually depends on the value of the longitudinal wavenumber k_l. This serves to perturb the results from the k-independent potential V_0 considered in Section 4.5. However, for any given value of k_l, the solution can still be obtained.

Degenerately Doped Heterojunctions. Earlier, we considered forming a two-dimensional electron gas by sandwiching a very thin layer of a smaller bandgap material (such as GaAs) between thick layers of a larger bandgap material (such as AlGaAs, more precisely denoted by $Al_\alpha Ga_{1-\alpha}As$). There is another method of forming a 2DEG that is of considerable interest, illustrated in Fig. 9.9.

In this method, the GaAs/$Al_\alpha Ga_{1-\alpha}As$ heterojunction is formed by n-type heavily doped (degenerate) $Al_\alpha Ga_{1-\alpha}As$, grown adjacent to *undoped* GaAs, typically using molecular-beam epitaxy. Silicon is often used as the dopant, forming donor atoms in the $Al_\alpha Ga_{1-\alpha}As$ material. A small amount of energy can liberate electrons from their donors, and these will be attracted to the GaAs material by the smaller bandgap. However, these electrons will still be attracted to their ionized donors in the $Al_\alpha Ga_{1-\alpha}As$ material, and thus will form a thin layer of electrons at the interface.

Furthermore, when there is an undoped layer between a heavily doped large bandgap material and a small bandgap material, a 2DEG forms in the undoped layer. This is known as *modulation doping*. Because the donors are separated from the electron gas layer, scattering from the donor atoms is avoided, resulting in very high mobility values.† Contributing to the high mobility is the good lattice match between AlGaAs and GaAs (Table V in

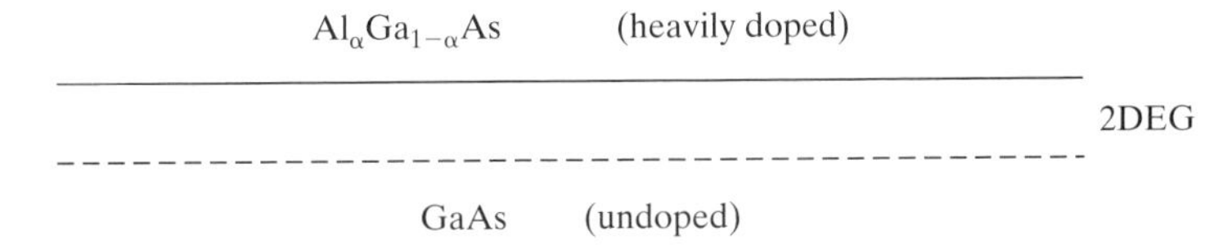

Figure 9.9 Degenerately doped heterojunction of $Al_\alpha Ga_{1-\alpha}As$ and GaAs.

†Mobility is discussed on page 320 , although, essentially, large (small) mobility means that electrons travel a relativity long (short) distance before colliding with the lattice or impurities.

Appendix B), forming an interface without many defects or much strain. For example, intrinsic GaAs at room temperature has an electron mobility of approximately $\mu_e = 8{,}600$ cm^2/V-s, whereas for intrinsic Si, $\mu_e = 1{,}350$ cm^2/V-s, and for copper, $\mu_e = 43.5$ cm^2/V-s. With a doping concentration of $N_D = 10^{18}$ cm^{-3}, still at room temperature, $\mu_e = 3{,}000$ cm^2/V-s for GaAs and $\mu_e = 300$ cm^2/V-s for Si. The decrease in mobility for the doped materials is due to scattering from the dopant atoms. In contrast, in modulation-doped AlGaAs/GaAs heterostructures, one can achieve $\mu_e = 7{,}000$ cm^2/V-s at room temperatures, and as high as $\mu_e = 10^6$–10^7 cm^2/V-s at low temperatures. Thus, modulation doping can offer a significant advantage over traditional doping methods. In fact, modulation doping is the basis for the *high electron mobility transistor* (HEMT), which is a common commercial structure. Other quantum well/heterostructure applications include resonant tunneling diodes, quantum well lasers, quantum well photodetectors, and optical modulators.

9.2 QUANTUM WIRES AND NANOWIRES

The preceding analysis assumed that electrons were confined along one coordinate, x, where the confinement was provided by the difference in bandgaps between different materials. With this assumption, we developed the concept of a two-dimensional electron gas and of energy subbands. The next logical step is to consider what happens if we confine electrons in a second direction, say along the y coordinate, as depicted in Fig. 9.10. The resulting (effectively one-dimensional) structure is called a quantum wire, as discussed previously in Section 4.7.2.

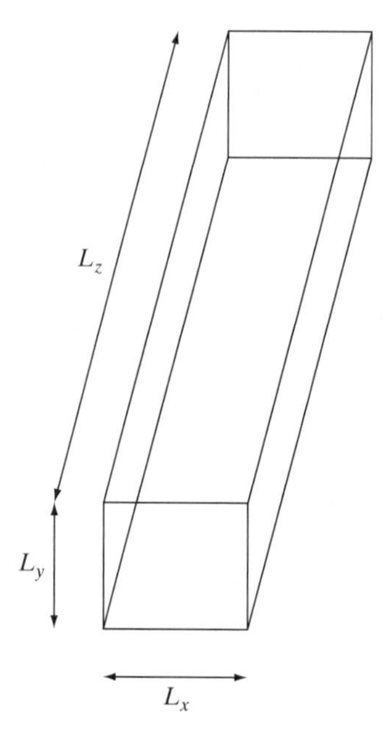

Figure 9.10 Quantum wire; $L_{x,y} \leq \lambda_F$, $L_z \gg \lambda_F$.

To analyze the quantum wire, we start with Schrödinger's equation,

$$\left(-\frac{\hbar^2}{2m^*}\nabla^2 + V(x, y)\right)\psi(\mathbf{r}) = E\psi(\mathbf{r}), \tag{9.46}$$

and express the wavefunction as

$$\psi(\mathbf{r}) = e^{ik_z z}\psi(x, y). \tag{9.47}$$

That is, the electron will be free to move along the z coordinate, i.e., along the wire, but will be confined in other directions. Substituting (9.47) into (9.46), we obtain

$$\left(-\frac{\hbar^2}{2m^*}\left(\frac{\partial^2}{\partial x^2} + \frac{\partial^2}{\partial y^2}\right) + \frac{\hbar^2 k_z^2}{2m^*} + V(x, y)\right)\psi(x, y) = E\psi(x, y). \tag{9.48}$$

Given a certain potential $V(x, y)$, we can solve (9.48) analytically if V has a particularly simple form, or numerically, or using some approximate method. Although the details may be complicated for general confining barriers, given our experience in the previous section, we expect to obtain an energy relation analogous to (9.42), having the form

$$E = E_{n_x,n_y} + \frac{\hbar^2}{2m^*}k_z^2, \tag{9.49}$$

where k_z is now the longitudinal wavenumber. This will be a continuous parameter since electrons are free along the z coordinate. The discrete indices n_x and n_y correspond to subband indices. The subband energy E_{n_x,n_y} will be given by

$$E_{n_x,n_y} = \frac{\hbar^2}{2m^*}\left(k_{x,n_x}^2 + k_{y,n_y}^2\right), \tag{9.50}$$

where k_{x,n_x} and k_{y,n_y} will be discrete, and will depend on the specific form of the potential $V(x, y)$. For example, for an infinite confining potential (i.e., a "hard" barrier) at $x = 0, L_x$ and $y = 0, L_y$, then

$$k_{x,n_x} = \frac{n_x\pi}{L_x}, \qquad k_{y,n_y} = \frac{n_y\pi}{L_y}, \tag{9.51}$$

with $n_x, n_y = 1, 2, 3, ...,$ such that

$$E_{n_x,n_y} = \frac{\hbar^2\pi^2}{2m^*}\left(\left(\frac{n_x}{L_x}\right)^2 + \left(\frac{n_y}{L_y}\right)^2\right). \tag{9.52}$$

We previously determined the density of states in one dimension to be (8.15),

$$N(E) = \frac{1}{\pi\hbar}\left(\frac{2m^*}{E - V_0}\right)^{1/2} \tag{9.53}$$

for $E > V_0$. As with the two-dimensional electron gas (quantum well case), for the one-dimensional quantum wire, each subband will have a density of states given by the one-dimensional result (9.53),

$$N_{n_x,n_y}(E) = \frac{1}{\pi\hbar}\left(\frac{2m^*}{E - E_{n_x,n_y}}\right)^{1/2}. \tag{9.54}$$

If the energy level E is low, such that only the first subband is filled, the preceding density of states holds, and the system is one dimensional. As energy increases and more subbands are filled, the system is quasi-one dimensional. In this case the density of states is found by summing over all subbands, resulting in

$$N(E) = \frac{1}{\pi\hbar}\sum_{n_x,n_y}\left(\frac{2m^*}{E - E_{n_x,n_y}}\right)^{1/2} H\left(E - E_{n_x,n_y}\right),$$

where H is the Heaviside function (9.32),

This quasi-one dimensional density of states is depicted in Fig. 9.11. The discontinuities in the density of states are known as *van Hove singularities*.

As a concrete example, the density of states for two different single-wall carbon nanotubes is given in Fig. 9.12. As discussed in Section 5.5 and Problem 5.24, the (9, 0) zigzag tube is metallic, and has a finite density of states at the Fermi level $E_F = 0$. The (10, 0) tube is semiconducting, and the density of states is zero at the Fermi level.

Transitions and Excitonic Effects. Optical transitions in quantum wires arise from allowed transitions (governed by appropriate selection rules) between energy bands of the wire, and can be associated with the locations of van Hove singularities. As with quantum well structures, excitonic effects are expected to be very important in quantum wires even at room temperature, due to larger binding energies associated with confinement. For example, although binding energies for bulk semiconductors are typically a very small

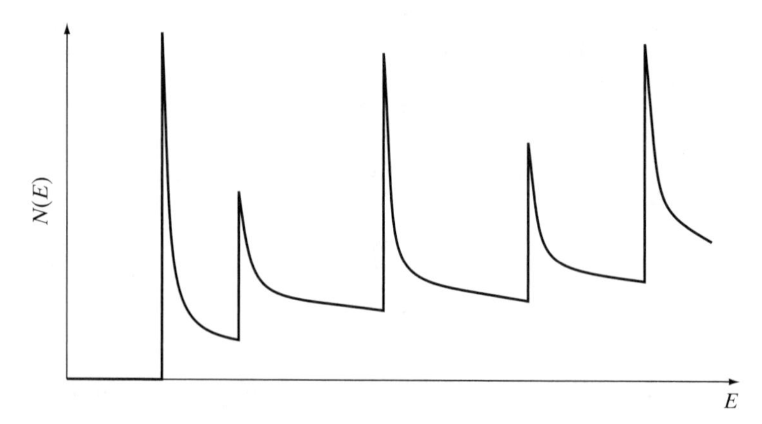

Figure 9.11 One-dimensional density of states.

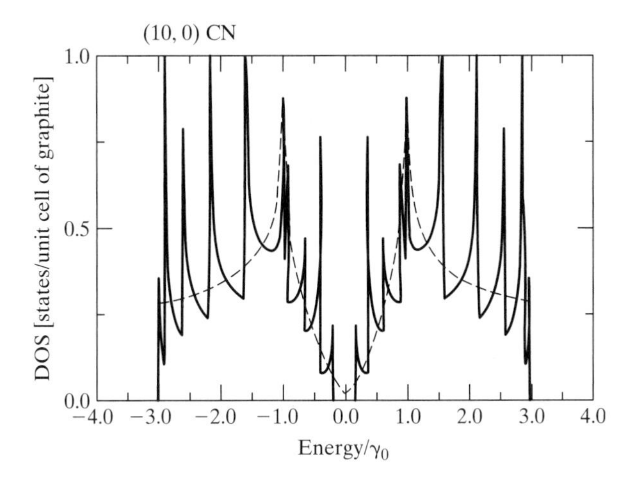

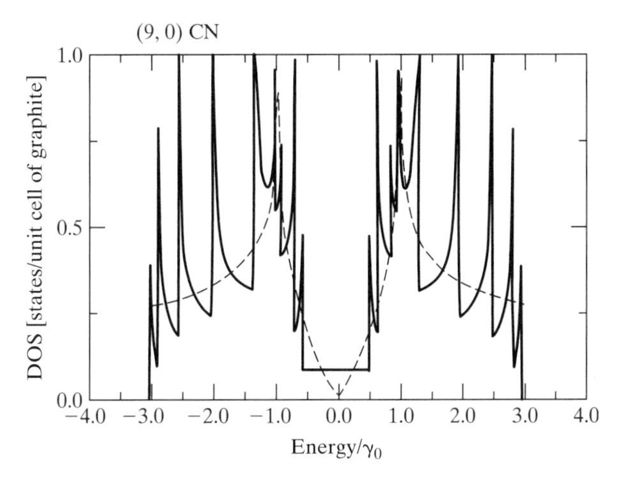

Figure 9.12 Density of states per unit cell for a (10, 0) (top) and (9, 0) (bottom) single-wall carbon nanotube. The (10, 0) tube is semiconducting, and the (9, 0) tube is metallic. $\gamma_0 = 2.5$ eV. The dashed lines are the density of states for graphene. (Based on a figure from Saito, R., et al., "Electronic Structure of Chiral Graphene Tubules," *Appl. Phys. Lett.* 60 (1992): 2204. © 1992, American Institute of Physics.)

fraction of the bandgap energy (and thus constitute a relatively small perturbation in optical properties (Problem 5.20)), exciton binding energies in semiconducting quantum wires can be a large fraction of the bandgap. In SWNTs, the binding energy can be close to half of the bandgap,† significantly altering optical properties. Although there is still some debate about the relative importance between excitonic and van Hove singularity effects,

†See, e.g., [21] and [22].

it seems clear that, especially for semiconducting nanotubes, excitonic effects are very important.

Last, it should be noted that for the quantum wire to act in a truly one-dimensional manner, the width (or radius for a round cross-section wire) should be on the order of the Fermi wavelength, or smaller. For the situation where the wire is thin but relatively wide, electrical transport will typically involve electrons in a single subband in the coordinate normal to the interface (i.e., in the "thin" direction, where energy level spacing is large), yet multiple subbands in the lateral (wide) direction due to more closely spaced energy levels. More generally, a wire that has radius values on the order of nanometers or tens of nanometers, but much larger than λ_F, might be called a *nanowire*, rather than a quantum wire. Quantum wires and nanowires will be further discussed in Chapter 10, where the concept of ballistic transport is introduced.

9.3 QUANTUM DOTS AND NANOPARTICLES

If we confine electrons in all three coordinates, forming a quantum dot (Section 4.7.3), electrons do not have a plane wave dependence in any direction. The three-dimensional Schrödinger's equation

$$\left(-\frac{\hbar^2}{2m^*}\nabla^2 + V(x, y, z)\right)\psi(\mathbf{r}) = E\psi(\mathbf{r}) \tag{9.55}$$

must be solved, and, not unexpectedly, the resulting energy will be fully quantized. For example, for the three-dimensional infinite potential well considered in Section 4.3.2, energy levels were obtained as (4.53),

$$E = E_{n_x,n_y,n_z} = \frac{\hbar^2\pi^2}{2m^*}\left(\left(\frac{n_x}{L_x}\right)^2 + \left(\frac{n_y}{L_y}\right)^2 + \left(\frac{n_z}{L_z}\right)^2\right). \tag{9.56}$$

In general, regardless of the specific potential $V(\mathbf{r})$, energy will take the form

$$E = E_{n_x,n_y,n_z} = \frac{\hbar^2}{2m^*}\left(k_{x,n_x}^2 + k_{y,n_y}^2 + k_{z,n_z}^2\right), \tag{9.57}$$

where k_{x,n_y}, k_{y,n_y}, and k_{z,n_z} will be discrete, and must be found from the specific boundary conditions of the problem.

As described previously, the density of states for a quantum dot is a series of delta functions (8.18). This discrete density of states leads to several fundamental differences compared with higher (than zero) dimensional systems, which have at least a piece wise continuous density of states. (A comparison of the density of states in various dimensions is shown in Fig. 9.13.)

For example, electron dynamics are obviously quite different in a quantum dot, since current cannot "flow." Thermal effects will be different in quantum dots than in bulk

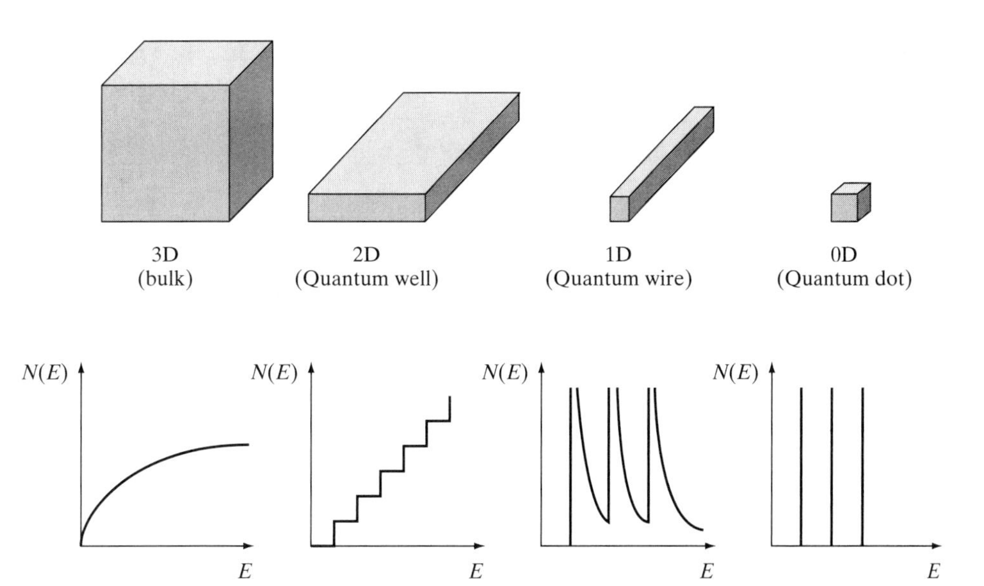

Figure 9.13 A comparison of the density of states in various dimensions.

materials (or even quantum wells), since thermal energy can only excite electrons to a limited number of widely separated states. For this same reason, the frequency spectrum of emitted optical radiation from energy level transitions (called the *luminescence linewidth*) in quantum dots is very narrow even at relatively high temperatures. This makes quantum dots attractive in laser applications, as biological markers (as discussed next), and in a host of other optical applications.

9.3.1 Applications of Semiconducting Quantum Dots

Some applications of quantum dots in electronics are discussed in Chapter 7, in particular, their use as charge islands in Coulomb blockade circuits. Here we will discuss an important application of quantum dots in biology and medicine: the use of quantum dots as *biological markers*. In this application, the main idea is to coat the dot with a material that causes it to bind selectively to a certain biological structure, such as a cancer cell, and then to reveal its presence by absorbing and emitting light (in a process known as *fluorescence*). Light is absorbed by the dot, raising electrons from lower to higher energy states. In general, there is some energy dissipation, and more importantly, the electrons will tend to fall back down to lower states, emitting a photon. The ability to tune precisely the energy levels in a quantum dot is of great importance.

As a rough approximation, the emission spectra of a quantum dot can be determined by the simple three-dimensional hard-wall particle-in-a-box model discussed in Section 4.3. For example, energy levels for a cubic dot having side length L are, from (4.54),

$$E_n = \frac{\hbar^2 \pi^2 n^2}{2m^* L^2}, \tag{9.58}$$

where $n^2 = n_x^2 + n_y^2 + n_z^2$ represents the triplet of integers $n = (n_x, n_y, n_z)$. Energy transitions from, say, the $n = (2, 1, 1)$ state to the $n = (1, 1, 1)$ state release the energy quantum

$$\Delta E_{2,1} = E_{2,1,1} - E_{1,1,1} = \frac{\hbar^2 \pi^2 (6)}{2m^* L^2} - \frac{\hbar^2 \pi^2 (3)}{2m^* L^2} = 3\frac{\hbar^2 \pi^2}{2m^* L^2}. \tag{9.59}$$

As an example, for a $10 \times 10 \times 10$ nm^3 GaAs dot ($m^* = 0.067 m_e$), $\Delta E_{2,1} = 0.168$ eV.

For a spherical dot of radius R, the details of obtaining the energy levels are a bit more involved than for the rectangular dot, although the resulting expression for the energy levels is very similar,

$$E_n = \frac{\hbar^2 \pi^2 n^2}{2m^* R^2}. \tag{9.60}$$

For an $R = 6.2$ nm radius sphere, which has the same volume as the $10 \times 10 \times 10$ nm^3 box considered previously, $E_{2,1} = 0.438$ eV.

We can see that the electron mass is important in determining energy spacings. In a typical metal, the electron mass is the familiar quantity $m_e = 9.1095 \times 10^{-31}$ kg, which would lead to $\Delta E_{2,1} = 0.0293$ eV for the $R = 6.2$ nm radius sphere. Therefore, it is much easier to distinguish the discrete nature of the energy levels in a semiconducting quantum dot than in a metallic dot.†

In light of the discussion in Section 9.1.2, it can be seen that (9.58) and (9.60) actually relate to intersubband transitions. Referring to Fig. 9.6, we find that direct-gap interband transitions in spherical dots from the nth level in the valence band well to the mth level in the conduction band well can occur for incident photon energies

$$\hbar\omega = E_g + E_n + E_m \tag{9.61}$$

$$= E_g + \frac{\hbar^2 \pi^2 n^2}{2m_h^* R^2} + \frac{\hbar^2 \pi^2 m^2}{2m_e^* R^2}, \tag{9.62}$$

where E_g is the band gap energy of the bulk semiconductor.

For transitions among the lowest states ($n = m = 1$), we have

$$\hbar\omega = E_g + \frac{\hbar^2 \pi^2}{2R^2}\left(\frac{1}{m_h^*} + \frac{1}{m_h^*}\right) \tag{9.63}$$

$$= E_g + \frac{\hbar^2 \pi^2}{2m_r^* R^2}, \tag{9.64}$$

where m_r^* is the reduced mass.

Furthermore, in contrast to bulk semiconductors, in quantum dots exciton effects often play a dominant role in determining optical properties of the dot at room temperature. In fact, optical transitions in quantum dots are usually associated with excitons (Section 5.4.5),

†We could say, equivalently, that this effect is due to the small Fermi wavelength in metals, and the relatively larger Fermi wavelength in semiconductors.

and an approximation expression known as the *Brus equation* models the transition energy in spherical dots,

$$E_g^{\text{dot}} = E_g + \frac{\hbar^2 \pi^2}{2m_r^* R^2} - \frac{1.8 q_e^2}{4\pi\varepsilon_r\varepsilon_0 R}. \tag{9.65}$$

The third term in (9.65) is related to the binding energy of the exciton (i.e., it is due to the Coulomb attraction between the electron and the hole), which is modified from the bulk case by the size of the dot.

In bulk semiconductors, the exciton radius is given by (5.56),

$$a_{ex} = \frac{4\pi\varepsilon_r\varepsilon_0\hbar^2}{m_r^* q_e^2} = \frac{\varepsilon_r m_e}{m_r^*} a_0 = \frac{\varepsilon_r m_e}{m_r^*} \left(0.53\ \text{Å}\right), \tag{9.66}$$

where a_0 is the Bohr radius. For quantum dots, the actual (highest probability) separation between the electron and hole is influenced by the size of the dot. In this case, we will consider a_{ex} to be the excition Bohr radius, which is often taken as the measure of quantum confinement in quantum dots. In particular,

- if $R \gg a_{ex}$, then confinement effects will generally not be very important. Otherwise,
- if $R > a_{ex}$, then we have the *weak confinement regime*, and
- If $R < a_{ex}$, then we have the *strong confinement regime*. In this case, the hydrogen model tends to break down, and the exciton is delocalized over the entire dot.

For semiconducting dots, Group II–VI semiconductors such as ZnSe, ZnS, and CdSe are often used, since these materials tend to have relatively large bandgaps and can be fabricated using a variety of methods. For self-assembly, in particular, CdSe/ZnSe systems naturally lead to quantum dots because of the large lattice mismatch between ZnSe and CdSe (about 7 percent).

As an example, consider a cadmium selenide (CdSe) quantum dot. Using $E_g^{\text{bulk}} = 1.74$ eV, $m_e^* = 0.13 m_e$, $m_h^* = 0.45 m_e$ ($m_r^* = 0.101 m_e$), and $\varepsilon_r = 9.4$ (choosing reasonable values from measurements), for transitions from the conduction to valence bandedges through the band gap, we obtain from (9.63), for an $R = 2.9$ nm dot, $E_g = 2.092$ eV. Using $E = hf$ and $c = \lambda f$, this energy corresponds to $\lambda = 593$ nm, indicating that yellow light (Table 9.1)

TABLE 9.1 COLORS AND ASSOCIATED ELECTROMAGNETIC WAVELENGTHS.

Color	λ (nm)
Red	780–622
Orange	622–597
Yellow	597–577
Green	577–492
Blue	492–455
Violet	455–390

would be absorbed by the dot. The experimental absorption peak is closer to 580 nm, although the Brus formula gives reasonably accurate (at least, better than order-of-magnitude) results. In this case, the exciton Bohr radius is

$$a_{ex} = \frac{\varepsilon_r m_e}{m_r^*} (0.53 \text{ Å}) = 4.93 \text{ nm}, \tag{9.67}$$

which would correspond to strong confinement.

The energy reradiated by the dot is less than that which excites the dot, so that radiated wavelengths of the fluorescence are longer. This difference is called the *Stokes shift*, which relates to relaxation of angular momentum in the dot, although the details won't be presented here. However, experiments show that an $R = 2.9$ nm dot emits at approximately $\lambda = 592$ nm, or yellow (and so, using the previously given experimental absorption peak of 580 nm, we find that the Stoke's shift is approximately $\Delta\lambda = 592$ nm $- 580$ nm $= 12$ nm). Note that having a Stokes shift is generally considered a positive attribute of a biological marker, since the illuminating energy and the re-radiated energy can be separated (filtered). This provides an advantage for quantum dots compared with fluorescent dyes, which radiate at nearly the same wavelength as their excitation. Illumination is often provided in the UV range, since the absorbance spectrum is relatively broad, whereupon the dot reradiates at the wavelength corresponding to its size and material composition. Thus, if different-sized dots are coated to bind with different biological features, with a common illumination, the various dots will fluoresce in different colors. As another example, if one breaks off a nanoscopic piece of a material (say, a rock) of a certain color, that piece will often be a different color from the original, due to the described quantum confinement effects. This has been used (unknowingly) in making stained glass for centuries.

Size Effects. As described previously, when material regions shrink to the nanoscale, the properties of the material change. This is due to a variety of reasons, such as the large ratio of surface-to-interior atoms, and the change in bandstructure due to having the periodic crystalline geometry of a material, which is effectively infinite in the bulk case, significantly truncated. The effect on bandgap energy is described previously.

Furthermore, it should be noted that even the lattice structure of a material may change for nanoscopic material samples. For example, both gold and silver have the fcc lattice in the bulk state, but for particles having a diameter of less than 5–8 nm, significant deviations from the fcc structure have been observed, which will obviously alter the material's properties.

Lastly, it can be appreciated that the role of semiconductor doping is also strongly affected when considering nanoscale material regions. Typical semiconductor doping levels for traditional device applications are on the order of 10^{16}–10^{18} impurity atoms per cm^3. This would result in, on average, 0.00001–0.001 atoms per nm^3. Thus, a quantum dot 10 nm on a side would occupy 10^3 nm^3, and would, therefore, contain approximately one extra electron or hole available for conduction.

Blinking and Spectral Diffusion. It has been observed that the emission spectrum of colloidal dots shifts randomly over time, which is known as *spectral diffusion*. This shift can occur over times as long as seconds, or even minutes. Although this phenomenon is not completely understood, it is thought to arise from the local environment of the dot, which can admit fluctuating electric fields. Spectral diffusion is not observed in self-assembled dots embedded in a host matrix.

Another phenomenon observed in colloidal quantum dots is known as *blinking*, which is, as its name suggests, a turning on and off of the dot's emissions. This phenomenon is also not completely understood (although it seems to be related to charging of the dot), and along with spectral diffusion, is generally considered a detrimental aspect. The off state of the dot can last from milliseconds to minutes.

Coated and Functionalized Quantum Dots. In addition to simple dot structures, often dots are coated with multilayered shells to tailor their electronic, chemical, or biological properties. A core material (often cadmium sulfide (CdS), CdSe, or cadmium telluride (CdTe)) and a core size/shape is chosen based on the desired emission wavelength; CdSe dots are currently often chosen for most of the visible spectrum. To the core is added a shell, consisting of a transparent material or layering of materials that can be attached to the core. The shell material (e.g., ZnS) serves as a protective coating for the core. Multilayer coatings can be added to stabilize the dots' emission spectra, to render the dots chemically inert,[†] and to provide a surface to attach biological structures for, e.g., selective binding.

For example, Fig. 9.14 shows cultured HeLa cells[‡] labeled with two different quantum dots, one that fluoresces red, and the other, green. The red and green dots are conjugated (bound) to different antibodies and label two different proteins of interest in cell biology.

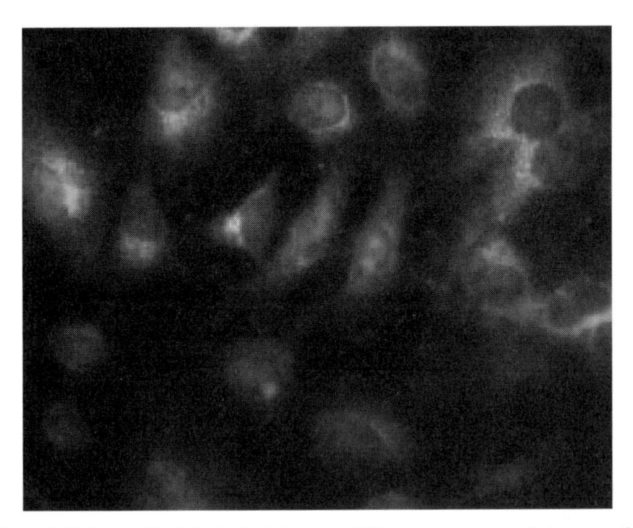

Figure 9.14 Cultured HeLa cells labeled with two different quantum dots, one that fluoresces red, and the other, green. The dots are conjugated (bound) to different antibodies, and label two different proteins of interest in cell biology. (Reprinted with permission from Quantum Dot Corporation.)

[†]Cadmium itself is a toxic heavy metal.

[‡]HeLa cells are cultured cervical cancer cells that are routinely used in the study of cancer.

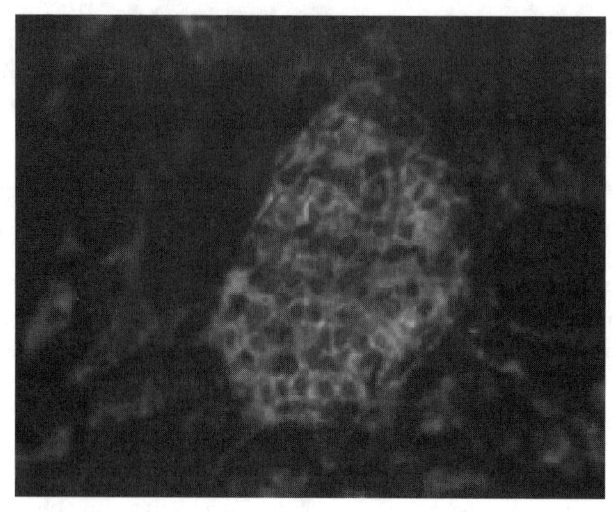

Figure 9.15 Two different quantum dots that fluoresce red and green are shown labeling proteins in a tissue section of mouse kidney. (Reprinted with permission from Quantum Dot Corporation.)

The intensity of the fluorescence and the distribution in the cells are the key features of this technique in cell biological research.

In a similar manner in Fig. 9.15, two quantum dots which fluoresce red and green are shown labeling proteins in a tissue section of mouse kidney. This application is typically used in clinical pathology and medicine for the diagnosis of disease states. An understanding of the amount and distribution of protein expression helps define the cancerous or noncancerous state of the cells in the tissue section.

In Fig. 9.16, quantum dots injected into a live mouse mark the location of a tumor.

In addition to quantum dots, quantum wires can also be functionalized. For example, nanowires can be coated with substances that will bind with certain molecules, which will, in turn, alter the conductance of the wire. Thus, such nanowires can be used as, for example, chemical sensors.

9.3.2 Plasmon Resonance and Metallic Nanoparticles

Quantum confinement effects can be seen in metallic nanoparticles if their size is extremely small. Due to the small Fermi wavelength of electrons in metals, quantum confinement effects generally only become important for metallic spheres having radius values far below 10 nm; perhaps on the scale of 1–2 nm or less. For metallic nanoparticles above this size range, optical properties are governed by *plasmon resonances*, which are collective modes of oscillation of the electron gas in the metal. This is a classical, rather than quantum, effect, and can be described by classical Maxwell's equations.

At the frequencies of plasmon resonances, the response of the material to illumination is particularly strong and easy to detect. It is important to note that these oscillations are not very sensitive to the size of the nanoparticle, contrasting the extreme size dependence

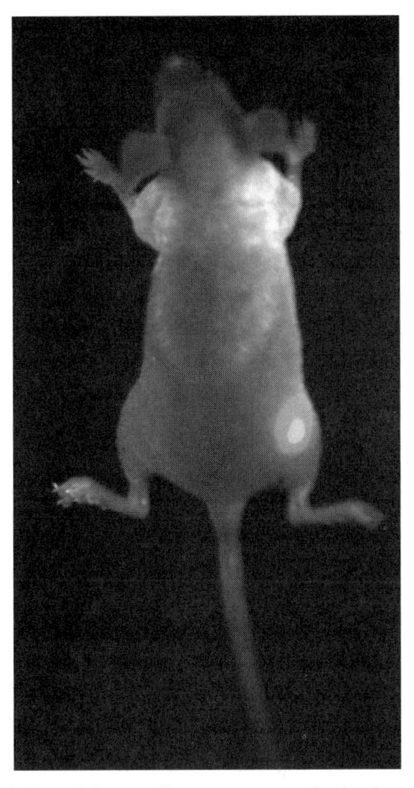

Figure 9.16 Quantum dots injected into a live mouse mark the location of a tumor. (From Seydel C., "Quantum Dots Get Wet," *Science* 4 April 2003, 300: 80 [DOI: 10.1126/Science.300.5616.80]. Courtesy Xiaohu Gao and Shuming Nie, Emory University School of Medicine.)

of confinement effects in semiconductor dots and very small metallic dots. For example, consider applying a static (non-time-varying) electric field to a small dielectric sphere of radius a in free space. It can be shown that the ratio of the field inside the sphere to the applied field is

$$\frac{3}{\varepsilon_r + 2}, \tag{9.68}$$

where ε_r is the relative permittivity of the dielectric. For low-frequency, time-varying fields where $\lambda \gg a$, with λ the electromagnetic wavelength, the same result holds.[†] Thus, if $\varepsilon_r = -2$, the field inside the sphere "blows-up," (i.e., in a real material the interior field takes on a very large value[‡]), independent of the specific radius of the sphere. That is, the effect is not due to fitting an integer number of half wavelengths inside the sphere, as is the case for most resonance phenomena. The effect is actually due to interactions between interior charge and charge induced on the surface of the metallic particle. Furthermore, a similar result holds for other structures, not only spheres. Metals in the optical range have negative

[†]Note that since a is on the order of nanometers, "low frequency" certainly includes optical frequencies.

[‡]All real materials have some damping, which prevents perfect oscillations from occurring.

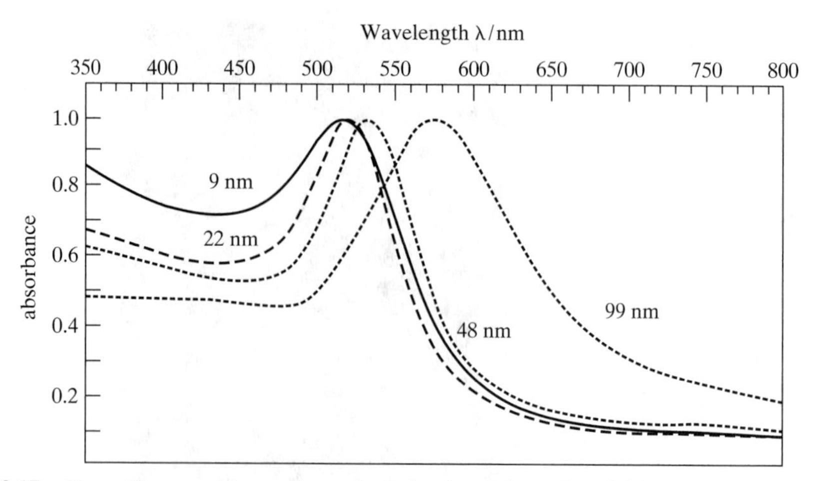

Figure 9.17 Size effects on the surface plasmon absorption of colloidal solutions of spherical gold nanoparticles with diameters varying between 9 and 99 nm. (Based on a figure from Link, S. and M. A. El-Sayed, "Spectral Properties and Relaxation Dynamics of Surface Plasmon Electronic Oscillations in Gold and Silver Nanodots and Nanorods," *J. Phys Chem B* 103 (1999): 8410–8426.)

relative permittivities, and therefore small metal particles tend to exhibit relatively size-insensitive plasmon resonances, rather than size-specific geometrical resonances, although the latter will also occur. In addition, particularly in hollow nanoshells of metal, these plasmon resonances can be tuned to various frequencies by controlling the relative thickness of the core and shell layers.

To demonstrate the relative size insensitivity of plasmon resonances, Fig. 9.17 shows the measured absorbance of a colloidal solution of spherical gold nanoparticles with diameters varying between 9 and 99 nm. It can be appreciated that as the diameter changes from 9 nm to 48 nm (a 433 percent change), the plasmon resonance energy only changes by approximately 0.06 eV (from 2.34 eV to 2.40 eV; a 2 percent change).

9.3.3 Functionalized Metallic Nanoparticles

In addition to semiconducting dots, metallic nanoparticles can also be functionalized. For example, Fig. 9.18(a) shows silver nanoparticles, and Fig. 9.18(b) shows gold nanoshells made by coating the silver particles and dissolving the silver. These metallic nanoparticles can be functionalized for diagnostic, or even therapeutic purposes. For example, cancer cells tend to have certain proteins on their surfaces that are not found, at least in the same concentration, in healthy cells. By conjugating (binding) a certain antibody to the gold nanoparticles, the nanoparticles will attach themselves preferentially to the cancer cells. Since gold nanoparticles can be detected using visible light (as mentioned previously, due to plasmon resonances), the cancer cells can be imaged. Furthermore, significant amounts of heat can be generated by a resonating nanoshell, and, by tuning the nanoshell to resonate

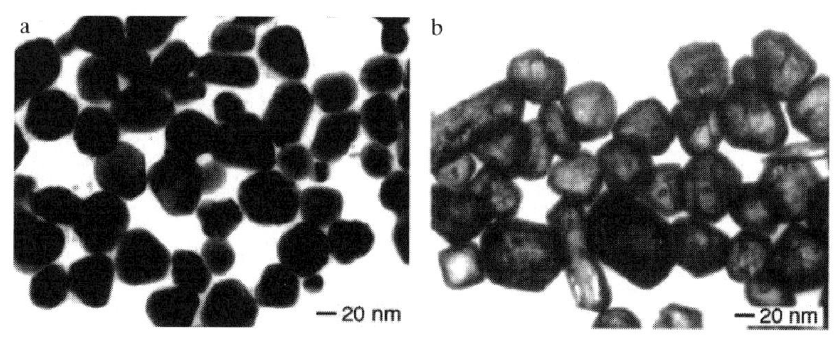

Figure 9.18 (a) TEM of silver nanoparticles. (b) TEM of gold nanoshells formed by reacting the silver nanoparticles of (a) with an aqueous HAuCl4 solution. The silver dissolves, leaving the gold shell remaining. (From Sun, Y., B. T. Mayers, and Y. Xia, "Template-Engaged Replacement Reaction: A One-Step Approach to the Large-Scale Synthesis of Metal Nanostructures with Hollow Interiors," *Nano Lett.* 2 (2002): 481. © 2002, American Chemical Society.)

in response to a particular frequency,† nanoshells embedded in cancer cells can deliver a therapeutic dose of heat, pinpointed to the cell itself. This only requires moderate amounts of energy illuminating the subject, such that excessive heating of nontargeted (i.e., healthy) cells can be avoided.‡

9.4 FABRICATION TECHNIQUES FOR NANOSTRUCTURES

The mainstay of commercial silicon technology fabrication is optical lithography. As discussed briefly in Chapter 1, it seems that optical lithography can be pushed to resolutions on the order of several tens of nanometers. Other technologies, like extreme UV lithography, immersion lithography, or electron beam lithography may become sufficiently mature for commercial applications in the near future. A very brief discussion of several important techniques will be provided next.

9.4.1 Lithography

The main idea of lithography was described in chapter 1, and here a brief recapitulation of electromagnetic lithography, which includes optical, EUV, and X-ray techniques, is provided.

1. Electromagnetic energy is directed at a photomask containing opaque and transparent regions that correspond to the desired pattern.

†Infrared frequencies are typically used, where transmission through tissue is adequate such that the illuminating field can penetrate to the targeted cell.

‡Heating of metal nanoparticles for medical applications has even been observed at frequencies far below the range of plasmon resonances. For example, in [23], a solution consisting of 10 nm gold nanoparticles and protein aggregates associated with Alzheimer's disease was heated using low-power (100 mW) 12 GHz electromagnetic radiation. It was estimated that each nanoparticle dissipated 10^{-14} J/s, which was more than enough to dissolve the protein aggregates.

2. Energy that passes through the photomask reaches a substrate coated with a photoresist, and sections of the photoresist that are illuminated by the energy undergo a chemical reaction due to the illumination.
3. Upon washing the structure in a solvent, sections of the resist are dissolved, forming a pattern on the resist. The regions of the resist that dissolve are either those sections that were exposed to the incident energy, or those that were shielded by opaque regions of the mask, depending on the type of photoresist (positive or negative) used.
4. Further processing steps are used to transfer the pattern from the resist to the substrate, or otherwise to change the substrate in the desired manner. These include
   (a) *Wet Etching*: Wet etching is the removal of material using acids, or any liquid solution that will dissolve the material in question. The patterned resist layer protects areas of material that are not to be removed. The depth of the etch can be controlled by the choice of etchant, and by limiting the etching time. However, undercutting of the resist is common in wet etching.
   (b) *Dry Etching*: Although there are several forms of dry etching, the easiest to understand is a method of bombarding the material by energetic ions inside a vacuum chamber. The ions collide with atoms in the material (called the target), resulting in momentum transfer, and causing these atoms to be ejected from the surface of the target. Again, the resist protects areas of the target that are not to be removed. In general, dry etching is more expensive than wet etching, although it has better resolution, and undercutting can be generally avoided. Another form of dry etching uses the chemical reactions between a gas and the substrate to remove material.
   (c) *Doping*: For semiconducting substrates, it is often necessary to dope the material. This can be accomplished, for instance, by accelerating a beam of dopant ions towards the substrate. The resist blocks the ions from reaching those regions of the substrate covered by the resist, and, thus, creates regions of doping in areas not covered by the resist. This is known as *ion implantation*.
   (d) *Material Deposition*: Material may be deposited (for example, metal) onto the wafer. There are several methods of deposition:
      i. *Sputtering* : Sputtering is basically the bombardment of a material (the target) by energetic ions inside a vacuum chamber. The ions collide with atoms in the target, resulting in momentum transfer, causing these atoms to be ejected from the surface of the target. These atoms then become deposited on the adjacent substrate, forming a thin film. Since sputtering is essentially a mechanical (rather than chemical) process, a wide range of materials can be deposited onto the substrate.
      ii. *Chemical vapor deposition* (CVD): In CVD, a substrate is placed inside a chamber with a number of gases. A chemical reaction between the gases produces a solid material that condenses on the substrate (and everywhere else), forming a thin film.

iii. *Evaporation*: In evaporation, a metal to be deposited and a substrate are placed inside a vacuum chamber. The metal is heated till it melts, and begins to evaporate. The evaporate then condenses on the substrate, forming a thin film. Heating of the metal occurs either by resistive effects in a sample holder, or by directing an electron beam at the metal.
iv. *Laser ablation*: In laser ablation, a pulsed laser beam vaporizes part of a target, and the evaporate forms a thin layer on a nearby substrate.
v. *Molecular beam epitaxy* (MBE): Molecular beam epitaxy† is an important technique that is somewhat similar to CVD. However, it can result in epitaxial layers of materials, including compound semiconductors. In MBE, elemental materials in separate chambers are evaporated in a high vacuum, and beams of the resulting molecules or atoms deposit material on a heated substrate. The deposited material builds up on the substrate with the same crystallographic orientation as the preceding atomic layer. This can form an atomically sharp interface between, say, GaAs and AlGaAs. Therefore, MBE in particular is an important technique in forming two-dimensional electron gases.

(e) After processing, the photoresist is removed (along with any material deposited on top of the photoresist, which is known as *lift-off*), leaving the desired structure.

For fabricating nanostructures, electron beam lithography, because of its nanometer scale resolution, has been a mainstay in laboratory settings. In this technique, electron beams serve as the illumination source, rather then electromagnetic waves, although the idea of having a resist, and doing various subsequent processing steps, is the same as for electromagnetic lithography. For research applications, it is common to perform electron beam lithography using an electron microscope. A figure of a quantum wire formed by electron beam lithography and ion etching is shown on page 125.

Many other variations of lithography exist, all involving forming some sort of template, and then removing or depositing material on the structure, after which the template is removed.

9.4.2 Nanoimprint Lithography

Nanoimprint lithography (NIL) consists of pressing a stamp or mold conforming to the desired pattern onto a thin film (perhaps a photocurable resist) on top of a wafer. The resist is heated to a temperature at which it is thermoplastic, becoming a viscous liquid. Thus, the film can flow, and be deformed into the shape of the mold, as depicted in Fig. 9.19. Various processing steps, including etching and material deposition, are then performed, as is done in the previously described lithography technologies. The mold itself is often produced by electron beam lithography.

Fig. 9.20 shows a T-structure made from the NIL process depicted in Fig. 9.19.

†Epitaxy is the growth of one material on another such that there is a crystallographic relationship between the two materials.

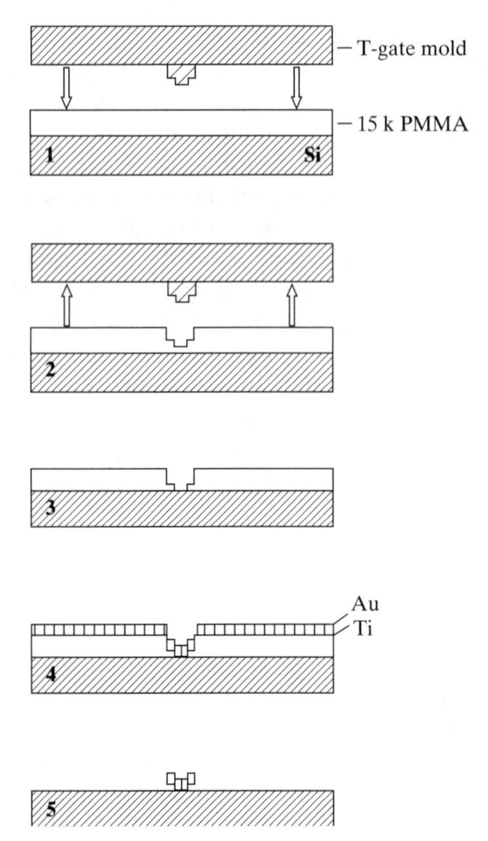

Figure 9.19 Depiction of NIL lithography to make a T-structure: the mold is pressed into the substrate, then retracted; metal is deposited; the substrate is dissolved and the extra metal is lifted off, leaving the desired shape. (Based on an image from Li, M., L. Chen, and S. Y. Chou, "Direct Three-Dimensional Patterning Using Nanoimprint Lithography," *Appl. Phys. Lett.* 78 (2001): 3322. © 2001, American Institute of Physics.)

9.4.3 Split-Gate Technology

Another method to form laterally confined nanostructures is the use of what are called *split gates*. This forms an electrically controllable structure that can be adjusted by an applied voltage. The idea is schematically depicted in Figs. 9.21 and 9.22.

One starts with a two-dimensional electron gas, formed, for instance, by a semiconductor heterostructure. As shown in Fig. 9.21, Schottky gate electrodes are deposited on top of the heterostructure. Left unbiased, the electron gas remains unperturbed. However, application of a negative potential to the electrodes depletes the electron gas below the gate, and so the gate electrodes form a narrow electron gas channel under the split, resulting in a quantum wire (into and out of the plane of the page), as shown in Fig. 9.22. Often the lateral confinement forming the wire is somewhat weak (this can also be true of wires formed by etching, due to, for example, undercutting), resulting in a "soft" barrier. This results in wire energies smaller than those given by (9.50).

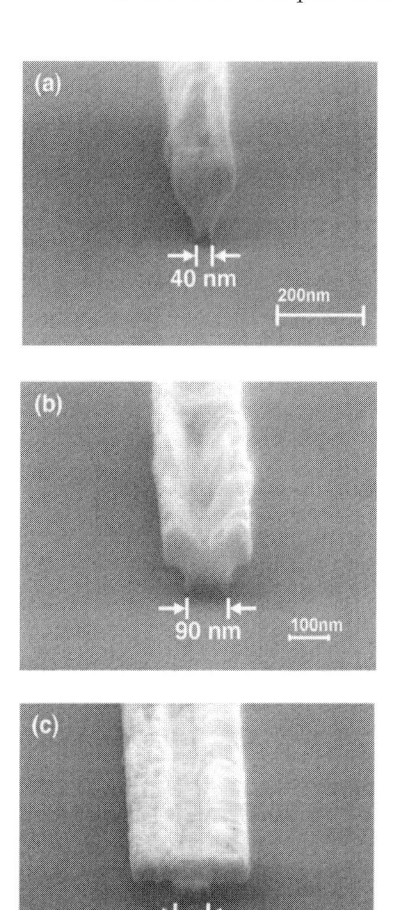

Figure 9.20 Resulting T-structure made from the NIL process depicted in Fig. 9.19. (Reused with permission from Li, M., L. Chen, and S. Y. Chou, "Direct Three-Dimensional Patterning Using Nanoimprint Lithography," *Appl. Phys. Lett.* 78 (2001): 3322. © 2001, American Institute of Physics.)

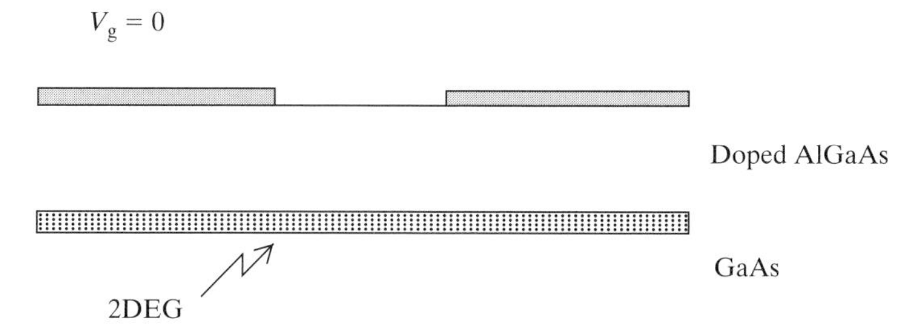

Figure 9.21 Split-gate electrodes over a two-dimensional electron gas.

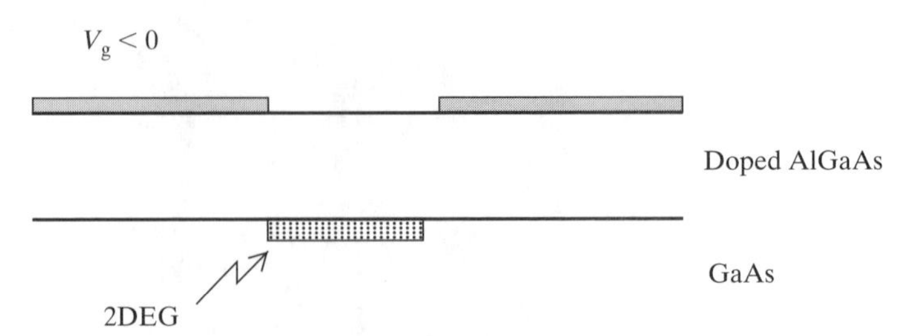

Figure 9.22 Split-gate electrodes over a two-dimensional electron gas. Negative gate bias depletes electrons below the gates, constraining the electron gas to a narrow region, forming a quantum wire.

Quantum dots can also be fabricated in a similar manner. In the case of a split-gate quantum dot, the electrode pattern constrains the 2DEG formed by a heterojunction in the shape of a hole in the electrode, as shown in Fig. 9.23.

9.4.4 Self-Assembly

Self-assembly techniques can be used to form a variety of nanostructures, and self-assembly is one of the ultimate goals in nanostructure fabrication. Many new techniques will undoubtedly be developed, although here we mention three common methods.

1. Lattice mismatch: The different crystallographic features of different materials can be used to advantage in self-assembly, especially in the formation of quantum dots

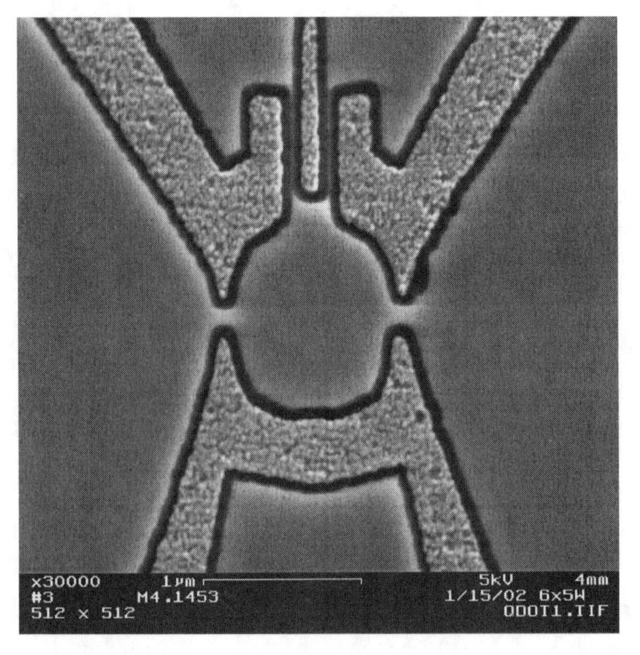

Figure 9.23 Quantum dot formed from split gates. (Courtesy of Jonathan Bird and Yuki Takagaki.)

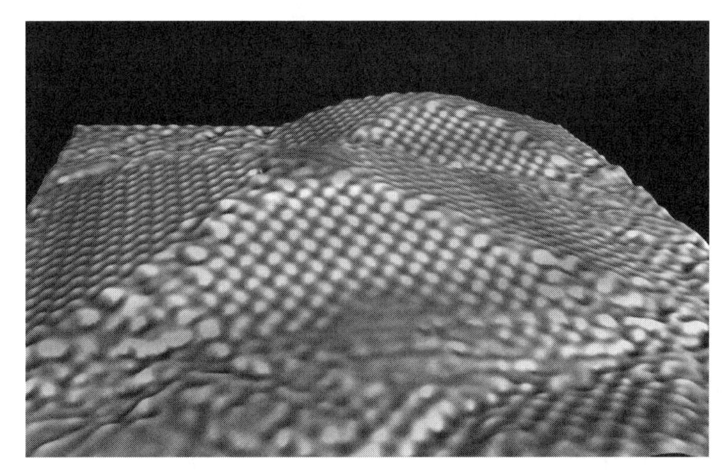

Figure 9.24 STM image of a self-assembled Ge islands on Si. (Reprinted with permission from Gilberto Medeiros-Ribeiro, HPL Palo Alto.)

and related nanostructures. For example, the lattice mismatch (and associated strain) between Ge and Si results in the formation of three-dimensional islands when Ge is deposited epitaxially on Si. The shape of the islands depend on the number of monolayers of Ge deposited on the Si surface—for six monolayers characteristic square pyramids tend to form, as shown in Fig. 9.24.

2. Wet chemistry methods: Quantum dots can also be synthesized by forming nano-sized precipitates from chemical solutions. Quantum dots so formed are called *colloidal quantum dots*.[†] This method has the advantage of producing large quantities of quantum dots in solution, which is often useful for biological or medical applications.
3. Molecular self-assembly: Molecular self-assembly is, as the name implies, the process of molecules organizing themselves into a desired structure. Often this takes place in liquids, but it can occur in a gas and other host mediums. Molecular self-assembly is obviously the goal in molecular electronics.

9.5 MAIN POINTS

In this chapter, we have considered models of quantum wells, quantum wires, and quantum dots, especially those structures fabricated from semiconducting materials. In particular, after studying this chapter, you should understand

- the idea of Fermi wavelength and exciton Bohr radius as important length scales in deciding when quantum confinement effects are important;
- the idea of a semiconductor heterostructure;
- hard-wall and other confinement models for forming a quantum well;

[†]A colloidal system (or colloid) is a collection of small particles, typically having sizes on the order of one to several hundred nanometers, dispersed in a host medium (often a liquid). The particles are generally small enough to be moved about by molecular collisions, and tend not to settle under gravity.

- energy subband structures in quantum wells and quantum wires;
- interband, intersubband, and exciton transitions in quantum wells, and the existence of selection rules;
- simple models of quantum dots, including the importance of exciton effects and the exciton radius;
- applications of quantum dots, including their use as biological markers;
- the idea of metal nanoparticles and plasmon resonance;
- fabrication methods, including various lithography techniques, split-gate techniques, and self-assembly.

9.6 PROBLEMS

1. A finite region of space is said to be effectively two-dimensional if $L_x \leq \lambda_F \ll L_y, L_z$. However, λ_F was computed from the three-dimensional result (9.3)

$$\lambda_F = \frac{2\pi}{\left(3N_e\pi^2\right)^{\frac{1}{3}}}, \tag{9.69}$$

where N_e is the electron density m^{-3}. Is it then permissible to use this formula to conclude that a structure is two or one dimensional? Shouldn't the two- or one-dimensional λ_F be used? Why or why not?

2. At the beginning of this chapter, it was stated that the Fermi wavelength is the important parameter in deciding if quantum confinement effects are important, and for determining the effective dimensionality of a system. Another method to characterize when quantum confinement effects are important is to say that if the difference in adjacent energy levels for a finite space is large compared with other energies in the system (thermal, etc.), then confinement effects are important. Considering a one-dimensional infinite-height potential well of width L, as considered in Section 4.3.1, show that confinement effects will be important compared with thermal energy when

$$L < \sqrt{\frac{3\hbar^2\pi^2}{m_e^* k_B T}}. \tag{9.70}$$

However, note that this result is independent of doping, and is, therefore, quite a rough approximation for semiconductors. For GaAs and Si, estimate the preceding relationship at room temperature.

3. As noted in Section 9.3.1, in quantum dots, one can take the exciton Bohr as the important length scale for deciding if quantum confinement effects are important. In what way is this a qualitatively similar idea to the use of the Fermi wavelength for the same purpose?

4. On page 288, it was shown that for GaAs at room temperature, one would need $L_x \ll$ 15 nm for two-dimensional subbands to be resolved. Determine the corresponding value for copper.

5. In a 2DEG, in order for energy subbands to be evident, k_BT should be much less than the difference between subbands, leading to (9.35). Consider the situation from the perspective of the de Broglie wavelength associated with thermal energy,

$$\lambda_e = \frac{h}{\sqrt{2m^*E}} = \frac{h}{\sqrt{2m^*k_BT}}, \tag{9.71}$$

and determine the size constraints on L_x to observe subbands at room temperature, and at $T = 4$ K, for GaAs. Comment on the relationship between the results of (9.35) and the results from (9.71).

6. Assume the well structure depicted in Fig. 9.3 on page 283. Since GaAs has a bandgap of $E_g \simeq 1.5$ eV, and $Al_\alpha Ga_{1-\alpha}As$ has a band gap of

$$E_g \simeq 1.426 + 1.247\alpha \tag{9.72}$$

for $\alpha < 0.45$, determine α if the height of the confinement barrier $\Delta E_C = E_C^{\text{AlGaAs}} - E_C^{\text{GaAs}}$ is 0.238 eV.

7. In Section 4.5.1, the quantum states in a finite-height potential well were determined, leading to the eigenvalue equation (4.83) for symmetric states, and (4.85) for antisymmetric states. In that example, the effective mass in each region was the same. Redo this problem assuming effective masses m_w^* and m_b^* for the well and barrier regions, respectively, and show that the resulting eigenvalue equations are

$$k_2 \tan(k_2L) = \frac{m_w^*}{m_b^*}k_1 \tag{9.73}$$

for the symmetric states, and

$$-k_2 \cot(k_2L) = \frac{m_w^*}{m_b^*}k_1 \tag{9.74}$$

for the antisymmetric states, where

$$k_2^2 = \frac{2m_w^*E}{\hbar^2}, \quad k_1^2 = \frac{2m_b^*(V_0 - E)}{\hbar^2}. \tag{9.75}$$

8. Comparing energy levels in one-dimensional quantum wells, for an infinite-height well of width $2L$ and effective mass m^*, energy states are given by (4.35) with L replaced by $2L$ and m_e replaced by m^*,

$$E_n = \frac{\hbar^2}{2m^*}\left(\frac{n\pi}{2L}\right)^2, \tag{9.76}$$

and for a finite-height well of width $2L$, the symmetric states are given by a numerical solution of

$$k_2 \tan(k_2L) = \frac{m_w^*}{m_b^*}k_1, \tag{9.77}$$

as described in Problem 9.7. Assume that $L = 5$ nm, $m_w^* = 0.067m_e$ (GaAs), and $m_b^* = 0.092m_e$ ($Al_{0.3}Ga_{0.7}As$; see Table VI in Appendix B). Compute E_1 and E_3 (the second symmetric state) for the infinite-height well, and compare it with the corresponding values obtained from the numerical solution for the finite-height well. (You will need to use a numerical root solver.) For the finite-height well, assume barrier height $V_0 = 0.3$ eV, which is $\Delta E_C = E_C^{\mathrm{Al}_\alpha\mathrm{Ga}_{1-\alpha}\mathrm{As}} - E_C^{\mathrm{GaAs}}$ for $\alpha = 0.3$. Comment on the appropriateness, at least for low energy states, of the much simpler infinite well model.

9. Assume a hard-wall model of a rectangular cross-section metallic quantum wire. If the wire is 1 nm thick and 10 nm wide, determine how many subbands are filled at 1 eV.

10. In a quantum wire, the available energy gets partitioned into various channels (subbands). Is this partitioning unique?

11. Calculate the energy levels in a quantum dot in the form of a cube, 5 nm on a side. Assume zero potential energy in the dot, and an infinitely high potential bounding the dot. Assume that $m^* = 0.045m_e$ in the dot material.

12. How small must a metal nanosphere be in order for the ground state energy to be 1 eV? Assume a hard-wall model.

13. Consider a hard-wall model of a cubical metal quantum dot, having side length $L = 3$ nm. What is the energy of a transition from the $(1, 1, 1)$ state to the $(2, 2, 1)$ state?

14. Consider a CdS quantum dot, having $E_g^{\mathrm{bulk}} = 2.4$ eV, $m_e^* = 0.21m_e$, $m_h^* = 0.8m_e$, and $\varepsilon_r = 5.6$. What is the expected peak absorbance energy (E_g^{dot}) and wavelength for a $R = 2$ nm dot? Repeat for a $R = 3$ nm dot.

15. Using the Internet, find a company that sells quantum dots for biological and/or medical applications, and write a one-half to one page summary of one of their products, and its applications.

16. There is significant debate about the possible adverse health effects of quantum dots in biological bodies. Much, but not all, of the concern revolves around the use of cadmium in quantum dot structures. Write a one page summary of the debate, including the benefits and drawbacks of using quantum dots in living bodies.

17. Determine quantum dot applications to lasers, and write a one-half to one page summary of the uses, pros, and cons, of using quantum dots as lasing materials.

Chapter

10

NANOWIRES, BALLISTIC TRANSPORT, AND SPIN TRANSPORT

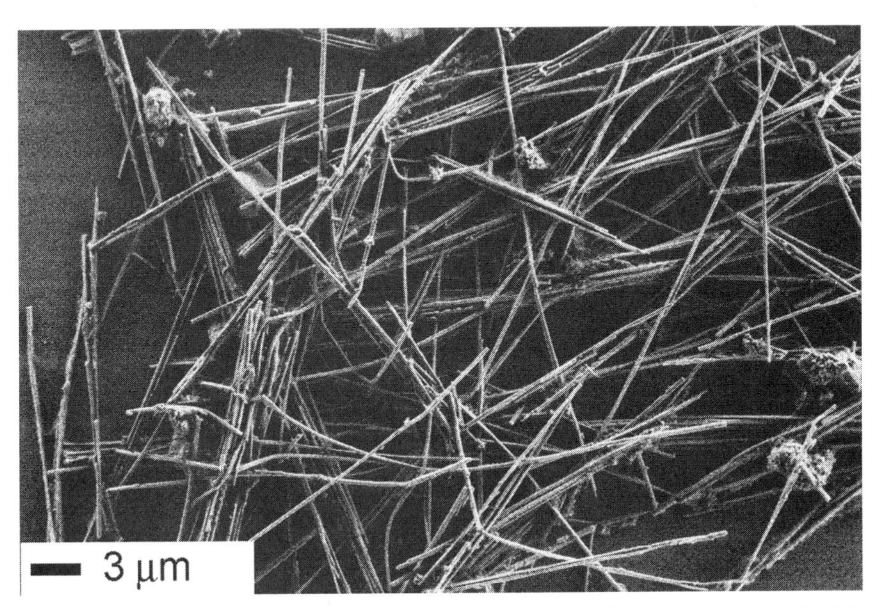

SEM image of 200 nm diameter nickel nanowires. Ann Bentley & Professor Art Ellis, Materials Science and Engineering Department, University of Wisconsin-Madison. Image courtesy the Nickel Institute.)

In an ordinary electrical circuit, conducting wires are used to interconnect electrical devices. For example, common residential wiring typically consists of 12 and 14 gauge† round cross-section wire, having diameters of 2.05 mm and 1.63 mm, respectively. Discrete electronic circuits built on a breadboard typically use 22 gauge wire, having a diameter of 0.64 mm. Integrated circuits use printed interconnects, having roughly rectangular cross sections, with typical widths on the order of 0.16 μm, and thicknesses on the order

†Americal Wire Gauge, AWG.

of 0.25 μm, although these dimensions continue to shrink along with transistor feature size.†

As electronics shrink, the "wires" used to interconnect devices obviously must also shrink in size. In this chapter, we consider what happens to the concept of conductivity when the dimensions of the conductor are very small. In this regard, there are two important dimensions to consider: the wire's radius, and its length. When the wire's radius is very small, on the order of the Fermi wavelength, we have a quantum wire, as discussed previously (Sections 4.7.2 and 9.2). Transport occurs axially along the wire, in well-quantized transverse energy subbands. We will discuss this topic further in this chapter.

However, we have not considered the case when the length of the conductor in the direction of propagation is "small." This topic not only is relevant to wire interconnects, but also is, for example, of interest in considering the channel of a FET, as channel lengths become a few nanometers. As we will see, this case leads to the concept of *ballistic transport*, which is a uniquely nanoscale phenomenon. The chapter concludes with a discussion of *spin transport*, and the possible applications of spin in forming devices.

10.1 CLASSICAL AND SEMICLASSICAL TRANSPORT

10.1.1 Classical Theory of Conduction—Free Electron Gas Model

Consider a solid material. The electrons (and the atoms themselves) are in a continual state of vibration, and temperature T is a measure of the energy associated with this motion. Considering that a particle has three degrees of freedom in which to move (x, y, and z), we see that Boltzmann's law states that each particle has thermal energy

$$E_T = \frac{3}{2} k_B T \tag{10.1}$$

joules, where k_B is Boltzmann's constant ($k_B = 1.38 \times 10^{-23}$ J/K). This is actually only true for gases, but is a reasonable approximation for many solids. For a particle having mass m and moving at an average velocity v in the absence of a potential field, the total energy (kinetic) is $mv^2/2$, and so

$$\frac{1}{2} m v^2 = \frac{3}{2} k_B T, \tag{10.2}$$

where v is the *mean thermal velocity* of electrons. Solving for velocity, we obtain

$$v = v_T = \sqrt{\frac{3 k_B T}{m}}. \tag{10.3}$$

At absolute zero,‡ $v_T = 0$, and at room temperature (293 K), for an electron this gives

$$v_T = 1.15 \times 10^5 \text{ m/s}. \tag{10.4}$$

†For integrated circuits, the thickness/width ratio is kept relatively large to keep interconnect resistance at a reasonable level.

‡As an aside, at $T = 0$ K, classical theory predicts that all motion stops, whereas quantum theory dictates some nonzero energy; atoms are in their ground state, analogous to the ground state particle-in-a-box result (4.53).

However, this motion is random and does not result in any net current, and so we'll denote this velocity as a scalar quantity.

Now consider applying a voltage for a time t across a solid containing free electrons. The resulting electric field accelerates the electrons in the direction opposite to that of the applied field ($\mathbf{F} = q_e\mathbf{E}$ and $q_e = -e = -1.6 \times 10^{-19}$ C), and, treating the electrons as classical particles, we find that Newton's law gives

$$\mathbf{F} = m_e\mathbf{a} = m_e\frac{d\mathbf{v}}{dt} = q_e\mathbf{E}, \tag{10.5}$$

where $\mathbf{E}$ is the vector electric field. This results in

$$\mathbf{v}(t) = \int_0^t \left(\frac{q_e}{m_e}\mathbf{E}\right) dt = \frac{q_e t}{m_e}\mathbf{E} = \mathbf{v}_d(t), \tag{10.6}$$

where we have assumed that $\mathbf{E}$ is independent of time. Note that this velocity is the component of velocity due to the applied electric field, which is called the *drift velocity*. Therefore, the total velocity of an electron in an applied field is the sum of the thermal and drift velocities,

$$\mathbf{v} = \mathbf{v}_T + \mathbf{v}_d, \tag{10.7}$$

where $\mathbf{v}_T$ is randomly directed.

However, the result (10.6) predicts that the electrons will increase their velocity indefinitely as time increases, which is obviously not true.[†] The problem is that collisions between the accelerated electrons and the material lattice (Chapter 5) have not been taken into account.[‡] It is these collisions that result in energy transfer from the applied field $\mathbf{E}$ to the material, via the electrons, that results in heating of the material. The mean time between collisions is τ, called the *relaxation time* (sometimes called the *momentum relaxation time*).

A good way to account for these collisions is to adjust the integral in (10.6), keeping in mind the model of electrons accelerating from $t = 0$ until they collide with something (a lattice imperfection, a thermal vibration, etc.) at $t = \tau$. Since the lattice atoms are much heavier than the electrons, it is assumed that each collision results in the electron completely losing its momentum[§] gained from the applied field. Therefore, in (10.6), we integrate from $t = 0$ to $t = \tau$, obtaining

$$\mathbf{v}_d = \frac{q_e\tau}{m_e}\mathbf{E}. \tag{10.8}$$

[†]For a particle in a vacuum this is closer to the truth, although relativistic effects limit the velocity.

[‡]As discussed in Chapter 5, an electron can move freely through a perfect periodic lattice, without scattering, and assuming that the electrons' energy is in an allowed energy band. However, that is a quantum result, and here we are discussing the classical model, and even in the quantum model, electrons will collide with lattice imperfections, or lattice vibrations (phonons). So, for simplicity, here we will just say that the electron collides with "the lattice."

[§]In a more precise treatment, the mean time between collisions and the momentum relaxation time are different, since a collision may not result in the complete loss of momentum.

We then have

$$\mathbf{v}_d = -\left(\frac{e\tau}{m_e}\right)\mathbf{E} = -\mu_e \mathbf{E}, \tag{10.9}$$

where μ_e is called the electrical *mobility* of the material, with units of m^2/V-s. Since mobility is proportional to the relaxation time, large mobility means that electrons travel a (relatively) long distance before colliding with the lattice.

We can consider τ as a temperature-dependent material parameter that can be measured, and, thus, $\mathbf{v}_d$ can be determined. As an example, if $\tau = 2.47 \times 10^{-14}$ s (an approximate value for copper at room temperature), and $|\mathbf{E}| = 1$ V/m, then $v_d = 4.35 \times 10^{-3}$ m/s, which is much less than the thermal velocity. Therefore, in the classical model, we view electron movement as being made up of rapid ($\sim 10^5$ m/s), random thermal motion, superimposed on much slower ($\sim 10^{-3}$ m/s) directed movement in the presence of an applied field or potential. Note, however, that despite the small value of drift velocity, electrical signals propagate as electromagnetic waves at the speed of light in the medium in question (e.g., $c \simeq 3 \times 10^8$ m/s for air). For highly conducting wires, the electromagnetic wave associated with the electrical signal carried by the wire propagates in the medium exterior to the wire (often air, or some dielectric insulation) even though in the wire itself, the electrons travel at a significantly slower velocity.

The total number of electrons crossing a unit plane each second is $N_e \mathbf{v}_d$, where N_e is the number of electrons per unit volume (i.e., the electron density, m^{-3}). The classical current density is obtained as

$$\mathbf{J} = q_e N_e \mathbf{v}_d = q_e N_e \left(\frac{q_e \tau}{m_e}\mathbf{E}\right) = \frac{q_e^2 \tau}{m_e} N_e \mathbf{E} \ \ \mathrm{A/m^2}, \tag{10.10}$$

often written as Ohm's law,

$$\mathbf{J} = \sigma \mathbf{E}, \tag{10.11}$$

where

$$\sigma = \frac{q_e^2 \tau}{m_e} N_e \tag{10.12}$$

is called the *conductivity* (S/m $= 1/\Omega$m). Large conductivity can result from large mobility, large electron density, or both. The conductivity (10.12) holds for a wide range of materials, although for semiconductors the effective mass should be used in place of m_e. In this case, mobility is defined by

$$\mu_e = \frac{e\tau_e}{m_e^*}, \qquad \mu_h = \frac{e\tau_h}{m_h^*} \tag{10.13}$$

for electrons and holes, respectively, where $\tau_{e,h}$ is the appropriate relaxation time for electrons and holes. Therefore,

$$\sigma = q_e^2 \left(\frac{\tau_e}{m_e^*} N_e + \frac{\tau_h}{m_h^*} N_h \right). \tag{10.14}$$

In copper, one of the best electrical conductors, at room temperature

$$\sigma \simeq 5.9 \times 10^7 \text{ S/m}, \tag{10.15}$$

using the electron density $N_e \simeq 8.45 \times 10^{28}$ /m^3 and $\tau = 2.47 \times 10^{-14}$ s. Therefore,

$$\mu_e \simeq \frac{\sigma}{eN_e} = 4.35 \times 10^{-3} \ \text{m}^2\text{V}^{-1}\text{s}^{-1}, \tag{10.16}$$

leading to

$$\mathbf{v}_d = -\mu_e \mathbf{E} \simeq -4.35 \times 10^{-3} \text{ m/s} \tag{10.17}$$

as previously stated for a one volt per meter electric field.

10.1.2 Semiclassical Theory of Electrical Conduction—Fermi Gas Model

The preceding *classical* model (called the *Drude model*) is simply a classical free electron gas model with the addition of collisions. Of course, it is better to use quantum physics principles. Rather than the thermal velocity, we should consider the Fermi velocity, which is the quantum velocity roughly analogous to the classical thermal velocity. By modifying Fig. 8.8 on page 273 to show momentum, rather than wavevector, we see that in the absence of an applied field, the Fermi surface associated with electrons in a bounded region of space is symmetric about the origin, as shown in Fig. 10.1.

All momentum states within the Fermi sphere are occupied, and those outside the Fermi sphere are empty.[†] Thus, for every occupied state $\mathbf{p}$, there is a state $-\mathbf{p}$, such that there is no net motion, analogous with the classical concept of thermal velocity being random.

From (8.47), the Fermi velocity is given by

$$v_F = \frac{\hbar k_F}{m^*} = \frac{\hbar \left(3N\pi^2\right)^{1/3}}{m^*} \tag{10.18}$$

[†]Of course, this is only strictly true at $T = 0$ K; for $T > 0$, the Fermi surface tends to become smeared out a bit, but the assumption of a sharp Fermi surface is a good approximation for even moderately large T, due to the behaviour of the Fermi-Dirac distribution.

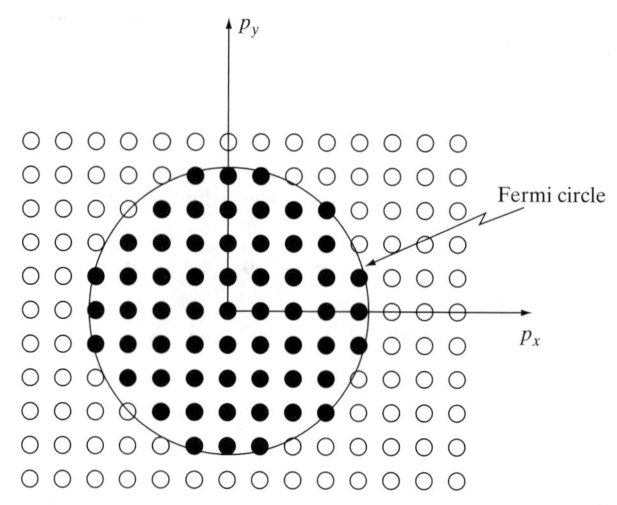

Figure 10.1 Electronic states in the momentum (p_x–p_y) plane for the case of a bounded region of space with no applied electric field. The states within the Fermi circle shown are occupied.

for the electron gas model of a metal, where N is the three-dimensional electron density. For metals, v_F is typically an order of magnitude larger than the thermal velocity. For example, for copper,

$$v_F = 1.57 \times 10^6 \text{ m/s}. \tag{10.19}$$

Therefore, the Fermi velocity is many orders of magnitude larger than the drift velocity, in general. As with the thermal velocity in the classical model, the motion due to the Fermi velocity is randomly directed. Therefore, electrons in a metal can often be considered to be moving randomly at the Fermi velocity in the absence of an applied field.

When we apply an electric field to a material, we will accelerate electrons, resulting in drift. Note that (10.8) for the drift velocity, and, hence, (10.10), is still approximately valid in this semiquantum model, even though they were derived using classical concepts. This can be understood because, as we are using τ here, it is a measured or empirical quantity, rather than being itself strictly derived from classical physics. Therefore, we obtain a somewhat empirical, although realistic model yielding good agreement with measurements.

In theory, τ in the classical model is the average relaxation time of all electrons (since in the free electron gas model, all electrons contribute to conduction), whereas in the quantum model, it is the relaxation time of electrons at the Fermi surface (i.e., the electrons most important for conduction). However, we don't need to be overly concerned with this detail here. The idea to keep in mind is that in going from the classical to the semiclassical model we are really replacing the thermal velocity with the Fermi velocity.[†] In either model, the net effect of applying an electric field, at least pertaining to the resulting current flow, is to consider the electrons to have directed movement at velocity $\mathbf{v}_d$.

[†]In addition, in the semiclassical treatment, rather than the collision-based model, we also consider Bloch's theorem, and the fact that electron scattering only occurs with impurities or lattice vibrations.

From the viewpoint of the Fermi surface, the energy or momentum gained by the electrons from the applied field can be obtained from†

$$\mathbf{F} = m^* \frac{d\mathbf{v}_d}{dt} = \frac{d\mathbf{p}}{dt} = \hbar \frac{d\mathbf{k}}{dt} = q_e \mathbf{E}, \tag{10.20}$$

and so the increase in momentum in a time increment dt is, using $\tau = dt$,

$$\delta \mathbf{p} = q_e \tau \mathbf{E}. \tag{10.21}$$

Thus, when an electric field is applied to the material, the Fermi surface is shifted from the origin, resulting in a nonzero net momentum, and electrical conduction, as shown in Fig. 10.2. That is, as shown in the figure, there are states having certain values of p_x that are not compensated for by states having $-p_x$, resulting in a net positive x-directed motion. When the applied field is removed, the Fermi surface shifts back (relaxes) to its previous position, symmetric about the origin.

This semiclassical theory of conduction, using the Fermi velocity rather than the thermal velocity, is called the *Fermi gas model*, in which electrons have total velocity

$$\mathbf{v} = \mathbf{v}_F + \mathbf{v}_d. \tag{10.22}$$

The formulas for conductivity, (10.12), and mobility are still used in this model.

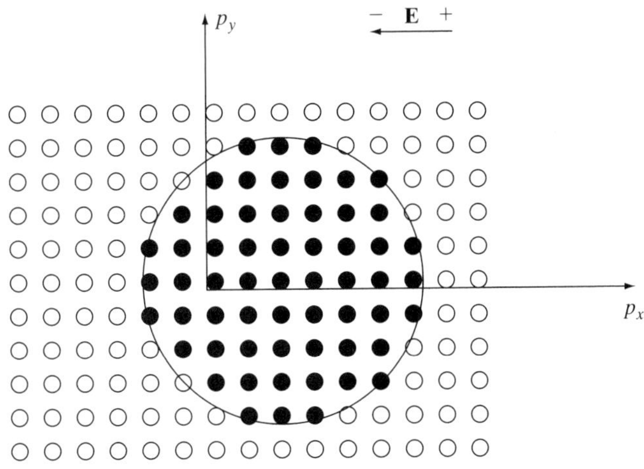

Figure 10.2 Electronic states in the momentum (p_x–p_y) plane for the case of an applied electric field, in which case the Fermi surface is shifted by the field. The states within the Fermi circle shown are occupied.

†As there are several different concepts of momentum described in the text, at the risk of being repetitive, it is worthwhile to be clear about what concept of momentum we are using. In (2.16), $\mathbf{p} = \hbar\mathbf{k}$ is the particle's actual momentum, and in Section 5.3.1, $\hbar\mathbf{k}$ is the crystal momentum, which is not the electron's physical momentum, but describes a Bloch electron's state. Here, because we are considering a Fermi gas model (no crystal lattice), $\mathbf{p} = \hbar\mathbf{k}$ represents the average physical momentum of electrons.

It is also convenient to define the *mean-free path* as the mean distance an electron travels before a collision with the lattice. This gives for copper

$$L_m = (v_d + v_F)\,\tau \simeq v_F\tau = 3.9 \times 10^{-8}\ \text{m}, \tag{10.23}$$

which indicates that the electron scatters off the lattice infrequently, since the mean free path is on the order of 100 lattice constants. (The lattice constant for copper is 3.61 Å.) Using the thermal velocity alone (i.e., the classical model), one obtains

$$L_m = 2.8 \times 10^{-9}\ \text{m}, \tag{10.24}$$

which is approximately eight lattice constants. Thus, using the thermal velocity, we would expect that the electron should scatter from the lattice quite frequently. This would also seem to make physical sense in the classical view, as we picture a dense sea of electrons moving through and colliding with the lattice. However, using the (better) Fermi velocity and the semiclassical model, we see that electrons actually scatter from the lattice infrequently. This is consistent with the fact that quantum mechanics predicts that an electron can pass through a perfect periodic lattice without scattering, where the effect of the lattice merely leads to the use of an effective mass, as described in Section 5.3.1. As discussed previously, scattering actually occurs not with a perfect lattice, but at the sites of lattice imperfections, or with lattice vibrations (phonons). Thus, the long mean free path, much longer than the lattice constant, makes perfect sense. Last, we should note that in a classical model, collisions between electrons and the lattice seem to be purely deleterious, giving rise to resistance. However, in the more realistic quantum model, as described in Section 5.3.1, collisions with impurities and vibrations are vital for electrical conduction to take place, thwarting Bloch oscillation.

Mobility (and obviously τ) is a strong function of temperature. As temperature decreases, mobility increases due to diminished phonon scattering. At sufficiently low temperatures, phonon scattering is suppressed, and scattering is mostly due to impurities. For crystals of relatively pure copper at liquid helium temperatures (4 K), it is possible to obtain $\tau \sim 10^{-9}$ s. This results in

$$L_m = v_F\tau = \frac{\hbar\left(3N\pi^2\right)^{1/3}}{m^*}\tau \sim 0.003\ \text{m}, \tag{10.25}$$

using $N = 8.45 \times 10^{28}\ \text{m}^{-3}$ for copper. In this case, L_m is on the order of eight million lattice constants! This is obviously in agreement with Bloch's theorem from Chapter 5.

10.1.3 Classical Resistance and Conductance

In electrical circuit theory, resistance is defined by Ohm's law in circuit form,

$$R = \frac{V}{I}, \tag{10.26}$$

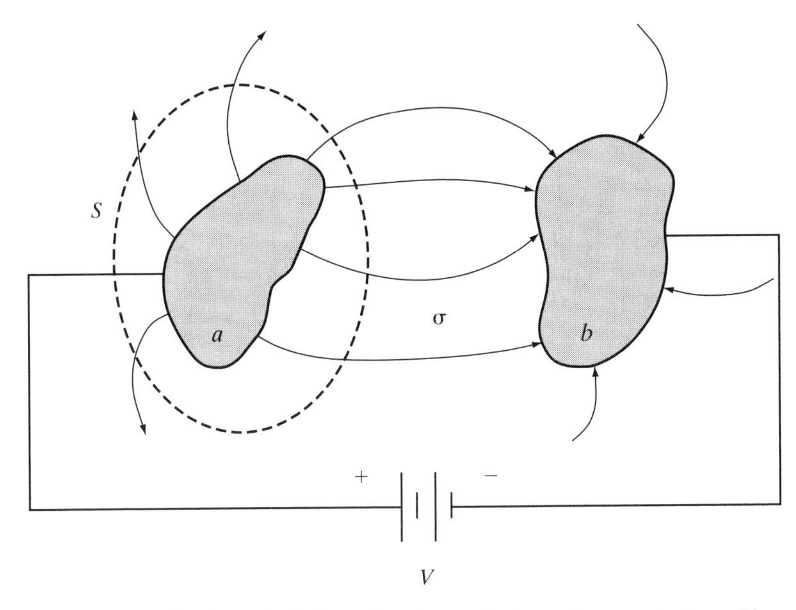

Figure 10.3 Geometry for the calculation of resistance between two conductors. The conductors are immersed in a material having conductivity σ. The arrows represent both electric field lines and electric current density, through $\mathbf{J} = \sigma\mathbf{E}$.

where V is the voltage across a circuit element and I is the resulting current that flows through the element. More generally, considering two arbitrarily shaped conductors immersed in a conducting medium characterized by conductivity σ, as shown in Fig. 10.3, the resistance between the conductors is

$$R = \frac{V}{I} = \frac{\int_a^b \mathbf{E}\cdot d\mathbf{l}}{\int\int_S \mathbf{J}\cdot d\mathbf{S}} = \frac{\int_a^b \mathbf{E}\cdot d\mathbf{l}}{\int\int_S \sigma\mathbf{E}\cdot d\mathbf{S}}. \tag{10.27}$$

The path a–b is any path from conductor a to conductor b, and S is any surface enclosing conductor a.

In particular, for a homogeneous rectangular solid having cross-sectional area $S = w_1 \times w_2$ and length L, we assume that the electric field $\mathbf{E}$ is uniform throughout the material and has magnitude $\mathcal{E}$, such that

$$R = \frac{\mathcal{E}L}{\sigma S\mathcal{E}} = \frac{L}{\sigma S}. \tag{10.28}$$

This is the three-dimensional result.

We are often interested in two-dimensional conductors, which are models of very thin metal strips, such as a printed circuit board interconnect. If $w_2 \to 0$, then (10.28) predicts that $R \to \infty$ (assuming that σ is finite). This result is true, however, we are often interested in the case of small yet finite values of w_2. For these situations, it is convenient to assume that a two-dimensional surface current density flows ($\mathbf{J}$ A/m), such that for a two-dimensional flat wire having length L and width w,

$$R = \frac{V}{I} = \frac{\int_0^L \mathbf{E} \cdot d\mathbf{l}}{\int_0^w \mathbf{J} \cdot d\mathbf{S}} = \frac{\mathcal{E} L}{\sigma_s \mathcal{E} w} = \frac{L}{\sigma_s w}. \tag{10.29}$$

In this case σ_s is the *sheet conductivity* in S (not S/m). In a real material, where w_2 (the thickness) is perhaps very small, but finite, (10.28) applies, or we can model the thin conductor as a two-dimensional conductor, in which case (10.29) applies, where $\sigma_s = \sigma \times w_2$.

10.1.4 Conductivity of Metallic Nanowires—The Influence of Wire Radius

In the next section we will examine what happens when wire length L becomes extremely small relative to the mean free path, resulting in an interesting phenomenon that is not encountered in the world of macroscopic conductors. However, before considering this case, it is worthwhile to make some comments about the influence of wire cross-section size on conductivity.

Consider a circular cross-section wire having radius a and length L, and assume that L is very large compared to the mean free path (and, therefore, certainly to the Fermi wavelength as well). When a is macroscopic, one uses the model of conductivity for a bulk material, such as (10.12). In this case, the resistance is given by (10.28). For example, assuming a copper ($\sigma = 5.9 \times 10^7$ S/m) wire having radius $a = 10$ mm, $R = 5.395 \times 10^{-5}$ ohms/m. Thus, one would need approximately 18, 536 m of wire to amount to a resistance of one ohm, which is why we can often ignore wire resistance in electrical circuits. However, if $a = 10$ μm, $R = 53.95$ ohms/m, amounting to 1 ohm in only 1.85 cm. If $a = 10$ nm, the resistance is huge, $R = 5.395 \times 10^7$ ohms/m.

However, it is important to note that for wires having radius values on the order of the mean free path or less, the conductivity value is changed from the case of a bulk material. Thus, the previously listed values of resistance for the 10 mm and 10 μm radius wires are reasonable, although the value for the 10 nm radius wire is not large enough. For example, copper has a mean free path of approximately 40 nm, and in this range, radius-dependent effects are usually manifest. In fact, one may consider that radius-dependent effects may occur even when the radius is approximately double this value, on the order of 80–100 nm. In the 1–20 nm radius range, the conductivity of the wire certainly will differ appreciably from the bulk value, and generally the conductivity significantly decreases as a is reduced. This is due to several effects, such as scattering from the wire's surface and from grain boundaries, not to mention the difficulty in fabricating high-quality, defect-free metals at small size scales. Thus, as a very rough rule of thumb, one can use the bulk value of conductivity for many good conductors when the radius value is above approximately $a = 80$–100 nm. Below this point, down to radius values of perhaps 5–10 nm (but above metallic quantum wire dimensions), one may expect to need to use a size-dependent value of conductivity, perhaps based on measurement. A relatively simple approximate formula for the resistivity (ρ; $\rho = \sigma^{-1}$) of rectangular cross-section wires is[†]

$$\rho = \rho_0 \left\{ \frac{1}{3\left[\frac{1}{3} - \frac{\alpha}{2} + \alpha^2 - \alpha^3 \ln\left(1 + \frac{1}{\alpha}\right)\right]} + \frac{3}{8} C (1 - p) \frac{1 + AR}{AR} \frac{L_m}{w} \right\}, \tag{10.30}$$

where

$$\alpha = \frac{L_m}{d} \frac{R_c}{1 - R_c}, \tag{10.31}$$

and where ρ_0 is the bulk resistivity, w is the wire width, AR is the aspect ratio (wire height divided by wire width), d is the average grain size (for relatively narrow wires this can be taken as the wire width), p is called the *specularity parameter* (relating to reflection from the wire surface), R_c is the grain boundary reflectivity coefficient, and C is a constant (taken to be 1.2 in this model). The first term is related to grain-boundary scattering, and the second term to wire-surface scattering. Both p and R_c can take values between 0 and 1, and typical values determined by fitting (10.30) to experimental results are $p = 0.3$–0.5 and $R_c = 0.2$–0.3. For example, using $p = 0.50$ and $R_c = 0.27$ we have $\sigma = 1.22 \times 10^7$ S/m for a 10×10 nm^2 copper wire (down from 5.9×10^7 S/m for the bulk value). The preceding model may work down to wire cross-sectional dimensions on the order of perhaps 5–10 nanometers, below which a quantum wire model that accounts for transverse quantization would be necessary.

However, as complicated as surface- and grain-boundary scattering are, other factors also determine the conductivity of a nanowire. For example, the I–V characteristic of a 30 nm radius, 2.4 μm long single-crystalline copper nanowire is shown in Fig. 10.4. In Fig. 10.4(a), the room temperature characteristics are shown, along with an SEM image of the wire contacting two Au electrodes. The resistance is approximately 10 times the value expected from (10.28) using σ for bulk copper. The difference could be due to electron scattering (as discussed earlier), large contact resistance between the electrodes and the wire, or surface oxidation. To consider the latter effect, in Fig. 10.4(b) the current–voltage characteristics of a 25 nm radius copper nanowire are shown for various times over a 12-hour period. The increasing oxidation of the copper resulted in the material becoming more like Cu_2O, and after complete oxidation, the wire acts like a p$-$type semiconductor.

Figure 10.5 shows the wire in Fig. 10.4(a) after complete oxidation, at both room temperature and 4.2 K. The Cu_2O wire creates a Schottky contact with each electrode, and so the I–V curve has a double-diode (back-to-back diode) behavior. Due to the large surface-to-volume ratio of nanostructures, these kinds of effects can be extremely deleterious to performance.

[†]See [27] and references therein.

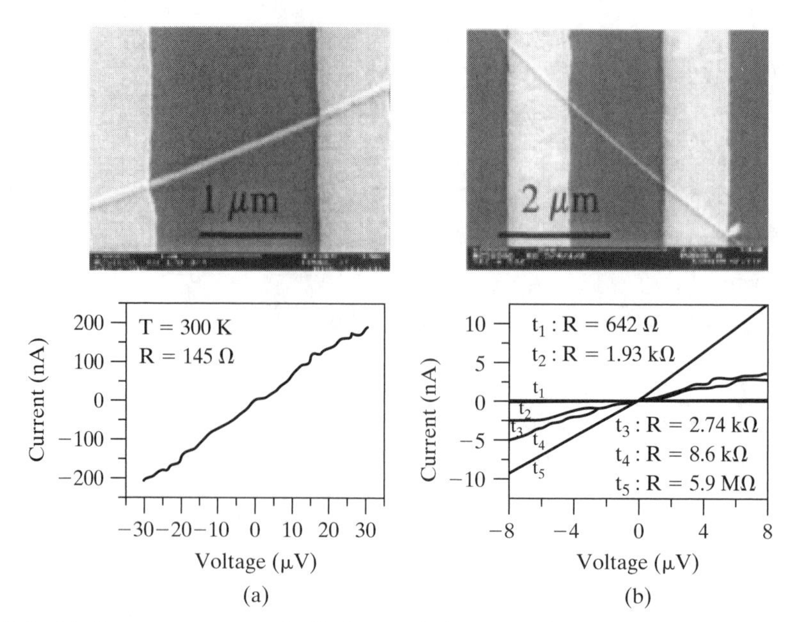

Figure 10.4 (a) Room temperature I–V characteristics of a 30 nm radius, 2.4 μm long single-crystalline copper nanowire. A SEM image of the wire contacting two Au electrodes is shown above the plot. (b) Current–voltage characteristics of a 25 nm radius copper nanowire for various times $t_1 < t_2 < t_3 < t_4 < t_5$ over a 12-hour period. The increasing oxidation of the copper resulted in the wire acting like a p− type semiconductor. (From Toimil Molares, M. E., et al., "Electrical Characterization of Electrochemically Grown Single Copper Nanowires," *Appl. Phys. Lett.* 82 (2003): 2139. © 2003, American Institute of Physics.)

10.2 BALLISTIC TRANSPORT

As the length L of a conduction path becomes very small, do the formulas (10.28) or (10.29) continue to hold? One may guess that they do not, since, for one thing, conductivity is a bulk parameter, and is derived assuming a large number of electrons (the electron gas model) and a large number of collisions between electrons and phonons, impurities, imperfections, etc. In particular, if L is reduced to become much less than the mean free path L_m, one would expect that no collisions would take place, rendering the collision-based model useless. This indeed occurs, and in this section, we consider the case when $L \ll L_m$.

It has been possible only recently to experimentally investigate resistance at the nanoscale. Much progress has been made in understanding the underlying physics of nanoscale and *mesoscopic*† transport. The overarching idea is that at very small length scales, electron transport occurs *ballistically*. It can be appreciated that ballistic transport will be important in many future nanoscopic devices.

†Mesoscopic refers to size scales between microscopic (atomic) and macroscopic (sizes of everyday objects). For example, carbon nanotubes are often mesoscopic, having nanometer radius values yet having anywhere from nm to cm lengths.

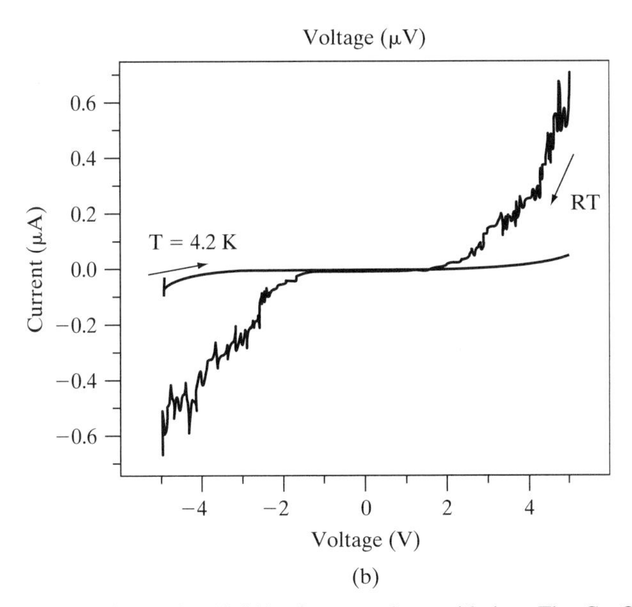

Figure 10.5 The Cu wire from Fig. 10.4(a) after complete oxidation. The Cu_2O wire acts like a p-type semiconductor, creating Schottky contacts with each electrode, and so the $I-V$ curve has a double-diode (back-to-back diode) behavior. (Based on a figure from Toimil Molares, M. E., et al., "Electrical Characterization of Electrochemically Grown Single Copper Nanowires," *Appl. Phys. Lett.* 82 (2003): 2139. © 2003, American Institute of Physics.)

10.2.1 Electron Collisions and Length Scales

There are two types of collisions to consider. First, an electron can collide with an object such that there is no change in energy. (Think of a ball bouncing off of a fixed surface.) This type of collision is called an *elastic collision*, and typically, collisions between electrons and fixed impurities are elastic. In the second type of collision, the energy of the electron changes (although total energy is conserved). This type of collision is called an *inelastic collision*, and typically results from collisions between electrons and phonons (quantized lattice vibrations), or between electrons and electrons.

Considering the preceding discussion, we'll define several length scales.

- L is the system length, in this case the length of the conductor in question.
- L_m is the mean free path defined previously. However, now we want to be explicit and define this to be the length that the electron can travel before having an *elastic* collision.
- L_ϕ is the length over which an electron can travel before having an *inelastic* collision. This is also called the *phase-coherence length*, since it is the length over which an electron wavefunction retains its coherence (i.e., retains its phase memory). Over

the phase-coherence length, the phase of the wavefunction evolves smoothly. From (3.135),

$$\Psi(\mathbf{r}, t) = \psi(\mathbf{r})\, e^{-iEt/\hbar}, \tag{10.32}$$

assuming a time-independent potential. We see that elastic collisions do not disrupt phase coherence, but that inelastic collisions destroy phase coherence. These inelastic collisions are called *dephasing events*. Among other things, these dephasing events will destroy interference effects, including electrons interfering with themselves. L_ϕ is usually on the order of tens to hundreds of nanometers at low temperatures.

As is typical of nanosystems, thermal effects play an important role in phase coherence. This is obviously the case for electron-phonon collisions, since phonon energy will be greater at higher temperatures; in the classical model, the lattice vibrates more at higher temperatures. Furthermore, due to thermal energy, at nonzero temperatures, an electron should be represented by a wavepacket, the energies of which will vary on the order of $k_B T$. As T increases, obviously, the spread of energies will increase, along with their associated phase differences, eventually leading to thermal decoherence even in the absence of particle scattering.† Decoherence of any kind is one of the reasons for systems to exhibit classical behavior, and is one of the most problematic issues facing the development of quantum computers. That is, the computation must be finished before the quantum state decoheres (with some exceptions), and, at present, this time is quite short.

For simplicity, we will divide electron transport into two regimes:‡

- For $L \gg L_\phi, L_m$, we have *classical transport*, which is the familiar macroscopic case previously described. Ohm's law applies, and momentum and phase relaxation occur frequently as charges move through the system. Because of this, we cannot solve Schrödinger's equation over the whole conductor length L. It is fortunate that semiclassical or even classical models generally work well in this case.
- For $L \ll L_m, L_\phi$, we have *ballistic transport*, which is our main interest here. Ballistic transport occurs over very small length scales, and is obviously coherent; the electron doesn't "hit" anything as it travels through the material, and, therefore, there is no momentum or phase relaxation. Thus, in a ballistic material, the electron's wavefunction can be obtained from Schrödinger's equation.§ One practical

†One can also view thermal decoherence as an interaction between the object and its environment, since a "warm" object can emit photons (classically, blackbody radiation). Recent experiments have shown that thermal decoherence will result if the object/particle emits photons capable of resolving its path. Just as in the case of the double-slit experiment discussed in Chapter 2, observing the particle's path, such as which slit it goes through in a two-slit experiment, results in the object exhibiting classical behavior.

‡There are several other transport categories between the two extremes of classical and ballistic transport, although these won't be considered here.

§In principle, if there are scattering events at specific positions in space, then Schrödinger's equation can be applied to each region between the scatterers, and the solutions connected. This is similar to what is done for heterostructures (i.e., applying Schrödinger's equation region by region, and then connecting the solutions by the boundary conditions), although the scattering problem presents a much more difficult task.

application of ballistic transport is to ultra-short-channel semiconducting FETs,[†] or carbon nanotube transistors. Short interconnects may exhibit ballistic transport properties, although material processing issues are important.

10.2.2 Ballistic Transport Model

In the following discussion, we consider a wire having small width (in two dimensions) or diameter/cross section (in three dimensions), on the order of[‡] λ_F, and length $L \ll L_m, L_\phi$, oriented along the x coordinate, extending from $x = 0$ to $x = L$. We assume that at both ends of the there are *reservoirs* of electrons; these reservoirs are simply macroscopic metal contacts that act as infinite sources and sinks for electrons, as shown in Fig. 10.6.

First, concerning the reservoirs themselves, from (4.54), for example, if the reservoir has dimensions $D \times D \times D$, then electrons in the reservoir have energy

$$E_n = \frac{\hbar^2 \pi^2}{2mD^2} n^2, \tag{10.33}$$

and for D sufficiently large the energy states in the reservoirs form essentially a continuum, as in a classical model. It is this continuum that is the infinite source and sink for electrons.

The Ballistic Channel and Subbands. Not surprisingly, we need to return to the quantum mechanical picture to consider ballistic transport between the reservoirs. From the discussion of subbands in Sections 4.7.2 and 9.2, electrons in the wire will have energy

$$\begin{aligned} E &= \frac{\hbar^2}{2m_e^*}\left(k_x^2 + k_{y,n_y}^2 + k_{z,n_z}^2\right) \\ &= E_{n_y,n_z} + \frac{\hbar^2}{2m_e^*} k_x^2, \end{aligned} \tag{10.34}$$

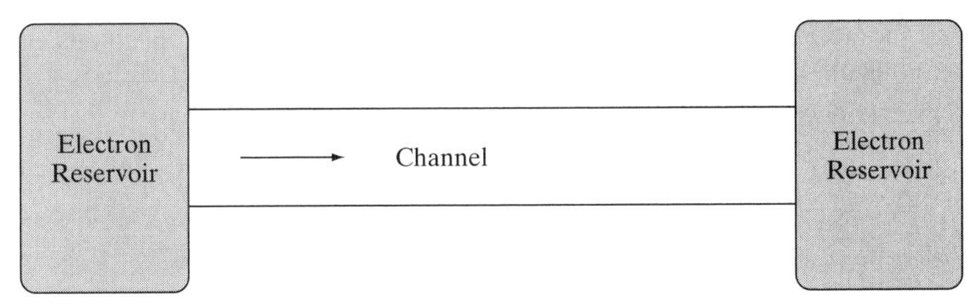

Figure 10.6 Ballistic channel connecting two electron reservoirs. The channel is parallel to the x-axis.

[†]For semiconducting channels, the specimen would need to be relatively pure, with low surface scattering, in order to achieve ballistic transport. Quasi-ballistic transport is more likely.

[‡]We assume a small cross section so as to have quantum confinement effects in the transverse direction; larger cross-section wires can be accommodated by a quasi-continuous, rather than discrete, energy level structure.

where k_x is the longitudinal wavenumber, which will be a continuous parameter (since electrons are free along the x coordinate), and the discrete indices n_y and n_z are subband indices. Subband energy levels are given by

$$E_{n_y,n_z} = \frac{\hbar^2}{2m_e^*}\left(k_{y,n_y}^2 + k_{z,n_z}^2\right), \tag{10.35}$$

where, for example, for an infinite confining potential (hard-wall case) and a rectangular cross-section ($w_1 \times w_2$) wire,

$$E_{n_y,n_z} = \frac{\hbar^2}{2m_e^*}\left(\left(\frac{n_y\pi}{w_1}\right)^2 + \left(\frac{n_z\pi}{w_2}\right)^2\right), \tag{10.36}$$

where $n_y, n_z = 1, 2, 3, \ldots$.

Let's assume that the quantum wire has a square cross section, such that $w_1 = w_2 = w$. Then, the hard-wall subband energy levels are given by

$$E_n = \frac{\hbar^2\pi^2}{2m_e^* w^2}n^2, \tag{10.37}$$

such that the number of subbands (also called the number of *electron channels*, or *modes*) at or below the Fermi energy is

$$N = \sqrt{\frac{E_F 2m_e^* w^2}{\hbar^2\pi^2}} = \frac{w}{\hbar\pi}\sqrt{E_F 2m_e^*} = \frac{wk_F}{\pi} = \frac{w}{\lambda_F/2}, \tag{10.38}$$

where we assumed the usual $E(k)$ relationship for electrons,

$$E = \frac{\hbar^2 k^2}{2m_e^*}. \tag{10.39}$$

Therefore, as the width of the wire increases, the number of electron channels increases, such that the wire gains a new channel (mode) each time the width becomes equal to an integral multiple of a half-Fermi wavelength. This is very similar to the case of electromagnetic wave propagation in a hollow conducting waveguide.

10.2.3 Quantum Resistance and Conductance

In order to further characterize ballistic transport, we need to determine the resistance of the channel. We assume that a potential V is applied across the two reservoirs in Fig. 10.6, positive on the left side, driving a current in the wire. The Fermi energy in the left reservoir is then $E_F - eV$, whereas in the right reservoir it is E_F. (We assume the left and right reservoirs are identical.) The electrons in the wire have wavefunction ψ,

$$\psi(x, y, z) = \psi_x(x)\,\psi_y(y)\,\psi_z(z), \tag{10.40}$$

with an associated probabilistic current density

$$\mathbf{J}(x, y, z) = \frac{\hbar}{m_e^*} \text{Im} \left(\psi^* \nabla \psi \right). \tag{10.41}$$

The current in the wire is, from (3.190),

$$\begin{aligned} I_x &= \int_y \int_z (-e) \, \mathbf{J}(x, y, z) \cdot \mathbf{a}_x \, dydz \\ &= -\frac{e\hbar}{m_e^*} \int_y \int_z \text{Im} \left(\psi^* \frac{\partial}{\partial x} \psi \right) dydz \\ &= -\frac{e\hbar}{m_e^*} \int_y \int_z \left| \psi_y(y) \right|^2 \left| \psi_z(z) \right|^2 \text{Im} \left(\psi_x^*(x) \frac{\partial \psi_x(x)}{\partial x} \right) dydz \\ &= -\frac{e\hbar}{m_e^*} \text{Im} \left(\psi_x^*(x) \frac{\partial \psi_x(x)}{\partial x} \right), \end{aligned} \tag{10.42}$$

where we used the normalization condition

$$\int_y \int_z \left| \psi_y(y) \right|^2 \left| \psi_z(z) \right|^2 \, dydz = 1. \tag{10.43}$$

Now, assume the wavefunction can be represented by a traveling state, indicating left-to-right (positive k) movement of the electron,

$$\psi_x(x) = \frac{1}{\sqrt{L}} e^{ikx}, \tag{10.44}$$

such that

$$I_x = -\frac{e\hbar}{m_e^* L} \text{Im} \left(e^{-ikx} (ik) \, e^{ikx} \right) = -\frac{e\hbar k}{m_e^* L}. \tag{10.45}$$

Since this is the current due to a certain state, and since two electrons can fill each state (accounting for spin), then, multiplying by two, we obtain

$$I_{x,k} = -\frac{2e\hbar k}{m_e^* L}, \tag{10.46}$$

where the subscript indicates both the direction and the state dependence.

The preceding quantity, (10.46), is the current in the x-direction carried by electrons in state k. However, we do not know if a certain state will be filled. Furthermore, the probability that the electron makes it into the channel from the left reservoir, and out of the channel into the right reservoir, must be incorporated. The desired result is obtained by summing over all possible states (k) and all possible electron channels (the N subbands). For each state, we multiply by the Fermi–Dirac probability of the state k in the left reservoir being occupied,

$f(E, E_F - eV, T)$, the energy-dependent transmission probability of passing through the channel, $T_n(E)$, and the current carried by each state, $I_{x,k}$. The resulting current flowing from left to right is

$$I_{L\to R} = \sum_{n=1}^{N}\sum_{k} f(E, E_F - eV, T)\, T_n(E)\left(-\frac{2e\hbar k}{m_e^* L}\right) \tag{10.47}$$

$$= \sum_{n=1}^{N}\frac{L}{2\pi}\int_{-\infty}^{\infty} f(E, E_F - eV, T)\, T_n(E)\left(-\frac{2e\hbar k}{m_e^* L}\right) dk$$

$$= -\frac{2e}{h}\sum_{n=1}^{N}\int_{-\infty}^{\infty} f(E, E_F - eV, T)\, T_n(E)\, dE,$$

where the factor $L/(2\pi)$ comes about in converting the sum over k to an integral.† Since the transmission probability $L \to R$ is the same as that for $R \to L$ conduction (assuming a symmetric potential barrier and identical reservoirs),

$$I_{R\to L} = -\frac{2e}{h}\sum_{n=1}^{N}\int_{-\infty}^{\infty} f(E, E_F, T)\, T_n(E)\, dE, \tag{10.48}$$

and the total current flowing is

$$I = I_{L\to R} - I_{R\to L} = -\frac{2e}{h}\sum_{n=1}^{N}\int_{-\infty}^{\infty}\{f(E, E_F - eV, T) - f(E, E_F, T)\}\, T_n(E)\, dE. \tag{10.49}$$

If eV is sufficiently small, then

$$f(E, E_F - eV, T) - f(E, E_F, T) \simeq \frac{\partial f(E, E_F, T)}{\partial E_F}(-eV) \tag{10.50}$$

$$= -\frac{\partial f(E, E_F, T)}{\partial E}(-eV),$$

resulting in

$$I = \frac{2e^2}{h} V \sum_{n=1}^{N}\int_{-\infty}^{\infty}\left(-\frac{\partial f(E, E_F, T)}{\partial E}\right) T_n(E)\, dE. \tag{10.51}$$

†For a free particle in one dimension, assuming periodic boundary conditions as discussed in Section 4.3.3, each state occupies a region of length $2\pi/L$ in $k-$space.

The current (10.51) leads to the temperature-dependent conductance

$$G(T) = \frac{I}{V} = \frac{2e^2}{h} \sum_{n=1}^{N} \int_{-\infty}^{\infty} \left(-\frac{\partial f(E, E_F, T)}{\partial E} \right) T_n(E) \, dE. \tag{10.52}$$

For very low temperatures, the Fermi–Dirac distribution resembles a step function,

$$\left(-\frac{\partial f(E)}{\partial E} \right) = \delta(E_F - E), \tag{10.53}$$

such that

$$G(T = 0) = \frac{2e^2}{h} \sum_{n=1}^{N} T_n(E_F). \tag{10.54}$$

As the width of the physical channel increases, there are more electronic conduction channels (subbands), as given by (10.38), increasing N.

From (10.54), if, for instance, there are N electronic channels, and if the transmission probability is one for each channel, then

$$G = \frac{2e^2}{h} N, \tag{10.55}$$

which is known as the *Landauer formula*. (Some of the preceding expressions could also be given this name.) Since N is the number of conduction channels, using (10.38) we have

$$G = \frac{2e^2}{h} \frac{2w}{\lambda_F} \tag{10.56}$$

for a wire with square cross section w^2. The resistance of each channel is

$$R_0 = \frac{h}{2e^2} = 12.9 \text{ k}\Omega, \tag{10.57}$$

which is known as the *resistance quantum* ($G_0 = 2e^2/h$ is called the *conductance quantum*), and the total resistance is

$$R = \frac{h}{2e^2 N} = \frac{12.9}{N} \text{ k}\Omega. \tag{10.58}$$

As the number of channels increases (i.e., as N increases by increasing the size of the physical channel's cross section), conductance increases and resistance decreases (analogous to adding additional wires to a wire bundle). The classical theory also predicts this behavior (e.g., (10.28)), although the quantum theory shows that this happens in discrete steps, as the number of electron channels increases. As N gets very large, the electron channels essentially form a continuum, and the quantum theory tends towards the classical

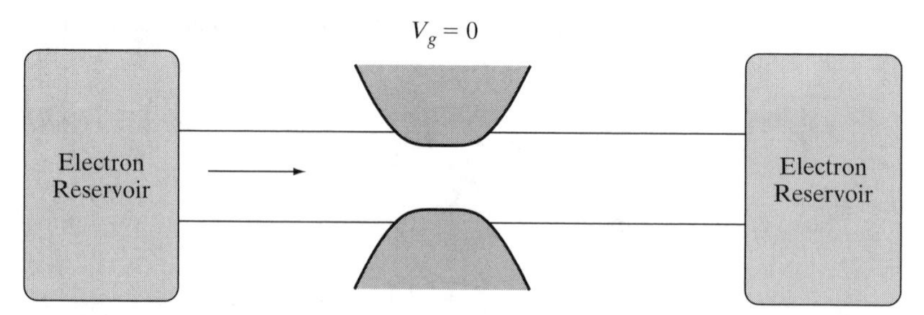

Figure 10.7 Two electron reservoirs connected by a ballistic channel. Channel width is controlled by the gate voltage applied at the gate electrodes. $V_g = 0$ and channel is relatively open.

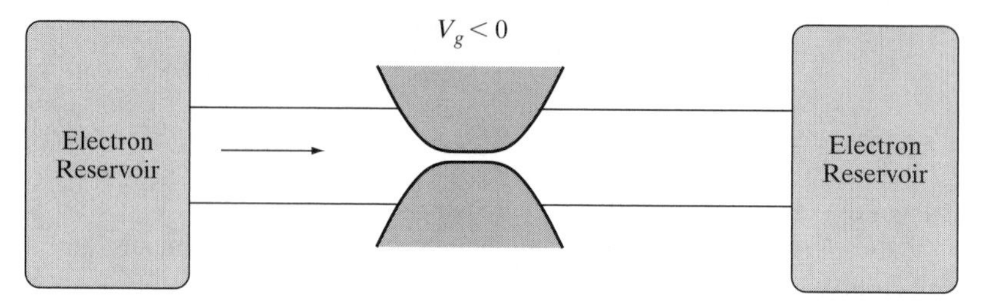

Figure 10.8 Same as Fig. 10.7, although with $V_g < 0$ so that the channel is constricted.

limit. However, in the presence of scattering, resistance quantization does not, in general, appear.†

The quantization of conductance is readily shown in low-temperature experiments. A schematic representation is shown in Figs. 10.7 and 10.8, where the left reservoir represents the source, and the right reservoir, the drain, and where a gate electrode is located near the center of the channel. The material between the source and drain is a two-dimensional electron gas (typically generated by a semiconductor heterojunction; see Section 9.1.1), such that when we apply a negative voltage to the gate, we deplete the carriers near to the gate, and constrict the channel. Such a structure is known as a *quantum point contact*.

When $V_g = 0$, the channel is open, and the number of electronic channels (subbands) are determined by the width of the physical channel (for example, using (10.36) in a hard-wall model). Say, for instance, that for a given maximum electron energy E, current is carried in N subbands. Then, the conductance will be $G = G_0 N$. As the gate voltage is increased in the negative direction, the physical channel is narrowed, as depicted in Fig. 10.8 resulting in larger spacing between subband energies by (10.36). In this case, fewer subbands

†It is worth noting that the general Landauer formula (10.54) applies to tunnel junctions as well, such as depicted in Fig. 6.2 on page 185 , resulting in

$$G = \frac{2e^2}{h} T\left(E_F\right), \tag{10.59}$$

where $T\left(E_F\right)$ is the transmission coefficient (6.15) obtained from solving Schrödinger's equation, evaluated at the Fermi energy.

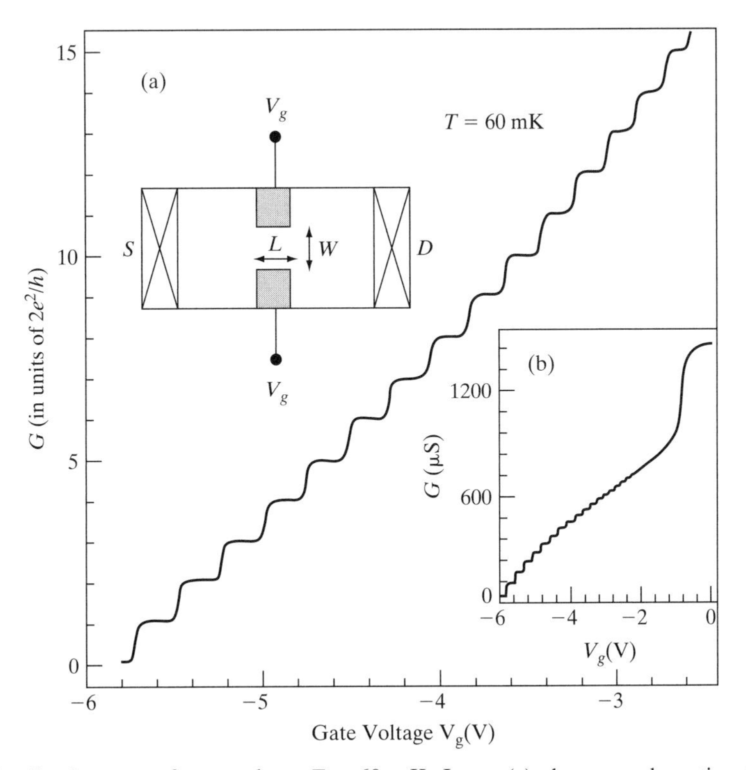

Figure 10.9 Conductance of a sample at $T = 60$ mK. Insert (a) shows a schematic of the split-gate device (not depicted is the electron gas connecting source and drain), and insert (b) shows the raw data. The width is 0.95 μm, and $L = 0.4$ μm. Although L and W are relatively large, ballistic transport occurs due to the low temperature, and commensurate long mean-free path. In this case, $\mu_e = 4.5 \times 10^6$ cm^2 V^{-1}s^{-1}. (Based on a figure from Thomas, K. J., et al., "Interaction Effects in a One-Dimensional Constriction," *Phys. Rev. B* 58 (1998): 4846. © 1998, American Physical Society.)

will be occupied, since only the subbands having $E_n \leq E$ will be populated (i.e., we hold E fixed while increasing the spacing between E_n levels). Thus, as the channel narrows, N decreases, and, therefore, the conductance decreases in integral units[†] of G_0. This forms the step-like behavior shown in Fig. 10.9.

Notice that as temperature increases, the observed quantization tends to vanish. This is due to thermal energy $k_B T$ becoming comparable to the subband energy spacing. For instance, at room temperature ($T = 293$ K), $k_B T \simeq 0.025$ eV. For a rectangular cross-section wire with infinite confining potential (hard-wall model), from (10.36) we have

$$E_{n_y,n_z} = \frac{\hbar^2}{2m}\left(\left(\frac{\pi n_y}{L_y}\right)^2 + \left(\frac{\pi n_z}{L_z}\right)^2\right). \tag{10.60}$$

[†]The observation of quantized conductance relies on assuming an *adiabatic* transition between the different regions, that is, that the transition is gradual enough not to engender scattering between subbands.

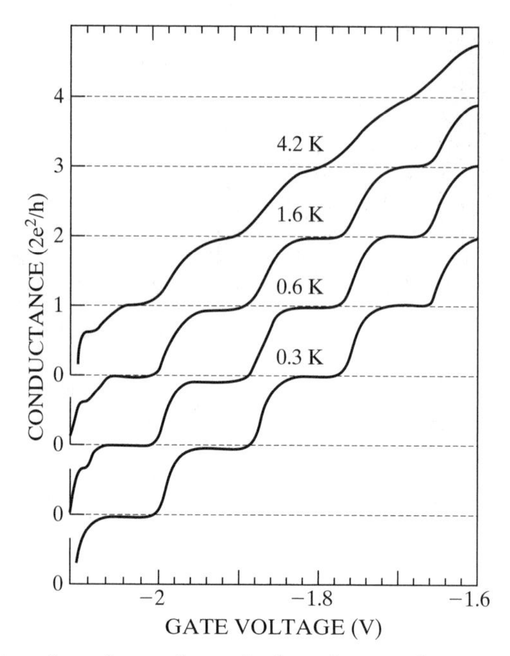

Figure 10.10 Temperature dependence of quantized conductance for a quantum point contact. The two-dimensional electron gas has a Fermi wavelength of 40 nm, and the width of the point contact is around 250 nm. (Based on Fig. 6 from van Wees, B.J., L.P. Kouwenhoven, E.M.M. Willems, C.J.P.M. Harmans, J.E. Mooij, H. van Houten, C.W.J. Beenakker, J.G. Williamson, and C.T. Foxon, "Quantum Ballistic and Adiabatic Electron Transport Studied with Quantum Point Contacts," *Phys. Rev. B* 43 (1991): 12431. © 1991, American Physical Society.)

Assuming $L_y = L_z = 10$ nm, then with $n_y, n_z = 1, 1$ and $n_y, n_z = 1, 2$ we have, $E_{1,1} = 0.0075$ eV and $E_{1,2} = 0.0188$ eV, such that $\Delta E = 0.0113$ eV, which is less then $k_B T$. Therefore, the subband quantization becomes "washed out." However, if $L_y = L_z = 1$ nm, then $E_{11} = 0.75$ eV and $E_{12} = 1.88$ eV, such that $\Delta E = 1.13$ eV, which is much greater than $k_B T$. In this case, the conductance quantization will be quite evident at room temperature. The temperature dependence of the conductance quantization of a quantum point contact is shown in Fig. 10.10, where quantization disappears quickly with increasing temperature due to the large size of the point contacts (around 250 nm).

Therefore, either temperature must be very low, or cross sections must be very small, for quantum effects to be noticed. Even for low temperatures, for paths longer than a few hundred nanometers, defects are likely to be present that destroy quantization.

Break Junctions. An interesting experiment involving break junctions can be performed to demonstrate the effect of conductance quantization in quantum wires. In a break junction, a wire having a relatively large radius is gradually pulled along the wire axis, and, as a result, the wire's radius becomes thinner as the wire stretches. Just before breaking, the wire's radius is on the order of an atom, and so at some point before breaking,

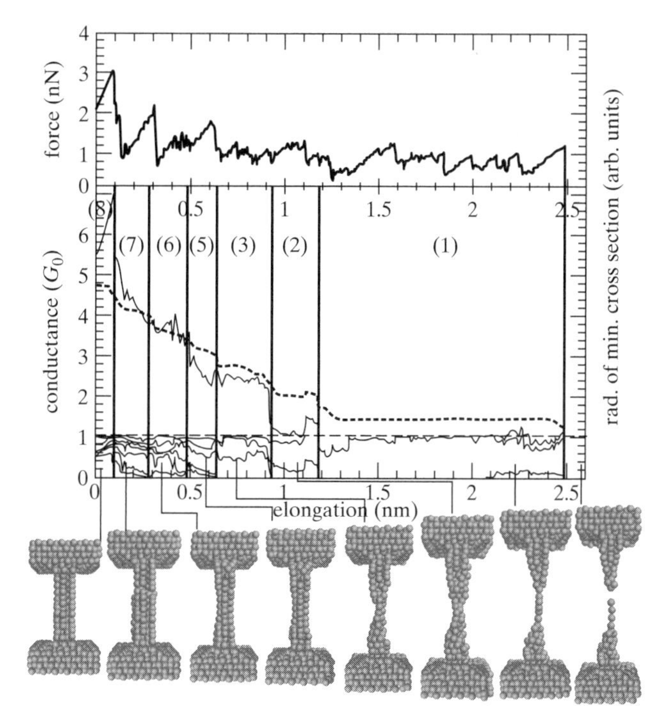

Figure 10.11 Theoretical simulation results from thinning a gold wire's radius by pulling the wire, at $T = 4.2$ K. The upper panel shows the strain forces as a function of the elongation of the contact. The middle panel shows the wire's conductance (solid line) and the radius of the minimum wire cross section (dashed line). The transmission coefficient of each channel is shown in the bottom panel. The vertical lines define regions with different numbers of open channels, ranging from 8 to 1. Breaking of the contact is depicted below the plot. (Based on a figure from Dreher, M.F., Pauly, J. Heurich, J.C. Cuevas, E. Scheer, and P. Nielaba, "Structure and Conductance Histogram of Atomic-Sized Au Contacts," *Phys. Rev. B* 72 (2005): 075435. © 2005, American Physical Society.)

one expects to see conductance quantization. These types of experiments have been widely performed, and quantization is indeed observed. In Fig. 10.11, the results of a molecular dynamics simulation are shown, and Fig. 10.12 provides measured results. In Fig. 10.11, elongation of the wire resulted in the formation of an atomic chain of gold atoms, reducing the number of channels to one.

Note that in these figures, the individual transmission coefficients of each channel are not generally unity, and so conductance values are not exactly integer multiples of G_0. Thus, the conductance (10.54) approximately applies to this case, rather than (10.55). For example, in Fig. 10.11, $G \simeq 2.4G_0$ when three channels exist, since the transmission coefficient of one of the channels is significantly below unity. Furthermore, although the number of modes decreases as the wire's radius decreases, at a given radius, the actual number of modes is a more complicated issue than given by the simple formula (10.38). This is due to the atomic structure of the wire when the wire's radius is reduced to only a few atoms.

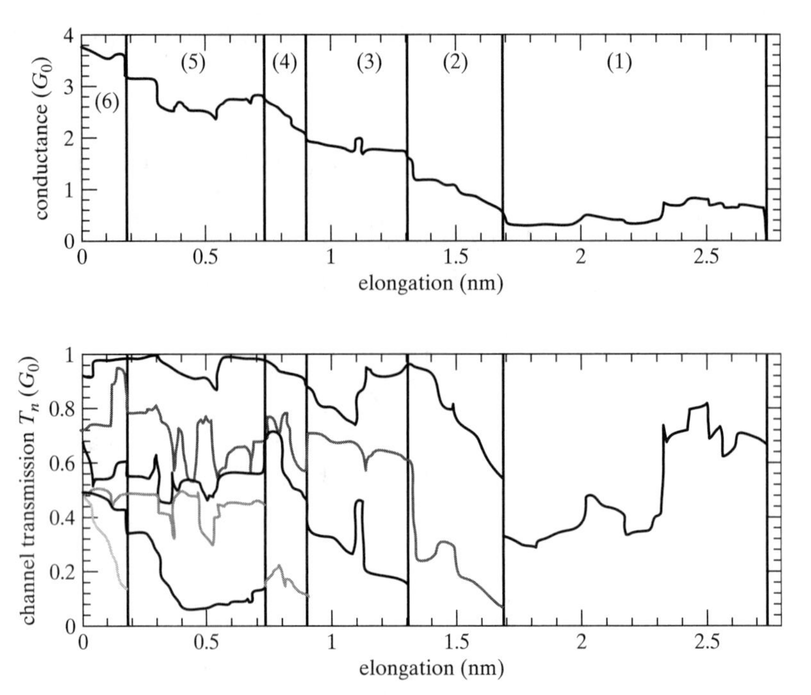

Figure 10.12 Measured total conductance of a gold wire (top panel) and the single-channel contributions (bottom panel) as a function of wire elongation. The vertical lines define regions with different numbers of channels, ranging from 6 to 1. Temperature was below 100 mK. (Based on a figure from Dreher, M.F., Pauly, J. Heurich, J.C. Cuevas, E. Scheer, and P. Nielaba, "Structure and Conductance Histogram of Atomic-Sized Au Contacts," *Phys. Rev. B* 72 (2005): 075435. © 2005, American Physical Society.)

10.2.4 Origin of the Quantum Resistance

Since $L \ll L_m, L_\phi$, there is generally no scattering along the conduction path, and we might have expected that the resistance of the channel would be zero (infinite channel conductance). That is, in our simple model, there are no collisions or interactions, and so there is no mechanism for energy dissipation. Therefore, there should not be any voltage dropped across the ballistic conductor, and, hence, resistance should be zero. However, the resistance obtained, (10.58), is not really the resistance of the channel, but of the contact between the physical channel and the electron reservoirs (similar to the junction between a large-diameter and small-diameter water hose). If the reservoirs are viewed as very wide material regions, one can think of them as having a large number of electronic channels. The small cross section physical channel has a small number of electronic channels, and so at the junction between the reservoir and the physical channel, there is a mismatch. Current passing through the reservoir, carried by the large number of electronic channels, must be redistributed among the small number of electronic channels available in the physically narrow channel, resulting in resistance. The resistance quantum, R_0, arises from perfect (infinitely wide) reservoirs in contact with a single electronic channel (i.e., a very narrow physical channel).[†] Indeed, one finds that the resistance of a ballistic channel is length independent, as long as $L \ll L_m, L_\phi$ is maintained. Furthermore, ballistic metal nanowires have been shown to be capable of carrying current densities much higher than bulk metals, due to the absence of heating in the ballistic channel itself.

10.3 CARBON NANOTUBES AND NANOWIRES

As discussed in Section 7.3.1, metallic carbon nanotubes are excellent conductors, and can exhibit d.c. ballistic transport over at least μm lengths. For an SWNT, when ballistic transport occurs, the resistance of the tube is length independent, and is, theoretically, approximately 6.45 kΩ. This resistance value results from having two propagation bands (called the $\pi-$bands) forming parallel propagation channels,[‡] where each channel has resistance equal to the resistance quantum, $R_0 \simeq 12.9$ kΩ. Thus,

$$G = \frac{2e^2}{h}N = \frac{4e^2}{h}, \tag{10.62}$$

$$R = \frac{h}{4e^2} = 6.45 \text{ k}\Omega. \tag{10.63}$$

Measurements on short tubes have verified this value, and values approaching (10.62) have been obtained in experiments on long tubes at lower temperatures. Note that, in contrast to solid metal quantum wires where the number of modes increase as the cross-sectional dimensions increase (e.g., (10.38)), for CNs the number of modes at the Fermi energy is two, independent of tube radius.

In practice, considerable care must be exercised to observe the resistance value 6.45 kΩ in carbon nanotubes, and, in fact, this is generally true in ballistic transport structures. The biggest experimental problem is to obtain sufficiently good (very low reflection) contacts to the tube, i.e., between the channel and the reservoirs. A multiwalled CN bridging two cobalt contacts is shown in Fig. 10.13.

In a metallic carbon nanotube the mean-free path is on the order of μm or more (often quoted values are in the range 1.3–1.7 μm) for low-bias voltages. The main source of resistivity for high-quality metallic CNs is scattering by acoustic phonons.[§] For lengths longer than the mean-free path, the resistance of a CN is length dependent. Several models

[†]Going in the other direction, we assume reflectionless contacts from the narrow physical channel to the large reservoirs.

[‡]If one wants to separate spin effects, carbon nanotubes can be considered to have four parallel channels (each of the two channels has two possible spin values), although in this case (10.55) becomes

$$G = \frac{e^2}{h}N \tag{10.61}$$

for an N channel system.

[§]Acoustic phonons are low-enegy phonons. The role of electron-phonon scattering in metallic carbon nanotubes is discussed in [22].

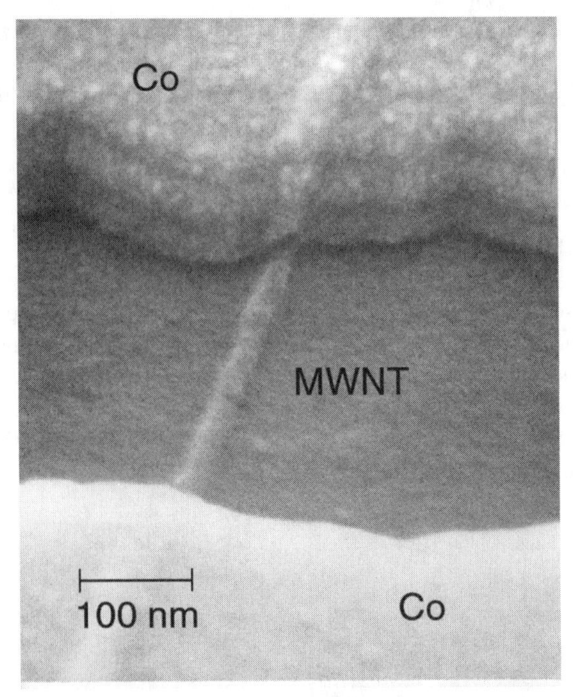

Figure 10.13 SEM image of a 15 nm radius MWNT bridging two cobalt contacts on an oxidized silicon wafer. The Co contacts lie on top of the MWNT; the image of the nanotube is seen through the Co layer due to the change in height of the evaporated film. This device was used for the study of spin transport (see the following section), although similar structures are used for d.c. and microwave charge transport characterization. (From Tsukagoshi, K., B.W. Alphenaar, and H. Ago, "Coherent Transport of Electron Spin in a Ferro Magnetically Contacted Carbon Nanotube," *Nature* 401 (1999): 572. Courtesy Macmillan Publishers Ltd. and Bruce Alphenaar.)

have been suggested, although in the low-bias regime (where voltages over the tube length are less than about one-tenth of a volt) a linear model,

$$R = \frac{h}{4e^2}\left(1 + \frac{L}{L_m}\right), \tag{10.64}$$

seems to work fairly at well least up to cm lengths. Figure 10.14 shows a plot of resistance versus tube length, where it can be seen that for longer tubes the resistance is approximately 6–7 $k\Omega$ per micron, in agreement with (10.64) if $L_m \simeq 1$ μm.

For high-bias voltages (above a few tenths of a volt) resistance becomes voltage dependent, and in typical SWNTs current tends to saturate around 25–30 μA. In high-quality tubes, this is mainly due to scattering with optical and zone-boundary phonons, which have energies on the order of 0.2 eV. This is because at high biases, electrons can gain enough energy to emit one of these phonons. In this case, the mean-free path can be as short as 10–30 nm. The length that an electron must travel to accelerate to this phonon

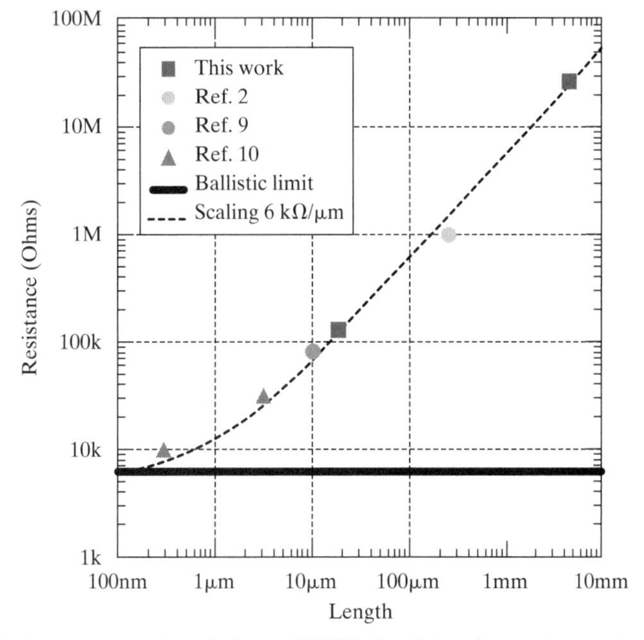

Figure 10.14 Resistance versus length for an SWNT. Symbols relate to the results from other work. All data are at room temperature. (Based on a figure from S. Li, Z.Y.C. Rutherglen, and P.J. Burke, "Electrical Properties of 0.4 cm Long Single-Walled Carbon Nanotubes," *Nano Lett.*, 4: 2003–2007, 2004. © 2004 American Chemical Society.)

range is†

$$L_h = \frac{E_{ph}}{e\,|\mathcal{E}|}, \tag{10.65}$$

where E_{ph} is the phonon energy and L_h is the high-bias mean-free path. Denoting the low-bias mean-free path by L_l, we see that *Matthiessen's rule* for combining mean free paths leads to

$$L_{mfp}^{-1} = L_l^{-1} + L_h^{-1}. \tag{10.66}$$

With this, CN resistance can be modeled by‡ (10.64), leading to

$$R = \frac{h}{4e^2}\left(1 + L\left(\frac{1}{L_l} + \frac{1}{L_h}\right)\right) \tag{10.67}$$

$$= \frac{h}{4e^2}\left(1 + \frac{L}{L_l}\right) + \frac{h}{4e}\frac{L\,|\mathcal{E}|}{E_{ph}} \tag{10.68}$$

$$= R_{\text{low-bias}} + \frac{V}{I_0}, \tag{10.69}$$

†This is obtained from energy being equal to force multiplied by distance.
‡See, e.g., [23] and [24].

where $V = L\,|\mathcal{E}|$ and $I_0 = 4eE_{ph}/h$. For $E_{ph} = 0.2$ eV, $I_0 = 30$ μA. It is easy to see that with this model current saturates, since

$$I = \frac{V}{R_{\text{low-bias}} + \frac{V}{I_0}}, \tag{10.70}$$

and for $V/I_0 \gg R_{\text{low-bias}}$,

$$I = \frac{V}{\frac{V}{I_0}} = I_0. \tag{10.71}$$

Although CNs can be grown that are nearly atomically perfect, defects do occur. One estimate is that high-quality, SWTNs have, on average,† one defect per 4 μm, although far fewer defects are possible. While even one defect per 4 μm is quite a small number (approximately one chemical bond out of a trillion is out of place, on a par with high-quality Si), the presence of a single defect can significantly influence electronic properties. Being a one-dimensional conductor, essentially current can't flow around the defect.‡ The conductance of a carbon nanotube in the presence of defects can vary substantially from the defect-free case, and defects obviously impact ballistic transport in CNs. Not surprisingly, defects are found more often in areas of tube curvature. Furthermore, defects are chemically active sites on the tubes, and seem to play a strong role in the sensitivity of CNs as gas sensors.

10.3.1 The Effect of Nanoscale Wire Radius on Wave Velocity and Loss

One important aspect of carbon nanotubes, and seemingly of all nanoradius conductors, is that the velocity of wave propagation along the structure is quite slow compared with macro- and even micron-scale conductors. Recall that although electrons move relatively slowly inside a conducting material (at approximately v_F, which for copper is 1.57×10^6 m/s $= 0.0052c$), for good conductors, electrical signals propagate outside the material as electromagnetic waves guided by the conductor. Therefore, wave velocities are approximately the speed of light in the medium outside of the wire. Assuming that $\mu_r = 1$, which is the usual case, we find that this velocity is given by $c/\sqrt{\varepsilon_r}$. Thus, for wires surrounded by free space, $v = c \simeq 3 \times 10^8$ m/s, and since ε_r for typical dielectric materials is in the range 1–10, signal propagation is usually close to c. This conclusion holds for macro- and micron-scale radius conducting wires and circuit traces.

However, as the radius of metal wires/interconnects decreases to the nanoscale, the signal velocity of the wave guided by the wire slows significantly. For example, Fig. 10.15 shows the normalized phase velocity of electromagnetic wave propagation, $s_r = v_p/c$, for an infinite carbon nanotube (radius $a = 2.712$ nm) and for various solid round copper wires

†See [13].

‡This is a general property of nanoscale devices; a very small number of defects, or a small amount of stray charge, can sometimes significantly affect device behavior.

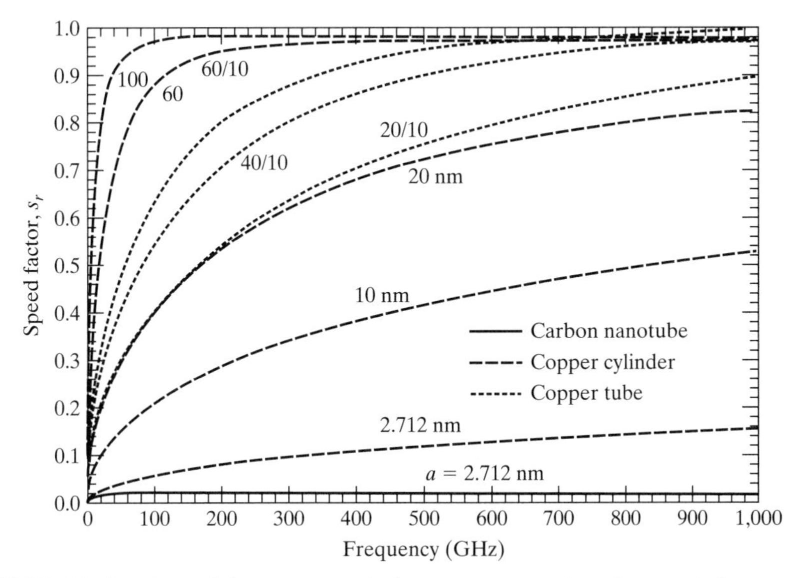

Figure 10.15 Predicted speed factor $s_r = v_p/c$ for wave propagation along an infinite carbon nanotube, and for various solid round copper wires and hollow copper tubes. For the CN and solid wires, a denotes the radius. For the tubular wires, results are labeled as a/d, where a is the outer radius and d is the tube wall thickness. (Based on a figure from Hanson, G.W. "Current on an Infinitely-Long Carbon Nanotube Antenna Excited by a Gap Generator," *IEEE Trans. Antennas Propagat.* 54 (2006): 76–81. © 2006, IEEE.)

and hollow copper tubes.† For the tubular wires, results are labeled as a/d, where a is the outer radius and d is the tube wall thickness. Results are for a single wire or tube in free space, i.e., no ground plane is present.

It can be seen that for relatively large radius copper wires, say, $a = 100$ nm or above, wave velocity is close to c, but for smaller radius wires and tubes, wave velocity is significantly decreased. Nevertheless, the carbon nanotube has the slowest wave velocity (although accounting for the decrease in copper conductivity at nanoradius values could change this conclusion). Theoretical analysis predicts that for waves guided by the CN, electromagnetic wave velocity is on the order of the Fermi velocity for electrons on the tube, $v_F \simeq 9.71 \times 10^5$ m/s. Different estimates yield the range $v_p \simeq 2v_F - 6v_F$, although all methods agree that wave velocity is quite slow, being several orders of magnitude below c. For example, assuming $v_p = 3v_F$, we have $v_p = 0.01c$. Given that interconnect delay presents a major bottleneck in chip performance, this slow speed is unfortunate (although, from Fig. 10.15, the same can be said about nanoradius copper interconnects). Signal propagation velocity can be increased by using larger-radius

†In Fig. 10.15, the results for the copper wires were obtained using the frequency-dependent conductivity of bulk copper. As described in Section 10.3.1, due to grain-boundary and surface scattering, impurities, and various fabrication aspects, the conductivity of nanoradius copper wires significantly decreases from the bulk value, and would tend to further slow wave propagation. However, the bulk value for copper conductivity is used here, since more realistic high-frequency values are not available.

metal wires, or by using, say, bundles of nanotubes, although this also brings about difficulties in connecting these larger-sized interconnects to devices having dimensions of a few nm.

However, signal speed is not the only figure of merit. Although the topic of IC interconnect performance will not be discussed in detail here, resistance and wire loss are obviously important. Considering again a single tube or wire in free space (as in Fig. 10.15) having $a = 2.712$ nm, it has been shown that a typical SWNT suffers roughly three times less power attenuation than a solid copper cylinder having the same radius,† even assuming the bulk value of conductivity for copper, which is unreasonably optimistic. Decreasing the conductivity by a factor of 10 from the bulk value, which is still probably not enough of a decrease to account for the small radius of the wire, we find that the copper wire has 10 times more power loss than the CN.

Of course, attenuation can be greatly reduced by using larger radius metal wires, say, $a \sim 100$ nm or more. Also, bundles on nanotubes can be used, reducing the resistance via (10.58), since each tube can be considered to add two propagation channels to the bundle. This not only decreases interconnect resistance and loss, but also increases speed, to the point that CN bundles can outperform larger-radius copper interconnects.‡ However, again it should be emphasized that to attach an interconnect to a nanoscopic circuit one may need to use interconnects with radius values on the order of several nm, and large-radius wires or bundles consisting of a large number of CNs may not be appropriate. Another consideration is that many nanoelectronic devices have high impedance (say, a tunneling device), and using a high impedance interconnect such as a single carbon nanotube may make impedance matching easier.

In summary, for nanoscale radius interconnects, a case can be made that carbon nanotubes seem to be superior to metal wires having similar radius values. This is due, in large part, to the difficulties in fabricating high-quality metal traces at the nanoscale (i.e., obtaining nanoradius metal interconnects that have material properties close to the bulk value). In contradistinction, high-quality CNs can be fabricated that are nearly defect free, and that perform close to their theoretical limits, although high-throughput volume manufacturing of electronic devices based on CNs is currently not possible.

10.4 TRANSPORT OF SPIN, AND SPINTRONICS

So far, the transport discussed in this chapter has been the transport of charge, and the electronics described in the preceding chapters have been based on charge. However, as described in Chapters 2 and 3, quantum particles also have spin. Unlike charge, spin is a purely quantum phenomenon, yet like charge, spin can be transported and/or used to store information. Devices based on spin (spin-based electronics, or *spintronics*) make up an important recent class of electronic devices. Current commercial applications include mass-storage devices such as hard drives, although logic and transistor applications, quantum computing technology, and a host of other applications are envisioned.

†See [12].
‡See [9], [24]

10.4.1 The Transport of Spin

A discussion of spin transport necessitates a discussion of magnetic materials. Although an everyday phenomenon, the theory of magnetism is extremely complicated, and in the following discussion, only a brief outline of the main ideas relevant to the transport of spin is provided.

The origin of magnetism is the orbital motion and spin of subatomic particles, with electrons being particularly important. It should be noted that all materials have some magnetic properties, and yet most materials are essentially nonmagnetic. That is, most materials have a very weak response to a magnetic field.

Materials can be divided into five basic groups according to the alignment of the magnetic moments of their atoms (due to orbital and spin momentum (Section 3.6)) in the presence of a d.c. magnetic field. Upon applying a magnetic field to a material, we find that the magnetic moments of the atoms tend to become oriented, inducing a net magnetization. Then, after the applied field is removed, this net magnetization may remain or disappear.

For example, in a *diamagnetic* material (DM), the induced magnetization disappears after the magnetic field is removed. That is, the net magnetic moment of an atom in a DM, due to orbital and spin momentum, is zero in the absence of an applied magnetic field. Application of a magnetic field produces a force (the Lorentz force, (3.212)) on the orbiting electrons, perturbing their angular momentum and creating a net magnetic moment. The resulting induced magnetic moments tend to oppose the direction of the applied field,[†] resulting in a relative permeability $\mu_r \simeq 0.9999$. Thus, diamagnetism is a very weak effect. Elements such as Ag and Pb are diamagnetic.

In a *paramagnetic* material (PM), atoms have a small net magnetic moment due to incomplete cancellation of the angular and spin momentums. Application of a magnetic field tends to align the magnetic moments in the direction of the applied field, resulting in $\mu_r \simeq 1.00001$. Thus, paramagnetism is also a very weak effect; for example, Al is paramagnetic. The overwhelming majority of materials are either diamagnetic or paramagnetic,[‡] and the effect is usually so weak that one often refers to these materials as nonmagnetic. This is why, unlike relative permittivity ε_r, which is almost always different than unity for materials, relative permeability $\mu_r \simeq 1$ is assumed for most materials. Note that a magnet will weakly attract PMs, and weakly repel DMs.

If the induced magnetization remains after the applied magnetic field is removed, the material is called a ferromagnetic material (FM). However, one does not necessary need to apply a magnetic field to align the magnetic moments in an FM—they may already be aligned. Ferromagnetism is the underlying principle of what one typically thinks of as a magnet. Ferromagnetic elements include Co, Fe, Ni, and their alloys, and these materials usually have[§] $\mu_r \gg 1$. Ferromagnetic materials become paramagnetic above a temperature known as the *Curie temperature*, which is above 600 K for Co, Fe, and Ni.

[†] In electromagnetics, this is known as *Lenz's law*.

[‡] All materials actually have a diamagnetic effect mainly from orbital motion of electrons, although often the diamagnetic effect is hidden by a slightly larger paramagnetic effect, or by a much larger ferromagnetic effect.

[§] However, this is an oversimplification, since relative permeability is usually nonlinear ($\mu = \mu(\mathbf{H})$, where $\mathbf{H}$ is the magnetic field intensity), and hysteric, and typically needs to be specified by a tensor since its value is different in different directions in the material.

Another class of materials is *antiferromagnetic*, such as chromium, Cr, below about 475 K, MnO, FeO, and others. These have ordered magnetic moments, although adjacent magnetic moments are antiparallel. A final category, *ferrimagnetic* materials (called *ferrites*, which are really compounds such as Fe_3O_4), have adjacent magnetic moments antiparallel but unequal. A depiction of the magnetic moments for the latter three material types is provided in Fig. 10.16.

In general, we are most interested in spin transport in FMs, or between ferromagnetic and nonmagnetic materials. In an FM, electronic current flow can be envisioned as occurring in two channels, one for spin-up electrons and the other for spin-down electrons. The density of states at the Fermi energy of the two channels differs in most ferromagnetics. Often the density of states for spin-up and spin-down electrons is nearly identical, but the states are shifted in energy, as depicted in Fig. 10.17.

The electrons having spin corresponding to the larger density of states at E_F are called the *majority carriers* (majority spin electrons), whereas the other spin electrons are called *minority carriers*. Each channel can be characterized by a conductivity ($\sigma_\uparrow$ for spin-up electrons, and $\sigma_\downarrow$ for spin-down electrons), and due to the asymmetry of the DOS for ferromagnetics, one channel will be dominant. That is, either $\sigma_\uparrow > \sigma_\downarrow$ or $\sigma_\uparrow < \sigma_\downarrow$ will occur. The most extreme example of asymmetry in the DOS is a *half-metallic ferromagnet*, which has only one spin channel (i.e., the density of states at the Fermi level is empty for one spin); an example is chromium dioxide, CrO_2. On the other hand, a nonmagnetic material has equal spin channels, as depicted in Fig. 10.17(a). The inherent asymmetry of spin channels in ferromagnetic materials means that they can be used as sources of spin-polarized currents.

Given the two-channel model for a ferromagnetic material, we can consider what happens when a spin-dominant electrical current passes from a ferromagnetic to a nonmagnetic material. (By nonmagnetic we actually mean a paramagnetic or diamagnetic material, obviously.) Consider the junction depicted in Fig. 10.18. On the FM side, there is strong

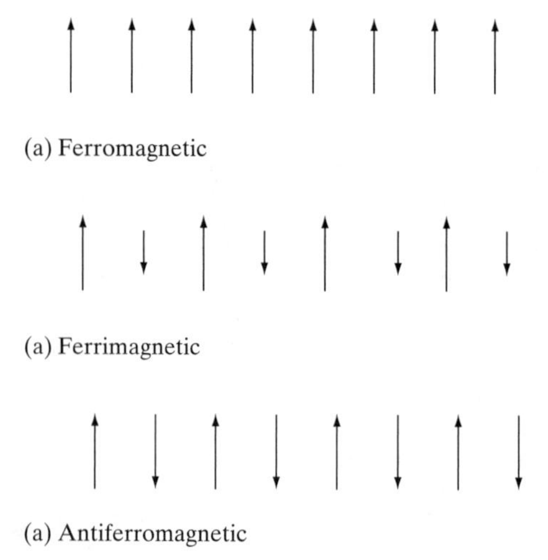

Figure 10.16 Depiction of the magnetic moments in (a) ferromagnetic, (b) ferrimagnetic, and (c) antiferromagnetic materials.

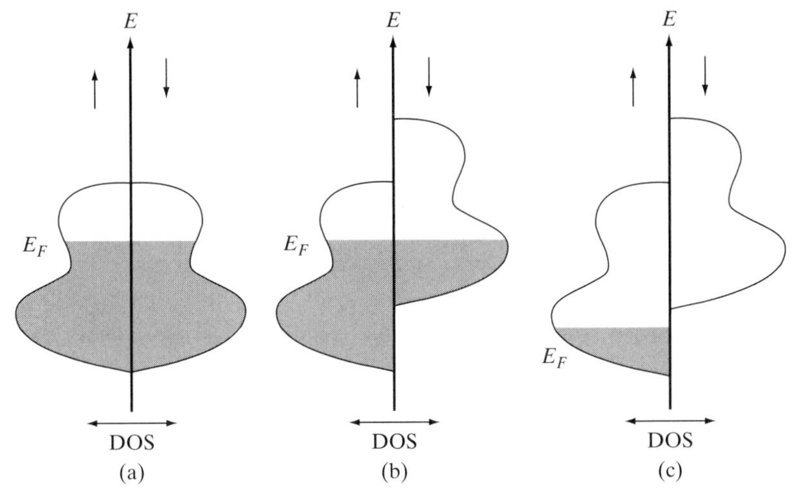

Figure 10.17 Density of states for spin-down electrons (indicated by the downward arrows), and for spin-up electrons (indicated by the upward arrows) for (a) a nonmagnetic material, (b) a ferromagnetic material, and (c) a half-metallic ferromagnet. A key parameter is the density of states at the Fermi energy.

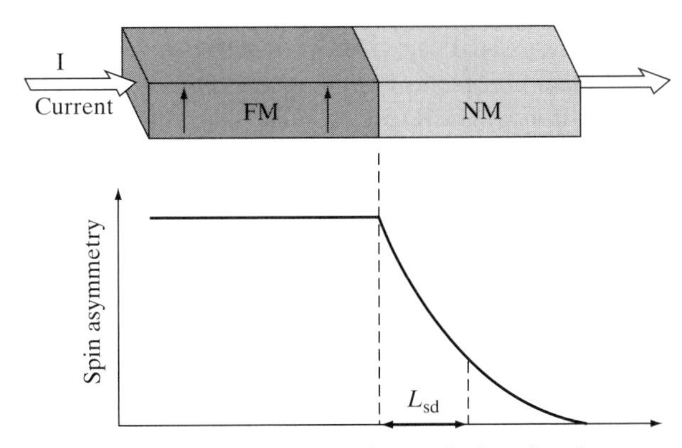

Figure 10.18 Depiction of spin asymmetry relaxation at the interface between a ferromagnetic and nonmagnetic material.

spin asymmetry. On the nonmagnetic material side, far from the interface, electrons will be carried in equal spin channels (assuming the nonmagnetic material is a conductor; if it is an insulator, there will be no current at all), and so there must be a region near to the interface where spin asymmetry decreases, as shown in the figure.

Spin is said to accumulate near the interface, and spin polarization decays exponentially away from the interface due to *spin-flip scattering events* on the scale of what is called the *spin diffusion length*, L_{sd}. Thus, spin-flip scattering events destroy the transport of spin, leading to a characteristic length. This is analogous to momentum scattering events that lead to the concept of the mean free path.

The spin diffusion length can be roughly estimated as follows. Consider that, upon entering the nonmagnetic material from the FM, any given electron will undergo a number of collisions (say, N) before flipping its spin. These collisions are the same as described previously in Section 10.1, so that the average distance between collisions is the mean-free path, L_m. Without going into the details of current diffusion, it can be shown that the average distance that spin penetrates into the nonmagnetic material is[†] $L_m\sqrt{N/3}$, which is the desired spin diffusion length L_{sd}. To eliminate N, consider that the total distance traveled by the electron is $NL_m = v_F\tau_{\uparrow\downarrow}$, where v_F is the Fermi velocity of electrons and $\tau_{\uparrow\downarrow}$ is called the *spin-flip time* (i.e., the characteristic time it takes an electron to flip its spin, analogous to the momentum relaxation time τ for charge transport). Therefore,

$$L_{sd} = \sqrt{\frac{v_F\tau_{\uparrow\downarrow}L_m}{3}}. \tag{10.72}$$

Typical values of L_{sd} are on the order of several hundred nm, with spin-flip times on the order of tens of picoseconds (ps) at room temperature. In comparison, the typical mean free path in metals is on the order of tens of nm (e.g., for copper at room temperature, $L_m \simeq 40$ nm), and momentum relaxation times are on the order of a hundredth of a ps (e.g., for copper, $\tau = .025$ ps). Thus, spin polarization is preserved for a relatively long time and over relatively large distances. However, impurities in the nonmagnetic material will increase collisions, and can greatly lessen the spin diffusion length.

The most common practical application of spin polarization involves what is called the *giant magnetoresistance* (GMR) effect. It can be most easily understood by considering two layers of the same FM, such as Co, separated by a very thin (on the order of nm) nonmagnetic conducting spacer material, such as Cu. The geometry is depicted in Fig. 10.19.

Each ferromagnetic layer has a magnetization vector **M**, and when these vectors are parallel,[‡] an electrical current can pass through the device. This is due to several reasons. First, since the spacer layer is thin (much thinner than the spin-diffusion length), spin can diffuse across the spacer to reach the other FM. Since that material's magnetization is parallel to the magnetization of the first layer, the density of states for those spin electrons is relatively high, and so there is low scattering. Thus, the flow of electrons can occur, resulting in a low electrical resistance. However, if the magnetization vector of the second FM is antiparallel with that of the first material, the majority-spin electrons see a small density of states in the second magnetic layer, and spin scattering is high. This is reflected by a much larger resistance to current flow,[§] depicted in Fig. 10.19. This resistance to current flow is called *magnetoresistance*, and the percentage ratio of the large and small resistance values is called the *GMR ratio*. Devices have been demonstrated with very high GMR ratios

[†]This comes from the theory of *random walks*, where the mean square distance that a particle moves in executing N steps of a random walk, with mean free path L_m, is NL_m^2.

[‡]We assume that the magnetization vectors can be set to be parallel or antiparallel by external magnetic fields.

[§]In the limiting case of two ideal half-metallic ferromagnets, if the magnetizations are antiparallel, resistance is infinite since the density of states in the last layer is empty for the electrons spin polarized from the first layer.

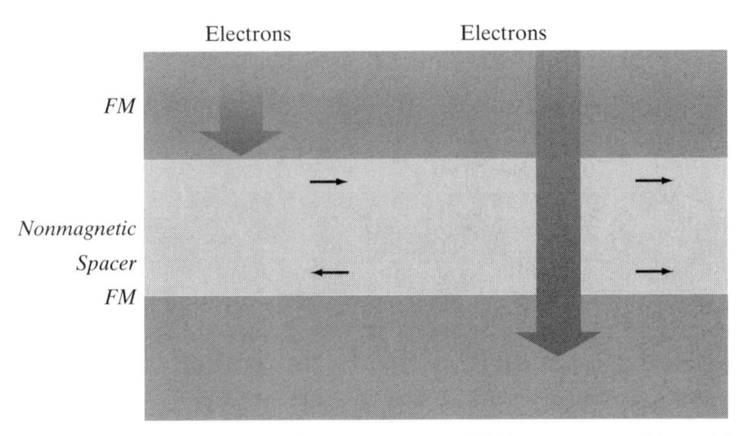

Figure 10.19 Depiction of the GMR effect. Two layers of FM are separated by a thin conducting (nonmagnetic) spacer. When the magnetization vectors in the two layers are parallel, an electrical current can easily pass through the device. If the magnetization vectors are antiparallel, then the resistance to current flow is much higher.

(hence the term "giant" in giant magnetoresistance), on the order of several hundred percent, although typical values are much less. Of course, this is a simplistic description, and various interface effects also play a role in the resistance obtained in such devices.

It should also be noted that current flow parallel to the layers is possible, and GMR can be obtained. In this case, however, the mean-free path is the important length parameter, and the GMR effect relies on the different mobilities of the spin-up and spin-down electrons, which are related to the two different densities of states. The non-magnetic layer is sufficiently thin (less than the mean-free path) so that current flow is spread across the multiple magnetic layers. If the layers have antiparallel magnetizations, each spin type will experience heavy scattering in one of the layers, resulting in relatively large resistance. However, if the layers have parallel magnetizations, one spin type will be heavily scattered in both layers, and the other spin type will experience little scattering, and a low resistance will result.

A related effect is when the spacer layer is insulating, and very thin. Electrons can tunnel through the insulating layer (spin is generally conserved in tunneling), and it can be considered that tunneling of spin-up and spin-down electrons are independent processes forming two independent spin tunneling channels. If the two ferromagnetic films are magnetized parallel to each other, the minority-spin electrons tunnel to minority states in the second ferromagnetic layer, and the majority-spin electrons tunnel to the majority states. If the two ferromagnetic layers have antiparallel magnetizations, the majority-spin electrons must tunnel to the minority states in the other layer. This forms a *magnetic tunnel junction*, the resistance of which (called the *tunneling magnetoresistance*) depends on the orientation of the two magnetization vectors and the spins of the incident electrons.

10.4.2 Spintronic Devices and Applications

A very important device based on the GMR effect is called the *spin valve*, depicted in Fig. 10.20.

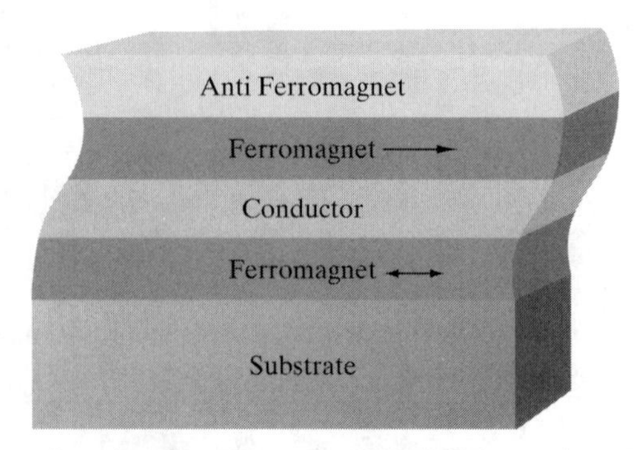

Figure 10.20 Depiction of a spin valve. (Courtesy of WTEC Inc. and Dr. James M. Daughton/NVE Corporation. Based on a graph by Professor Eshel Ben-Jacob.)

In this case, one of the ferromagnetic layers has a fixed (pinned) magnetization direction, due to the presence of the antiferromagnetic layer, and the other ferromagnetic layer magnetization is free to rotate upon application of a magnetic field (depicted by the double-headed arrow). An applied field, perhaps due to a magnetic bit on a hard drive, rotates the magnetization vector of the free layer, and when the two magnetization vectors are aligned, a minimum in device resistance is achieved. When the applied magnetic field results in antiparallel magnetization vectors, resistance of the device is greatly increased. Because of the strong GMR effect, GMR-based devices can be used as extremely sensitive magnetic read heads, allowing the storage capacity of a hard disk to increase considerably. Since the introduction of GMR read heads in 1997, hard disk density has increased greatly each year. Nearly all hard disk drives currently incorporate GMR read heads.

Mention must also be made of spin-based transistors. Considering the three layer structures already examined (i.e., ferromagnetic-nonmagnetic-ferromagnetic), it is not hard to envision adding a gate contact to the middle layer and forming a three-terminal transistor-like device, either in bipolar or field effect form. The middle layer may be made from a variety of materials, from normal metals to semiconductors, and applications of other materials, such as ferromagnetic semiconductors, are being activity considered. Furthermore, spin transistors have even been proposed using nonmagnetic semiconducting heterostructures, although in all cases there are many technological issues to be dealt with before spin-based transistors could be commercially viable. Other applications in spintronics include solid-state nonvolatile memories (since in many cases, spin states are retained after power is removed) and quantum information processing and quantum computation.

10.5 MAIN POINTS

In this chapter, we have considered classical and ballistic transport in materials, and the transport of spin. In particular, after studying this chapter, you should understand

- the classical theory of electrical conduction, including the concepts of conductivity and mobility, and the role of scattering;
- the semiclassical (Fermi) model of conductivity, and the idea of resistance;

- the idea of ballistic transport, including the various length scales (mean-free path, decoherence length, etc.) that define different transport regimes;
- the quantization of conductance and resistance;
- ballistic transport on CNs;
- size-dependent effects in nanowires;
- the basic ideas of spintronics, including the GMR effect and the operating principle of spin valves.

10.6 PROBLEMS

1. Compare the thermal velocity at room temperature with the Fermi velocity for a one-dimensional conductor having $N = 2 \times 10^7$ electrons/cm.
2. Determine the drift velocity for an electron in a material having momentum relaxation time $\tau = 2 \times 10^{-13}$ s, for an applied electric field $|\mathbf{E}| = 100$ V/m. If there are 3.4×10^{22} electrons per cm^3, what is the conductivity?
3. Another method to obtain electron drift velocity follows from the usual treatment of elementary differential equations modeling mass-spring systems; we can account for the effect of collisions by including a damping term in (10.5), resulting in

$$m_e \frac{d\mathbf{v}_d}{dt} + \frac{m_e \mathbf{v}_d}{\tau} = q_e \mathbf{E}. \tag{10.73}$$

   Solve (10.73) to determine the frequency-dependent version of the conductivity (10.12), assuming time-harmonic conditions.

   Comparing with the usual frictional term in a damped mass-spring system, we find that m_e/τ is analogous to a coefficient of friction. It is easy to see that (10.8) is a solution of (10.73).
4. Consider a two-dimensional electron gas in GaAs, with $n = 10^{12}$ cm^{-2}, and mobility 8, 200 cm^2/Vs. Determine the mean-free path from the semi-classical model, using $0.067 m_e$ as the effective mass of electrons in GaAs.
5. We can consider a thin metal sheet as being two-dimensional if the thickness of the sheet, d, is sufficiently small such that we can ignore electron movement perpendicular to the plane of the sheet. One way to assess this issue is to compare the two-dimensional Fermi energy, $E_F^{(2d)}$, given by (8.39), with the energy ΔE required to excite an electron above the ground state in a one-dimensional hard-wall quantum well of size d. Using (4.35) we find that,

$$\Delta E = E_2 - E_1 = \frac{\hbar^2 \pi^2}{2 m_e d^2} \left(2^2 - 1^2\right)^2. \tag{10.74}$$

   If $\Delta E \gg E_F$, the metal sheet is approximately two dimensional. This comparison is meaningful since the most important electrons will have energy equal to the Fermi energy, and if movement perpendicular to the plane of the sheet is to take place, the electrons must gain enough energy to move to a higher (perpendicular) state.

(a) Consider copper having $N_e \simeq 8.45 \times 10^{28}$ /m^3, such that the two dimensional density of electrons is approximately $N_e^{2d} = N_e^{2/3} = 1.93 \times 10^{19}$ /m^2. If the sheet thickness is $d = 0.2$ nm, is the sheet approximately two dimensional?

(b) Repeat (a) for a sheet with $d = 0.7$ nm.

6. One can approximately ignore quantization, and treat a layer of material as being three dimensional, if the (three-dimensional) Fermi wavelength is small compared with the sheet thickness d. Assuming we require

$$\lambda_F \leq 10d \tag{10.75}$$

in order to ignore quantization, determine the required sheet thickness that a copper sheet should have in order to be able to treat it as three dimensional. Repeat for n-type doped Si, assuming $N \simeq N_d = 1.5 \times 10^{16}$ cm^{-3}.

7. Do problems 10.5 and 10.6 provide self-consistent criteria for when a thin sheet can be considered to be two dimensional?

8. Using (10.30) with $p = 0.49$, $R_c = 0.27$, and d equal to wire width w, plot ρ as a function of wire width for a square cross-section copper wire.

9. A conducting wire carrying a signal is surrounded by teflon ($\varepsilon \simeq 2.5\varepsilon_0$). If the signal generates an a.c. 10 V/m electric field, thereby accelerating electrons, what is the approximate velocity of signal propagation?

10. Figures 10.1 (page 322) and 10.2 (page 323) show the Fermi surface in momentum space for a material. Draw a similar picture for a one-dimensional conductor.

11. Consider two concentric spherical conducting shells. The inner shell has radius a, the outer shell has radius b, and the material between the shells has conductivity σ. If the outer shell is grounded, and the inner shell is held at potential V_0, then elementary electrostatics shows that the resulting electric field in the space between the shells is†

$$\mathbf{E}(r) = \mathbf{r}\frac{V_0}{\frac{1}{a} - \frac{1}{b}}\frac{1}{r^2}. \tag{10.78}$$

Using this electric field, determine the resistance R seen between the two shells using (10.27).

†To derive (10.78), solve Laplace's equation,

$$\nabla^2 V = \frac{1}{r^2}\frac{\partial}{\partial r}\left(r^2\frac{\partial V}{\partial r}\right) = 0 \tag{10.76}$$

subject to $V(a) = V_0$ and $V(b) = 0$, then compute the electric field as

$$\mathcal{E} = -\nabla V = -\mathbf{r}\frac{\partial V}{\partial r}. \tag{10.77}$$

12. In Chapter 5, it was stated that collisions are necessary to frustrate Bloch oscillations, otherwise a d.c. voltage would result in an a.c. current. In this chapter, we have learned that if no collisions take place because $L \ll L_m, L_\phi$, d.c. ballistic transport will occur. These two facts seem to be in contradiction. Why are they not?

13. Consider a 20 nm length of a ballistic conductor, carrying $N = 4$ electron modes. Determine the current that will flow if a 0.3 V potential difference is applied across the length of the conductor. Assume low temperature, and that $T_n = 1$.

14. Consider a 5 nm length of a ballistic conductor, having square cross section of side 1 nm. If $E_F = 3.5$ eV, and $m^* = m_e$, determine the current that will flow if a 0.2 V potential difference is applied across the length of the conductor. Assume low temperature.

15. Consider a 50 nm length of a ballistic conductor, having square cross section of side 0.8 nm. If $E_F = 2.5$ eV, $m^* = m_e$, determine the current that will flow if a 0.2 V potential difference is applied across the length of the conductor. Assume low temperature.

16. Describe the difference between the physical channel and electronic channels in a quantum wire.

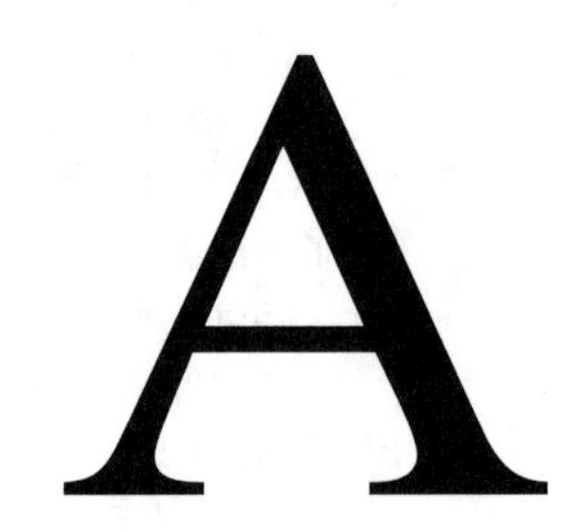

Appendix A

SYMBOLS AND ACRONYMS

List of symbols used in text.

Symbol	Meaning
a	lattice constant
a_0	Bohr radius
$\mathbf{a}$	acceleration
$\mathbf{a}_x$, $\mathbf{a}_y$, $\mathbf{a}_z$	rectangular coordinate unit vectors
$\mathbf{B}$	magnetic flux density
c	speed of light in vacuum
C	capacitance
e	proton charge
$\mathbf{E}$	electric field intensity
$\mathcal{E}$	electric field magnitude
E	energy
E_F	Fermi energy
E_g	gap energy
E_c	conduction bandedge energy (or charging energy)
E_v	valence bandedge energy
$\mathbf{F}$	force
f	frequency
G	conductance
h	Planck's constant
$\hbar$	Planck's reduced constant
H	Hamiltonian
H_n	Hermite polynomials
i	imaginary unit
I	current (or optical intensity)
$\mathbf{J}$	current density
$\mathbf{k}$, k	wavevector, wavenumber
k_F	Fermi wavenumber
k_B	Boltzmann's constant
L_m	mean-free path
L_ϕ	phase-coherence length
L_{sd}	spin-diffusion length
m	mass
m_e	electron rest mass
m_e^*	electron effective mass
m_h^*	hole effective mass
m^*	general effective mass
m_p	proton rest mass
μ_0	permeability of free space
μ	permeability (or chemical potential)
$\mu_{e(h)}$	electron (hole) mobility
Ψ, ψ	quantum mechanical state function
ω	radian frequency

$[\cdot,\cdot]$	commutator	$\mathbf{v}_g,\ v_g$	group velocity
$\langle\cdot\rangle$	expectation	$\mathbf{v}_p,\ v_p$	phase velocity
$N(E)$	density of states	$\mathbf{v}_T,\ v_T$	thermal velocity
$\mathcal{O}$	observable	$\mathbf{v}_d,\ v_d$	drift velocity
$\widehat{o}$	operator	V	potential energy (or voltage)
$\mathbf{p},\ p$	momentum	W	work
P	probability (or power)	$e\phi$	work function
q_e	electron charge	$e\chi$	electron affinity
Q	charge	ε_0	permittivity of free space
$\mathbf{r}\ (r)$	position vector (or distance)	ε_r	relative permittivity
R	resistance	ε_n	Neumann's number
R_0	resistance quantum	ρ	probability density function (or charge density)
R_t	tunnel resistance		
t	time	λ	wavelength
T	temperature (or period or tunneling probability)	λ_F	Fermi wavelength
		λ_n	eigenvalue
$\mathbf{v},\ v$	velocity	σ	conductivity
v_F	Fermi velocity	$\tau_{\uparrow\downarrow}$	spin-flip time

List of acronyms used in text.

AFM	atomic force microscope
CMOS	complementary MOS
CN	carbon nanotube
CVD	chemical vapor deposition
DOS	density of states
DM	diamagnetic material
EUV	extreme ultraviolet
FET	field-effect transistor
FM	ferromagnetic material
HEMT	high electron mobility transistor
IC	integrated circuit
NIL	nanoimprint lithography
MOS	metal–oxide semiconductor
MOSFET	metal–oxide-semiconductor field-effect transistor
PM	paramagnetic material
QD	quantum dot
SEM	scanning electron microscope
SET	single-electron tunneling
SET	single-electron transistor
STM	scanning tunneling microscope
TEM	transmission electron microscope
TJ	tunnel junction

Appendix B

Physical Properties of Materials

I. Fundamental constants.

Quantity	Symbol	Value
Electron charge	q_e	-1.602×10^{-19} C
Proton charge	e	1.602×10^{-19} C
Speed of light in vacuum	c	2.998×10^{8} m/s
Planck's constant	h	6.626×10^{-34} J s
Planck's reduced constant	$\hbar \; (= h/2\pi)$	1.055×10^{-34} J s
Boltzmann's constant	k_B	1.381×10^{-23} J/K
Permittivity of free space	ε_0	8.854×10^{-12} F/m
Permeability of free space	μ_0	$4\pi \times 10^{-7}$ H/m
Electron rest mass	m_e	9.110×10^{-31} kg
Proton rest mass	m_p	1.673×10^{-27} kg

II. Fermi energy, free electron density, and conductivity for some elements.

Element	E_F (eV)	N_e^{3d} ($\times 10^{28}/m^3$)	σ ($\times 10^7/\Omega m$)
Cs (Cesium)	1.58	0.91	0.50
Rb (Rubidium)	1.85	1.15	0.80
K (Potassium)	2.12	1.40	1.39
Na (Sodium)	3.23	2.65	2.11
Ba (Barium)	3.65	3.20	0.26
Sr (Strontium)	3.95	3.56	0.47
Ca (Calcium)	4.68	4.60	2.78
Li (Lithium)	4.72	4.70	1.07
Ag (Silver)	5.48	5.85	6.21
Au (Gold)	5.51	5.90	4.55
Cu (Copper)	7.00	8.45	5.88
Mg (Magnesium)	7.13	8.60	2.33
Cd (Cadmium)	7.46	9.28	1.38
In (Indium)	8.60	11.49	1.14
Zn (Zinc)	9.39	13.10	1.69
Pb (Lead)	9.37	13.20	0.48
Ga (Gallium)	10.35	15.30	0.67
Al (Aluminum)	11.63	18.06	3.65
Be (Beryllium)	14.14	24.20	3.08

All values for room temperature except for Na, K, Rb, and Cs at 5 K and Li at 78 K.

III. Bandgap properties of several important semiconductors.*

Crystal	Gap type	E_g (eV) @ 0 K	E_g (eV) @ 300 K
Si	I	1.17	1.11
Ge	I	0.74	0.66
GaAs	D	1.52	1.43
InP	D	1.42	1.27
AlAs	I	2.23	2.16
CdS	D	2.58	2.42
CdSe	D	1.84	1.74
CdTe	D	1.61	1.44
ZnS	—	3.91	3.60

*For gap type, I indicates an indirect bandgap semiconductor, and D a direct gap semiconductor. (Table repeated from Section 5.4.4.) Data from Kittel, C. (1986). *Introduction to Solid State Physics*, 6th ed., New York: Wiley.

IV. Effective mass.

(m_e^* is the electron effective mass, and m_h^* is the hole effective mass.)

Semiconductor	Effective m_e^*/m_e	Mass m_h^*/m_e
Ge	0.12	0.04
		0.28
		0.08
Si	0.26	0.50
		0.24
GaAs	0.067	0.50
		0.08
GaP	0.35	0.50
CdSe	0.13	0.45
InP	0.073	0.40
		0.08
InSb	0.015	0.39
		0.02

V. Lattice constant for some compound (III-V and II-VI) semiconductors.

(The lattice constant (a) for Si is 0.543 nm, and for Ge, 0.566 nm.)

Material	a (nm)	Material	a (nm)	Material	a (nm)
ZnS	0.541	CdS	0.583	ZnTe	0.610
AlP	0.545	HgS	0.585	GaSb	0.610
GaP	0.545	InP	0.587	AlSb	0.614
GaAs	0.565	CdSe	0.605	HgTe	0.646
AlAs	0.566	InAs	0.606	CdTe	0.648
ZnSe	0.567	HgSe	0.608	InSb	0.648

VI. Properties of two semiconductor alloys.

$Ga_{1-x}Al_xAs$	$Cd_{1-x}Mn_xTe$
$E_g = (1.426 + 1.247x)$ eV	$E_g = (1.606 + 1.587x)$ eV
$m_e^* = (0.067 + 0.083x)\, m_e$	$m_e^* = (0.11 + 0.067x)\, m_e$
$m_{hh}^* = (0.62 + 0.14x)\, m_e$	$m_{hh}^* = \left(0.60 + 0.21x + 0.15x^2\right) m_e$
$\varepsilon_r = 13.18$	

VII. Static relative dielectric constant ε_r ($\varepsilon = \varepsilon_r \varepsilon_0$).

Material	ε_r
Si	11.7
Ge	15.8
GaAs	13.13
AlAs	10.1
AlSb	10.3
InP	12.37
InSb	17.88
CdSe	9.4
CdTe	7.2

VIII. Element symbols and names.

Symbol	Element	Symbol	Element	Symbol	Element
Ag	Silver	Ge	Germanium	Pd	Palladium
Al	Aluminum	H	Hydrogen	Pt	Platinum
Ar	Argon	He	Helium	Pu	Plutonium
As	Arsenic	Hg	Mercury	Ra	Radium
Au	Gold	I	Iodine	Rn	Radon
B	Boron	In	Indium	S	Sulfur
Ba	Barium	Ir	Iridium	Sb	Antimony
Be	Beryllium	K	Potassium	Sc	Scandium
Bi	Bismuth	Kr	Krypton	Se	Selenium
Br	Bromine	La	Lanthanum	Si	Silicon
C	Carbon	Li	Lithium	Sn	Tin
Ca	Calcium	Mg	Magnesium	Sr	Strontium
Cd	Cadmium	Mn	Manganese	Ti	Titanium
Ce	Cerium	Mo	Molybdenum	Tl	Thallium
Cl	Chlorine	N	Nitrogen	U	Uranium
Co	Cobalt	Na	Sodium	W	Tungsten
Cr	Chromium	Ne	Neon	Xe	Xenon
Cs	Cesium	Nb	Niobium	Y	Yttrium
Cu	Copper	Ni	Nickel	Zn	Zinc
F	Fluorine	O	Oxygen	Zr	Zirconium
Fe	Iron	P	Phosphorus		
Ga	Gallium	Pb	Lead		

Appendix

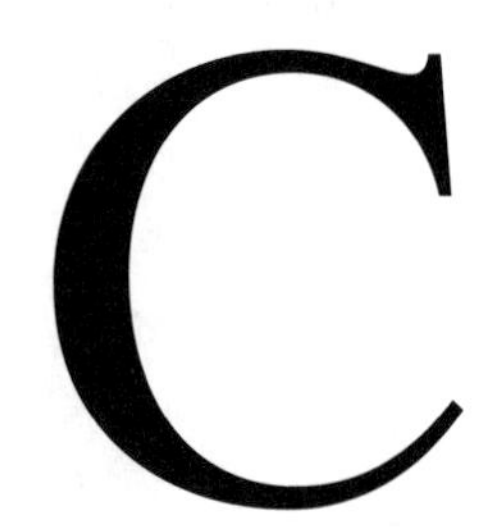

Conventional MOSFETs

As described in the text, MOSFETs are the workhorses of the electronics industry, a fact that is unlikely to change soon. Therefore, in this appendix, the basic device operation of a conventional n-type enhancement mode MOSFET is briefly reviewed. Short-channel and hot-electron effects, device scaling, and other issues are ignored, and the emphasis is simply on gaining a physical understanding of device operation.

Consider the usual n-type MOSFET structure shown in Fig. C.1, where s, d, and g indicate the source, drain, and gate electrodes, respectively. There is also often a substrate contact at the bottom of the device (called the body contact), which is typically connected to the source. Here we consider a simplified structure without this contact. The circuit symbol of the n-channel FET is shown in Fig. C.2. The oxide layer is conventionally formed by oxidizing the silicon substrate, forming an SiO_2 insulating barrier between the gate electrode and the rest of the device, effectively blocking d.c. current. However, if the oxide is too thin (on the order of 1–2 nm), tunneling can occur. High dielectric constant insulators have been developed for MOS devices that can replace the SiO_2 layer, allowing for a thicker layer resulting in less tunneling.

As shown in Fig. C.1, there is no conduction channel between the source and the drain (both n-type Si) when the gate voltage V_{gs} is zero. (Throughout this appendix, we assume that the source is grounded.) Thus, $I_{ds} = 0$ irrespective of V_{ds}. When a positive gate voltage $V_{gs} > 0$ is applied, and has sufficient magnitude,[†] an inversion layer is formed under the gate, connecting the source and drain. (This is called an *enhancement type* MOSFET.) The inversion layer is formed since positive voltage V_{gs} pushes away holes, and attracts electrons under the gate, forming, in effect, an n-type channel, as shown in Fig. C.3. This

[†] In order to induce enough of a channel we need $V_{gs} > V_t$, where V_t is called the *threshold voltage*, which is typically on the order of a volt.

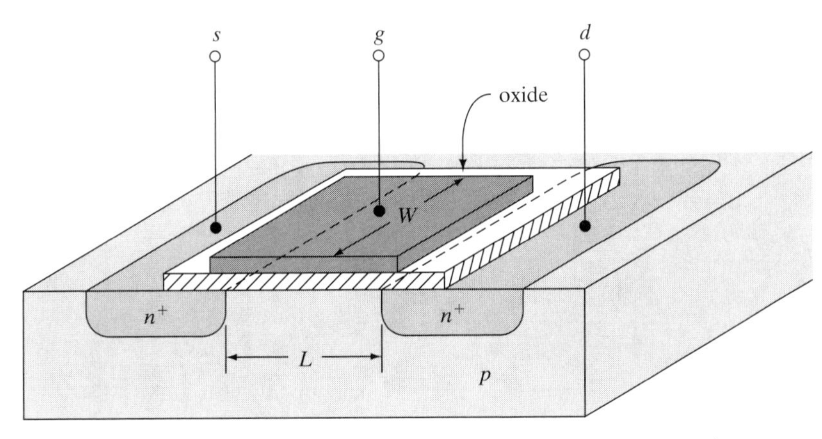

Figure C.1 Physical structure of an n-type MOSFET. The regions denoted as n^+ are heavily doped n-type Si, and p denotes the p-type Si substrate. For the gate electrode, typically poly crystalline silicon (*polysilicon*) is used for a variety of technical reasons, including leading to a lower threshold voltage.

allows current flow in the induced channel from drain to source upon applying a potential[†] $V_{ds} > 0$.

From an energy barrier viewpoint, there are barriers in two directions. A vertical energy barrier is between the gate and the substrate. (The barrier is basically the oxide.) A horizontal barrier is between the source and the drain. (The barrier is essentially the p-type substrate between the two n-type contacts.) Both the gate–oxide–substrate junction and the source-substrate-drain junction have energy profiles similar to the metal–insulator–metal junction depicted in Fig. 6.8 on page 193, although band bending actually occurs in the semiconducting regions. We will assume that the gate–oxide–substrate barrier is sufficiently high such that no current can flow from the gate into the device. However, the gate voltage plays a critical role. As V_{gs} is increased, forming an n-type channel as described earlier, the source–drain–substrate barrier is pushed down (Section 5.4.3), eventually below the filled states of the source and drain, and conduction can take place.

For an n-type MOSFET, device characteristics can be summarized as follows:

1. *Cut-off region*: When $V_{gs} < V_t$ the device is off. There is no conduction channel between the drain and the source, and, therefore,

$$I_{ds} = 0,$$

although weak current can arise due to what is called *subthreshold leakage* relating to minority carrier transport.

[†] The name "source" and "drain" is due to the fact that electrons flow from the source to the drain when $V_{ds} > 0$, current flow being defined in the opposite direction.

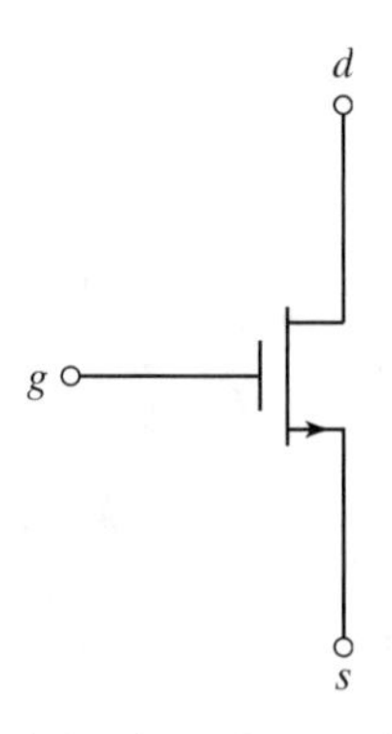

Figure C.2 Circuit symbol of an n-channel MOSFET.

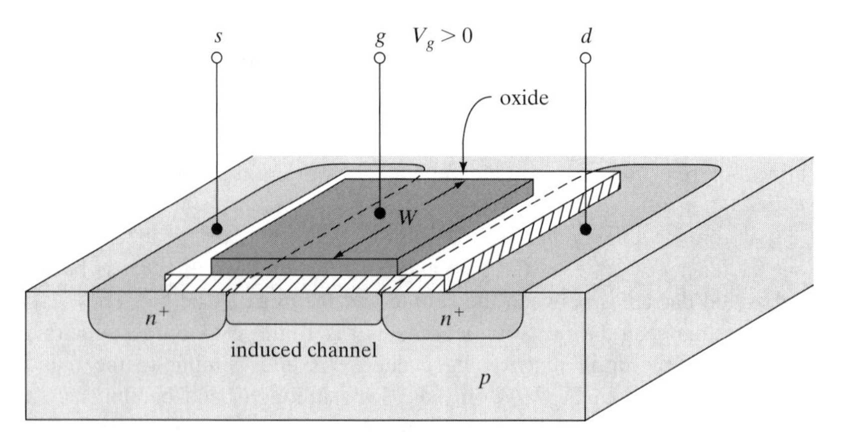

Figure C.3 MOSFET (n-type) with positive gate voltage, creating a conducting channel between source and drain.

2. *Triode or linear region*: When $V_{gs} > V_t$ and $V_{ds} < V_{gs} - V_t$ the device is on. (The inversion channel has been created.) In this region the MOSFET acts like a resistor controlled by the gate voltage. Drain to source current is

$$I_d = \frac{\mu_n C_{\mathrm{ox}}}{2} \frac{W}{L} \left(2 \left(V_{gs} - V_t\right) V_{ds} - V_{ds}^2\right). \tag{C.1}$$

where μ_n is the electron mobility (Section 10.1), $C_{\mathrm{ox}n}$ is the gate capacitance per unit area ($C_{\mathrm{ox}} = \varepsilon_{\mathrm{ox}}/t_{\mathrm{ox}}$, where $\varepsilon_{\mathrm{ox}}$ is the oxide permittivity and t_{ox} is the oxide thickness), W is the gate width (i.e., the dimension into the paper), and L is the gate length (essentially, from the n^+ source to the n^+ drain).

Holding V_{gs} constant, as V_{ds} is increased the voltage along the channel increases from the source to drain. Therefore, the voltage from the gate to points along the channel decreases as one moves from the source to the drain. The channel depth (the "thickness" of the channel) is dependent on this voltage, and, thus, the channel near

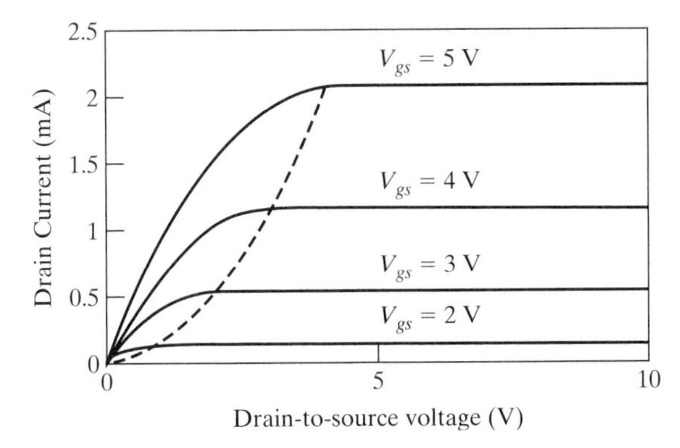

Figure C.4 Drain current versus drain-source voltage in an n-type MOSFET. The dashed line seperates the linear and saturation regions, and the different curves denote different positive gate voltages.

the drain becomes narrower than near the source. (The channel becomes tapered.) As V_{ds} is further increased, the voltage between the gate and the channel near the drain is reduced to V_t (when $V_{ds} = V_{gs} - V_t$), and the channel depth near the drain is reduced to almost zero. This point is called *pinch-off*, and beyond pinch-off, a further increase in the drain-source voltage will not significantly affect the drain current.† This is called the saturation region of operation.

3. *Saturation region*: When $V_{gs} > V_t$ and $V_{ds} > V_{gs} - V_t$ the device is on (the inversion channel has been created), but the channel is pinched off. The drain to source current is

$$I_d = \frac{\mu_n C_{\mathrm{ox}}}{2} \frac{W}{L} \left(V_{gs} - V_t\right)^2 . \tag{C.2}$$

Fig. C.4 shows the I-V characteristics of a typical device.

†Even if the channel is pinched off well before the drain, so that it would seem like there is no longer a channel for current flow, current can still flow to the drain due to the high value of the electric field at that end of the device. That is, the relatively large drain voltage causes a large electric field between the pinched-off end of the channel and the drain. Electrons are transported across this depletion region by the strong electric field. Current is relatively insensitive to further changes in the drain voltage since this affects primarily the width of the depletion region.

Appendix

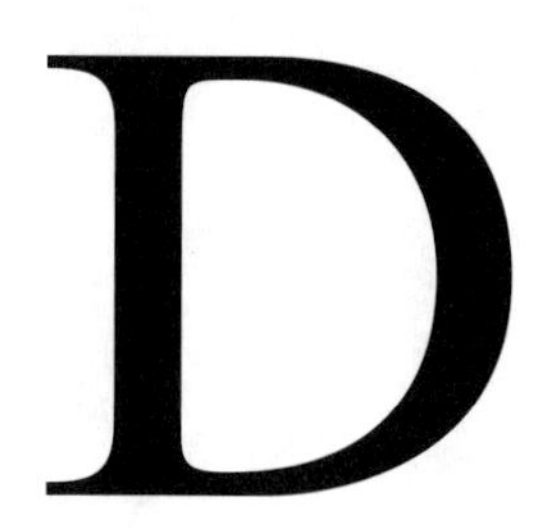

ANSWERS TO PROBLEMS

Chapter 2: Classical Particles, Classical Waves, and Quantum Particles

1. For the photon, $E = 1.91$ eV. For the electron, $E = 3.56 \times 10^{-6}$ eV.

2. Yes, these wavelengths are the same.

3. $E_{\text{elec}} = 2.48 \times 10^{-13}$ eV, $E_{\text{oven}} = 9.92 \times 10^{-6}$ eV, $E_{\text{uv}} = 124.2$ eV.

4. (a) Each photon carries $E_p = 1.963$ eV.
(b) $N = 3.18 \times 10^{15}$ photons/s.

5. $N = 3.18 \times 10^{16}$ photons/s. The number of photons scales linearly with power.

6. (a) $\lambda = 9.065 \times 10^{-13}$ m.
(b) $\lambda = 8.631 \times 10^{-13}$ m.
(c) $\lambda = 7.274 \times 10^{-8}$ m.
(d) $\lambda = 4.969 \times 10^{-38}$ m.

7. $\lambda = 1.098 \times 10^{-34}$ m.

8. $m = 1.647 \times 10^{-35}$ kg.

9. $p = 1.035 \times 10^{-27}$ Js/m=kg m/s.

10. For the photon,

$$\lambda = 310.17 \text{ nm},$$

$$f = 9.672 \text{ Hz},$$

$$p = 2.136 \times 10^{-27} \text{ kg m/s}.$$

For the electron,

$$\lambda = 0.613 \text{ nm},$$

$$f = 9.672 \text{ Hz},$$

$$p = 1.080 \times 10^{-24} \text{ kg m/s}$$

For the proton,

$$\lambda = 0.0143 \text{ nm},$$

$$f = 9.672 \text{ Hz},$$

$$p = 4.630 \times 10^{-23} \text{ kg m/s}$$

11. $\lambda = 1.001$ nm.

12. (a) $\Delta v \geq 2.637 \times 10^{-30}$ m/s.
(b) $\Delta v \geq 2.637 \times 10^{-34}$ m/s.
(c) $\Delta v \geq 2.637 \times 10^{-36}$ m/s.

13. For the electron, $\Delta x \geq 5.789 \times 10^{-3}$ m. For the baseball, $\Delta x \geq 3.52 \times 10^{-32}$ m.

14. $\Delta v \geq 0.0115$ m/s.

15. $\Delta v \geq 5788.5$ m/s.

Chapter 3: Quantum Mechanics of Electrons

4. $u = \sqrt{2/a}\sin(n\pi x/a)$, $\lambda = (n\pi/a)^2$.

5. $u = \sqrt{\varepsilon_n/a}\cos(n\pi x/a)$, $\lambda = (n\pi/a)^2$, where

$$\varepsilon_n \equiv \begin{cases} 1, & n = 0 \\ 2, & n \neq 0 \end{cases}.$$

7. $\Delta E \geq 3.291 \times 10^{-8}$ eV.

8. 2. $\langle p_x \rangle = 0$.

9. $a_m = \int \psi_m^*(\mathbf{r})\,\Psi(\mathbf{r}, 0)\,dr^3$.

11. 0, $x = \pm L/4$.

12. 100 percent, 0 percent.

13. 1. $|a|^2 + |b|^2 = 1$.
2. $P(E_2) = |b|^2$.

14. $\mathbf{J} = (\hbar k/m)\left(1 - |R|^2\right)$ on the left, and $\mathbf{J} = (\hbar q/m)\,|T|^2$ on the right.

15. $\mathbf{J} = 0$.

16. $k = 7.244 \times 10^9$ m^{-1}, $\omega = 3.039 \times 10^{15}$, $\mathbf{J} = \widehat{\mathbf{z}}\, 3.355 \times 10^{10}$ A/m^2.

Chapter 4: Free and Confined Electrons

1. $\psi(z, t) = Ae^{i(8.872\times 10^9)z}e^{-i\left(\frac{3|q_e|}{\hbar}\right)t}$.

2. $\psi(z, t) = Ae^{i(8.638\times 10^8)z}e^{-i\left(\frac{3|q_e|}{\hbar}\right)t}$, $V = 2.971$ eV.

4. $n = 3.65 \times 10^{10}$, $E_{2,1,1} - E_{1,1,1} = 1.8 \times 10^{-39}$ J $= 1.124 \times 10^{-20}$ eV.

5. $n = 3.65$, $E_{2,1,1} - E_{1,1,1} = 1.8 \times 10^{-19}$ J $= 1.12$ eV.

6. For an electron, the space should be on the order of 0.776 nm, or smaller. For a proton, the space should be on the order of 0.0181 nm.

7. (a) $v = 1.3 \times 10^{-36}$ m/s.
(b) $n = 1.3 \times 10^{34}$.

8. $P = 1/2$.

9. $v_e = 3.161 \times 10^5$ m/s.

10.

$$E_3 - E_1 = 0.752 \text{ eV},$$

$$E_3 - E_2 = 0.470 \text{ eV},$$

$$E_2 - E_1 = 0.282 \text{ eV},$$

11. (a) $A = \sqrt{30/L^5}$.
(b) $c_n = 4\sqrt{15}\,(1 - \cos n\pi) \,/\, \left(n^3\pi^3\right)$.
(c)

$$P(E_n) = |c_n|^2.$$

$$P(E_1) = 0.99856 = 99.8 \text{ percent}.$$

$$P(E_3) = 0.00137 = 0.137 \text{ percent}.$$

$$P(E_5) = 6.391 \times 10^{-5} = 0.0064 \text{ percent}.$$

12. 1. $A = \sqrt{630/L^9}$.

2. $$c_n = \begin{cases} 0, & n = 0, 2, 4, \ldots \\ \sqrt{\frac{1,260}{L^{10}}}\left(\frac{1}{\pi^5 n^5}\left(48L^5 - 4\pi^2 L^5 n^2\right)\right), & n = 1, 3, 5, \ldots. \end{cases}$$

3. $$P(E_n) = |c_n|^2.$$

$$P(E_1) = 0.9770 = 97.7 \text{ percent}.$$

$$P(E_3) = 0.0215 = 2.15 \text{ percent}.$$

$$P(E_5) = 0.00122 = 0.122 \text{ percent}.$$

13. $31.963/L^2$ eV, where L is in nm.

14. $\mu = 0.135$ eV.

17. For the infinite-height well, $E_1 = 0.0235$ eV, $E_3 = 0.21152$ eV. For the finite-height well, as a representative result, at $V_0 = 0.5$ eV, for E_1 we have $E_{\text{numerical}}^{\text{finite-height}} = 0.01812$ eV, and for E_3 we have $E_{\text{numerical}}^{\text{finite-height}} = 0.16085$ eV.

18. $\langle E \rangle = E_1$.

19. $\langle r \rangle = 6a_0$.

20. $\langle r \rangle = 5a_0$.

21. $\left(v_g\right)_z = 2.953 \times 10^5$ m/s.

22. $L \leq \lambda_e = 0.708$ nm.

Chapter 5: Electrons Subject to a Periodic Potential—Band Theory of Solids

1. We need a material cube having side length 2.5 nm.
2. The bandedges are at $E = 0.217$ eV and $E = 0.459$ eV.
4. $\mathbf{J}(\mathbf{r}, t) = \widehat{\mathbf{x}}\frac{p}{m}u^2$.
5. $m^* = (3/2)\, m$.
6. $m^* = -2m/\cos k$, $v_g = -\hbar \sin k/2m$.
7. $x(t) = \frac{2A}{\mathcal{E}q_e}\left(\cos\left(q_e\mathcal{E}at/\hbar\right) - 1\right)$.
8. $m^* = 5m_e/2$.
10. 5.32 eV.
11. 0.347 eV.
12. $\lambda = 174.67$ nm.
13. Metal.
14. Insulator, Semiconductor.
17. $\lambda = 721.3$ nm.
18. $E_{pn} = 0.20$ eV, $k_{pn} = k_a$.
19. (a) $\Delta E = 10.2$ eV, $\lambda = 121.6$ nm, between visible and UV.
(b) $\Delta E = 0.306$ eV, $\lambda = 4054.5$ nm, far infrared.
(c) $\Delta E = 0.0319$ eV, $\lambda = 3.89 \times 10^{-5}$ m, microwave.
(d) $\Delta E = 3.188$ eV, $\lambda = 389.2$ nm, visible, violet.
(e) $\Delta E = 13.51$ eV, $\lambda = 91.83$ nm, between visible and UV.
(f) $\Delta E = 13.6$ eV, $\lambda = 91.23$ nm, between visible and UV.
20. (a) $E = 1.426$ eV, (b) —, (c) magnitude of applied field greater than 5.5×10^5 V/m.
22. $r_{(19,0)} = 0.7437$ nm, $r_{(10,10)} = 0.678$ nm. For $(n, 0)$, $n = 9$.

Chapter 6: Tunnel Junctions and Applications of Tunneling

3. $V_0 = 6.858$ eV.
4. (a) $2e^{-9.48\times 10^{35}}$.
6.

$$T = \left|\frac{2k_1}{k_1 + k_2}\right|^2, \qquad R = \left|\frac{k_1 - k_2}{k_1 + k_2}\right|. \tag{D.1}$$

9. $T = 0.791$. For $a = 2$ nm, $T = 0.515$, for $a = 5$ nm, $T = 0.293$, and for $a = 10$ nm, $T = 0.901$.
10. $T \simeq e^{-2a\sqrt{\frac{2m^*}{\hbar^2}(V_0 - E)}}$.

Chapter 7: Coulomb Blockade and the Single-Electron Transistor

1. $\tau = 5.0 \times 10^{-14}$.
2. $T \ll 1,855$ K. For $C = 1.2$ pF, $T \ll 7.73 \times 10^{-4}$ K.
4. $V > -Nq_e/2C$.
6. $T = |q_e|/I_s$.
8. $V_s > 13.42$ V.
11. $V_s > 1.41$ V.
14. (a) $\lambda = 0.548$ nm.
(b) $L \leq \lambda$.
15. $|\mathbf{F}| = 4.617 \times 10^{-12}$ N, at 10 μm, $|\mathbf{F}| = 4.617 \times 10^{-18}$ N.

Chapter 8: Particle Statistics and Density of States

4. (a)

$$E_F = \left(\frac{hN_f}{3\sqrt{2m}}\right)^{3/2}.$$

(b) The Fermi wavelength is the de Broglie wavelength at the Fermi energy.
6. $\lambda_F^{(2d)} = 0.571$ nm.
7.

$$\lambda_F = 0.359 \text{ for aluminum} \tag{D.2}$$

$$= 0.399 \text{ nm for zinc.} \tag{D.3}$$

8. $n \simeq N_d = 10^{14}$ cm^{-3}.

Chapter 9: Models of Semiconductor Quantum Wells, Quantum Wires, and Quantum Dots

2. For GaAs, $L < 36$ nm, and for Si, $L < 18$ nm.
4. $L_x \ll 3.86$ nm.
6. $\alpha = 0.25$.
8. For the infinite-height well, $E_1 = 0.056$ eV, $E_3 = 0.505$ eV. For the finite-height well, $E_1 = 0.036$ eV and $E_3 = 0.275$ eV .
9. The first two states will be filled.
11. $E_{n_x,n_y,n_z} = 0.334\left(n_x^2 + n_y^2 + n_z^2\right)$ eV.
12. $R = 0.613$ nm.
13. $E_{1,1,1} - E_{2,2,1} = -0.251$ eV.
14.

$$E_g^{2 \text{ nm dot}} = 2.735 \text{ eV}, \lambda = 453.67 \text{ nm (blue-violet).}$$

$$E_g^{3 \text{ nm dot}} = 2.497 \text{ eV}, \lambda = \frac{hc}{2.497\,|q_e|\,10^{-9}} = 496.81 \text{ nm (orange-yellow).}$$

Chapter 10: Nanowires, Ballistic Transport, and Spin Transport

1. $v_T = 6.66 \times 10^4$ m/s, $v_F = 3.635 \times 10^5$ m/s.

2. $|\mathbf{v}_d| = 3.517$ m/s, $\sigma = 1.916 \times 10^8$ S/m.

3.
$$\sigma = \frac{q_e^2 \tau N_e}{m_e (i\omega\tau + 1)}.$$

4. $L_{\text{mfp}} = 135.3$ nm.

6. $d \geq 0.046$ nm for copper, $d \geq 8.235$ nm for silicon.

9. $v = 1.897 \times 10^8$ m/s.

11.
$$R = \frac{1}{\sigma 4\pi} \left(\frac{1}{a} - \frac{1}{b} \right).$$

13. $I = 92.98$ μA.

14. $I = 46.5 \times 10^{-6}$ A.

15. $I = 30.99 \times 10^{-6}$ A.

References

CHAPTER 1

[1] Drexler, E. (1986). *Engines of Creation: The Coming Era of Nanotechnology*, New York: Anchor.

[2] Goddard, W. A., D. W. Brenner, S. E. Lyshevski, and G. J. Iafrate, eds. (2003). *Handbook of Nanoscience, Engineering, and Technology*, Boca Raton: CRC Press.

[3] Wolf, E. L. (2004). *Nanophysics and Nanotechnology*, Weinheim: Wiley-VCH.

[4] Poole, C. P., and F. J. Owens (2003). *Introduction to Nanotechnology*, Hoboken: Wiley.

[5] Wilson, M., K. Kannangara, G. Smith, M. Simmons, and B. Raguse (2002). *Nanotechnology: Basic Science and Emerging Technologies*, Boca Raton: Chapman & Hall/CRC Press.

[6] Goser, K., P. Glösekötter, and J. Dienstuhl (2004). *Nanoelectronics and Nanosystems—From Transistors to Molecular and Quantum Devices*, Berlin: Springer-Verlag.

[7] Turton, R. (1995). *The Quantum Dot: A Journey into the Future of Microelectronics*, Oxford: Oxford University Press.

[8] Wong, B. P., A. Mittal, Y. Cao, and G. Starr (2005). *Nano-CMOS Circuit and Physical Design*, New York: Wiley.

[9] Nalwa, H.S., ed. (2004). *Encyclopedia of Nanoscience and Nanotechnology*, Stevenson Ranch, California: American Scientific Publishers.

CHAPTER 2

[1] Kroemer, H. (1994). *Quantum Mechanics for Engineering, Materials Science, and Applied Physics*, Englewood Cliffs, New Jersey: Prentice Hall.

[2] Weiss, R. J. (1996). *A Brief History of Light*, Singapore: World Scientific.

[3] Shankar, R. (1994). *Principles of Quantum Mechanics*, 2d. ed., New York: Kluwer Academic.

[4] Greiner, W. (1989). *Quantum Mechanics, An Introduction*, New York: Springer-Verlag.

[5] Baggott, J. (1992). *The Meaning of Quantum Theory*, Oxford: Oxford University Press.

[6] Hey, T., and P. Walters (2003). *The New Quantum Universe*, Cambridge: Cambridge University Press.

[7] Wieder, S. (1973). *The Foundations of Quantum Theory*, New York: Academic Press.

[8] Hund, F. (1974). *The History of Quantum Theory*, New York: Harper & Row.

CHAPTER 3

[1] Singh, J. (1999). *Modern Physics for Engineers*, New York: Wiley-Interscience.

[2] Kroemer, H. (1994). *Quantum Mechanics for Engineering, Materials Science, and Applied Physics*, Englewood Cliffs, New Jersey: Prentice Hall.

[3] Shankar, R. (1994). *Principles of Quantum Mechanics*, 2d. ed., New York: Kluwer Academic.

[4] Greiner, W. (1989). *Quantum Mechanics, An Introduction*, New York: Springer-Verlag.

[5] Ferry, D. K. (2001). *Quantum Mechanics, An Introduction for Device Physicists and Electrical Engineers*, Bristol: Institute of Physics Publishing.

[6] Harrison, W. A. (2000). *Applied Quantum Mechanics*, Singapore: World Scientific.

[7] Fromhold, A. T. (1981). *Quantum Mechanics for Applied Physics and Engineering*, New York: Academic Press.

[8] Marinescu, D. C., and G. M. Marinescu (2005). *Approaching Quantum Computing*, Upper Saddle River, New Jersey: Pearson Prentice Hall.

[9] Baggott, J. (1992). *The Meaning of Quantum Theory*, Oxford: Oxford University Press.

[10] Hey, T., and P. Walters (2003). *The New Quantum Universe*, Cambridge: Cambridge University Press.

[11] Tang, C. L. (2005). *Fundamentals of Quantum Mechanics for Solid State Electronics and Optics*, Cambridge: Cambridge University Press.

CHAPTER 4

[1] Kroemer, H. (1994). *Quantum Mechanics for Engineering, Materials Science, and Applied Physics*, Englewood Cliffs, New Jersey: Prentice Hall.

[2] Shankar, R. (1994). *Principles of Quantum Mechanics*, 2d. ed., New York: Kluwer Academic.

[3] Greiner, W. (1989). *Quantum Mechanics, An Introduction*, New York: Springer-Verlag.

[4] Solymar, L., and D. Walsh (1984). *Lectures on the Electrical Properties of Materials*, 3d. ed., Oxford: Oxford University Press.

[5] Schmid, G. (2004). *Nanoparticles, from Theory to Application*, Weinheim: Wiley-VCH.

[6] Singh, J. (1999). *Modern Physics for Engineers*, New York: Wiley-Interscience.

[7] Ferry, D. K. (2001). *Quantum Mechanics, An Introduction for Device Physicists and Electrical Engineers*, Bristol: Institute of Physics Publishing.

[8] Harrison, W. A. (2000). *Applied Quantum Mechanics*,Singapore: World Scientific.

[9] Fromhold, A. T. (1981). *Quantum Mechanics for Applied Physics and Engineering*, New York: Academic Press.

[10] Ferry, D. K., and J. P. Bird (2001). *Electronic Materials and Devices*, San Diego: Academic Press.

CHAPTER 5

[1] Solymar, L., and D. Walsh (1984). *Lectures on the Electrical Properties of Materials*, 3d. ed., Oxford: Oxford University Press.

[2] Sutton, A. P. (1993). *Electronic Structure of Materials*, Oxford: Clarendon Press.

[3] Bube, R. H. (1988). *Electrons in Solids, An Introductory Survey*, 2d. Ed., Boston: Academic Press.

[4] Singh, J. (1999). *Modern Physics for Engineers*, New York: Wiley-Interscience.

[5] Rogalski, M. S., and S. B. Palmer (2000). *Solid State Physics*, Boca Raton: CRC Press.

[6] Kittel, C. (1986). *Introduction to Solid State Physics*, 6th ed., New York: Wiley.

[7] Yang, E. S. (1978). *Fundamentals of Semiconductor Devices*, New York: McGraw-Hill.

[8] Ferry, D. K., and J. P. Bird (2001). *Electronic Materials and Devices*, San Diego: Academic Press.

[9] Saito, R., G. Dresselhaus, and M. S. Dresselhaus (1998). *Physical Properties of Carbon Nanotubes*, Singapore: World Scientific.

[10] Iijima, I., (1991). "Helical microtubules of graphitic carbon," *Nature*, 354 (1991): 56–58.

[11] Yao, Z., C. Dekker, and P. Avouris (2001). "Electrical transport through single-wall carbon nanotubes," in *Carbon Nanotubes; Topics in Applied Physics*, ed, M. S. Dresselhaus, G. Dresselhaus, and P. Avouris, Berlin: Springer-Verlag.

[12] Chen, Z. (2004). "Nanotubes for Nanoelectronics," in *Encyclopedia of Nanoscience and Nanotechnology*, ed. H. S. Nalwa, Stevenson Ranch, California: American Scientific Publishers.

[13] Heinze, S., J. Tersoff, R. Martel, V. Derycke, J. Appenzeller, and P. Avouris, "Carbon Nanotubes as Schottky Barrier Transistors," *Phys. Rev. Lett.*, 89 (2002): 106801.

[14] Fox, M. (2001). *Optical Properties of Solids*, Oxford: Oxford University Press.

[15] Singleton, J. (2001). *Band Theory and Electronic Properties of Solids*, Oxford: Oxford University Press.

[16] Ouyang, M., J.-L. Huang, C. L. Cheung, and C. M. Lieber, "Energy Gaps in 'Metallic' Single-Walled Carbon Nanotubes," *Science* 292 (2001): 702.

CHAPTER 6

[1] Singh, J. (1999). *Modern Physics for Engineers*, New York: Wiley-Interscience.

[2] Takeda, E., C. Y. Yang, and A. Miura-Hamada (1995). *Hot-Carrier Effects in MOS Devices*, San Diego: Academic Press.

[3] Kroemer, H. (1994). *Quantum Mechanics for Engineering, Materials Science, and Applied Physics*, Englewood Cliffs, New Jersey: Prentice Hall.

[4] Shankar, R. (1994). *Principles of Quantum Mechanics*, 2d. ed., New York: Kluwer Academic.

[5] Greiner, W. (1989). *Quantum Mechanics, An Introduction*, New York: Springer-Verlag.

[6] Ferry, D. K. (2001). *Quantum Mechanics, An Introduction for Device Physicists and Electrical Engineers*, Bristol: Institute of Physics Publishing.

[7] Harrison, W. A. (2000). *Applied Quantum Mechanics*,Singapore: World Scientific.

[8] Fromhold, A. T. (1981). *Quantum Mechanics for Applied Physics and Engineering*, New York: Academic Press.

[9] Goser, K., P. Glösekötter, and J. Dienstuhl (2004). *Nanoelectronics and Nanosystems—From Transistors to Molecular and Quantum Devices*, Berlin: Springer-Verlag.

[10] Lenzlinger, M., and E. H. Snow, "Fowler-Nordheim Tunneling into Thermally Grown SiO_2," *J. Appl. Phys.* 40 (1969): 278–283.

[11] Shakir, M. I., M. Nadeem, S. A. Shahid, and N. M. Mohamed, "Carbon Nanotube Electric Field Emitters and Applications," *Nanotechnology* 17 (2006): R41–R56.

[12] Groening, O. O. M. Kuettel, C. Emmenegger, P. Groening, and L. Schlapbach, "Field Emission Properties of Carbon Nanotubes," *J. Vac. Sci. Technol. B* 18 (2000): 665–678.

[13] Bonard, J. M., K. A. Dean, B. F. Coll, and C. Klinke, "Field Emission of Individual Carbon Nanotubes in the Scanning Electron Microscope," *Phys. Rev. Lett.* 89 (2002): 197602.

[14] Brodie, I., and C. Spindt, "Vacuum microelectronics," *Adv. Electron. Electron Phys.* 83 (1992): 1–106.

[15] Quan, W.-Y., D. M. Kim, and M. K. Cho, "Unified Compact Theory of Tunneling Gate Current in Metal–Oxide–Semiconductor Structures: Quantum and Image Force Barrier Lowering," *J. Appl. Phys.* 92 (2002): 3724.

CHAPTER 7

[1] Ferry, D. K. (2001). *Quantum Mechanics, An Introduction for Device Physicists and Electrical Engineers*, Bristol: Institute of Physics Publishing.

[2] Ferry, D. K. and S. M. Goodnick (1999). *Transport in Nanostructures*, Cambridge: Cambridge University Press.

[3] Grabert, H., and Devoret, M. H., eds. (1992). *Single Charge Tunneling—Coulomb Blockade Phenomena in Nanostructures*, New York: Plenum Press.

[4] Bird, J. P., ed. (2003). *Electron Transport in Quantum Dots*, New York: Springer.

[5] Singh, J. (1999). *Modern Physics for Engineers*, New York: Wiley-Interscience.

[6] Kroemer, H. (1994). *Quantum Mechanics for Engineering, Materials Science, and Applied Physics*, Englewood Cliffs, New Jersey: Prentice Hall.

[7] Shankar, R. (1994). *Principles of Quantum Mechanics*, 2d. ed., New York: Kluwer Academic.

[8] Irwin, J. D. (2002). *Basic Engineering Circuit Analysis*, 7th ed., New York: Wiley.

[9] Sze, S. M. (1969). *Physics of Semiconductor Devices*, New York: Wiley.

[10] Likharev, K. K., "SET: Coulomb Blockade Devices," *Nano et Micro Technologies* 3 (2003): 71–114.

[11] Yao, Z., C. Dekker, and P. Avouris (2001). "Electrical transport through single-wall carbon nanotubes," in *Carbon Nanotubes; Topics in Applied Physics*, ed. M. S. Dresselhaus, G. Dresselhaus, and P. Avouris, Berlin: Springer-Verlag.

[12] Datta, S. (1995). *Electronic Transport in Mesoscopic Systems*, Cambridge: Cambridge University Press.

[13] Datta, S. (2005). *Quantum Transport—Atom to Transistor*, Cambridge: Cambridge University Press.

[14] Goser, K., P. Glösekötter, and J. Dienstuhl (2004). *Nanoelectronics and Nanosystems—From Transistors to Molecular and Quantum Devices*, Berlin: Springer-Verlag.

[15] Avouris, P., J. Appenzeller, R. Martel, and S. J. Wind, "Carbon Nanotube Electronics," *Proceedings of the IEEE* 91 (2003): 1772–1784.

[16] Rakitin, A., C. Papadopoulos, and J. M. Xu, "Carbon Nanotube Self-Doping: Calculation of the Hole Carrier Concentration," *Phys. Rev. B* 67 (2003): 033411.

[17] Mantooth, B. A., and P. S. Weiss, "Fabrication, Assembly, and Characterization of Molecular Electronic Components," *IEEE Proceedings* 91 (2003): 1785–1802.

[18] Chen, Z. (2004). "Nanotubes for Nanoelectronics," in *Encyclopedia of Nanoscience and Nanotechnology*, ed. H. S. Nalwa, Stevenson Ranch, California: American Scientific Publishers.

[19] Heinze, S., J. Tersoff, R. Martel, V. Derycke, J. Appenzeller, and P. Avouris, "Carbon Nanotubes as Schottky Barrier Transistors," *Phys. Rev. Lett.* 89 (2002): 106801.

[20] Shea, H. R., R. Martel, T. Hertel, T. Schmidt, and P. Avouris, "Manipulation of Carbon Nanotubes and Properties of Nanotube Field-Effect Transistors and Rings," *Microelectronic Engineering* 46 (1991): 101–104.

[21] Kastner, M. A., "The Single-Electron Transistor," *Rev. Mod. Phys.* 64 (1992): 849.

[22] Wilkins, R., E. Ben-Jacob, and R. C. Jaklevic, "Scanning-Tunneling-Microscope Observations of Coulomb Blockade and Oxide Polarization in Small Metal Droplets," *Phys. Rev. Lett.* 63 (1989): 801.

CHAPTER 8

[1] Solymar, L., and D. Walsh (1984). *Lectures on the Electrical Properties of Materials*, 3d. ed., Oxford: Oxford University Press.

[2] Sutton, A. P. (1993). *Electronic Structure of Materials*, Oxford: Clarendon Press.

[3] Bube, R. H. (1988). *Electrons in Solids, An Introductory Survey*, 2d. Ed., Boston: Academic Press.

[4] Singh, J. (1999). *Modern Physics for Engineers*, New York: Wiley-Interscience.

[5] Rogalski, M. S., and S. B. Palmer (2000). *Solid State Physics*, Boca Raton: CRC Press.

[6] Kittel, C. (1986). *Introduction to Solid State Physics*, 6th ed., New York: Wiley.

[7] Ferry, D. K. (2001). *Quantum Mechanics, An Introduction for Device Physicists and Electrical Engineers*, Bristol: Institute of Physics Publishing.

CHAPTER 9

[1] Solymar, L., and D. Walsh (1984). *Lectures on the Electrical Properties of Materials*, 3d. ed., Oxford: Oxford University Press.

[2] Sutton, A. P. (1993). *Electronic Structure of Materials*, Oxford: Clarendon Press.

[3] Bube, R. H. (1988). *Electrons in Solids, An Introductory Survey*, 2d. ed., Boston: Academic Press.

[4] Singh, J. (1999). *Modern Physics for Engineers*, New York: Wiley-Interscience.

[5] Rogalski, M. S., and S. B. Palmer (2000). *Solid State Physics*, Boca Raton: CRC Press.

[6] Kittel, C. (1986). *Introduction to Solid State Physics*, 6th ed., New York: Wiley.

[7] Ferry, D. K. (2001). *Quantum Mechanics, An Introduction for Device Physicists and Electrical Engineers*, Bristol: Institute of Physics Publishing.

[8] Ferry, D. K., and S. M. Goodnick (1999). *Transport in Nanostructures*, Cambridge: Cambridge University Press.

[9] Mitin, V. V., V. A. Kochelap, and M. A. Stroscio (1999). *Quantum Heterostructures*, Cambridge: Cambridge University Press.

[10] Davies, J. H. (1998). *The Physics of Low-Dimensional Semiconductors*, Cambridge: Cambridge University Press.

[11] Schmid, G., ed. (2004). *Nanoparticles, from Theory to Application*, Weinheim: Wiley-VCH.

[12] Harrison, P. (2005). *Quantum Wells, Wires, and Dots*, 2d. ed., New York: Wiley.

[13] Madou, M. J. (2002). *Fundamentals of Microfabrication: The Science of Miniaturization*, 2d. ed., Boca Raton: CRC Press.

[14] Campbell, S. A. (2001). *The Science and Engineering of Microelectronic Fabrication*, 2d. ed., Oxford: Oxford University Press.

[15] Nalwa, H. S., ed. (2004). *Encyclopedia of Nanoscience and Nanotechnology*, Stevenson Ranch, California: American Scientific Publishers.

[16] Li, M., L. Chen, and S. Y. Chou, 2001 "Direct three-dimensional patterning using nanoimprint lithography," *Appl. Phys. Lett.* 78 (2001): 3322.

[17] Fox, M. (2001). *Optical Properties of Solids*, Oxford: Oxford University Press.

[18] Chui, H. C., E. L. Martinet, M. M. Fejer, and J. S. Harris, "Short Wavelength Intersubband Transitions in InGaAs/AlGaAs Quantum Wells Grown on GaAs," *Appl. Phys. Lett.* 64 (1994): 736.

[19] Link, S., and M. A. El-Sayed, "Spectral Properties and Relaxation Dynamics of Surface Plasmon Electronic Oscillations in Gold and Silver Nanodots and Nanorods," *J. Phys. Chem. B* 103 (1999): 8410–8426.

[20] Sun, Y., B. T. Mayers, and Y. Xia, "Template-Engaged Replacement Reaction: A One-Step Approach to the Large-Scale Synthesis of Metal Nanostructures with Hollow Interiors," *Nano Lett.* 2 (2002): 481.

[21] Dukovic, G., F. Wang, D. Song, M. Y. Sfeir, T. F. Heinz, and L. E. Brus, "Structural Dependence of Excitonic Optical Transitions and Band-Gap Energies in Carbon Nanotubes," *Nano Lett.* 5 (2005): 2314.

[22] Wang, F., G. Dukovic, L. E. Brus, and T. F. Heinz, "The Optical Resonances in Carbon Nanotubes Arise from Excitons," *Science* 308 (2005): 838.

[23] Kogan, M. J., N. G. Bastus, R. Amigo, D. Grillo-Bosch, E. Araya, A. Turiel, A. Labarta, E. Giralt, and V. F. Puntes, "Nanoparticle-Mediated Local and Remote Manipulation of Protein Aggregation," *Nano Lett.* 6 (2006): 110.

[24] Fox, A. M., "Optoelectronics in Quantum Well Structures," *Contemp. Phys.* 37 (1996): 111–125.

CHAPTER 10

[1] Solymar, L., and D. Walsh (1984). *Lectures on the Electrical Properties of Materials*, 3d. ed., Oxford: Oxford University Press.

[2] Ferry, D. K., and S. M. Goodnick (1999). *Transport in Nanostructures*, Cambridge: Cambridge University Press.

[3] Datta, S. (1995). *Electronic Transport in Mesoscopic Systems*, Cambridge: Cambridge University Press.

[4] Plonus, M. A. (1978). *Applied Electromagnetics*, New York: McGraw-Hill.

[5] Sutton, A. P. (1993). *Electronic Structure of Materials*, Oxford: Clarendon Press.

[6] Bube, R. H. (1988). *Electrons in Solids, An Introductory Survey*, 2d. ed., Boston: Academic Press.

[7] Kittel, C. (1986). *Introduction to Solid State Physics*, 6th ed., New York: Wiley.

[8] Naeemi, A., R. Sarvari, and J. D. Meindl, "Performance Comparison Between Carbon Nanotube and Copper Interconnects for Gigascale Integration (GSI)," *IEEE Electron Device Lett.* 26 (2005): 84–86.

[9] Srivastava, N. and K. Baberjee, "Performance Analysis of Carbon Nanotube Interconnects for VLSI Applications," *IEEE Int. Conf. on Computer-Aided Design (ICCAD)* (2005): 383–390.

[10] Ye, H., Z. Gu, T. Yu, and D. H. Gracias, "Integrating Nanowires with Substrates Using Directed Assembly and Nanoscale Soldering," *IEEE Trans. Nanotech.* 54 (2006): 62-66.

[11] Li, S., Z. Yu, C. Rutherglan, and P. J. Burke, "Electrical properties of 0.4 cm long single-walled carbon nanotubes," *Nano Lett.* 4 (2004): 2003–2007.

[12] Hanson, G. W., "Current on an Infinitely-Long Carbon Nanotube Antenna Excited by a Gap Generator, " *IEEE Trans. Antennas Propagat.* 54 (2006): 76–81.

[13] Fan, Y., B. R. Goldsmith, and P. G. Collins, "Identifying and counting point defects in carbon nanotubes," *Nature Materials* 4 (2005): 906–911.

[14] Chen, Z. (2004). "Nanotubes for Nanoelectronics," in *Encyclopedia of Nanoscience and Nanotechnology*, ed. H. S. Nalwa, Stevenson Ranch, California: American Scientific Publishers.

[15] Wolf, S. A., A. Y. Chtchelkanova, and D. M. Treger "Spintronics—A retrospective and perspective," *IBM J. Res. Dev.* 50 (2006): 101–110.

[16] Gregg, J. F., I. Petej, E. Jouguelet, and C. Dennis, "Spin electronics—a review," *J. Phys. D: Appl. Phys.* 35 (2002): R121–R155.

[17] Schmidt, G., "Concepts for spin injection into semiconductors—a review," *J. Phys. D: Appl. Phys.* 38 (2005): R107–R122.

[18] Poole, C. P., and F. J. Owens (2003). *Introduction to Nanotechnology*, Hoboken: Wiley.

[19] Goser, K., P. Glösekötter, and J. Dienstuhl (2004). *Nanoelectronics and Nanosystems—From Transistors to Molecular and Quantum Devices*, Berlin: Springer-Verlag.

[20] Molaresa, M. E. T., E. M. Höhberger, C. Schaeflein, R. H. Blick, R. Neumann, and C. Trautmann, "Electrical characterization of electrochemically grown single copper nanowires," *Appl. Phys. Lett.* 82 (2003): 2139.

[21] Dreher, M., F. Pauly, J. Heurich, J. C. Cuevas, E. Scheer, and P. Nielaba, "Structure and conductance histogram of atomic-sized Au contacts," *Phys. Rev. B* 72 (2005): 075435.

[22] Park, J.-Y., S. Rosenblatt, Y. Yaish, V. Sazonova, H. Üstunel, S. Braig, T. A. Arias, P. W. Brouwer, and P. L. McEuen, "Electron–Phonon Scattering in Metallic Single-Walled Carbon Nanotubes," *Nano Lett.* 4 (2004): 517.

[23] Naeemi, A., and J. D. Meindl, "Impact of Electron-Phonon Scattering on the Performance of Carbon Nanotube Interconnects for GSI," *IEEE Electron Device Lett.*, 26 (2005): 476–478.

[24] Salahuddin, S., M. Lundstrom, and S. Datta, "Transport Effects on Signal Propagation in Quantum Wires," *IEEE Trans. Electron Devices* 52 (2005): 1734–1742.

[25] van Wees, B. J., L. P. Kouwenhoven, E. M. M. Willems, C. J. P. M. Harmans, J. E. Mooij, H. van Houten, C. W. J. Beenakker, J. G. Williamson, and C. T. Foxon "Quantum ballistic and adiabatic electron transport studied with quantum point contacts," *Phys. Rev. B* 43 (1991): 12431.

[26] Thomas, K. J., J. T. Nicholls, N. J. Appleyard, M. Y. Simmons, M. Pepper, D. R. Mace, W. R. Tribe, and D. A. Ritchie "Interaction effects in a one-dimensional constriction," *Phys. Rev. B* 58 (1998): 4846.

[27] Steinhögl, W., G. Schindler, G. Steinlesberger, M. Traving, and M. Engelhardt, "Comprehensive study of the resistivity of copper wires with lateral dimensions of 100 nm and smaller," *J. Appl. Phys.* 97 (2005): 023706.

INDEX

T

U

V

W

Z